T0231806

SOFTWARE FOR ENGINEERING CONTROL OF LANDSLIDE
AND TUNNELLING HAZARDS

Dedicated to readers and users

Software for Engineering Control of Landslide and Tunnelling Hazards

BHAWANI SINGH
Indian Institute of Technology, Roorkee, India

R.K. GOEL
Central Mining Research Institute, Roorkee, India

A.A. BALKEMA PUBLISHERS / LISSE / ABINGDON / EXTON (PA) / TOKYO

Library of Congress Cataloging-in-Publication Data

Singh, Bhawani.
   Software for engineering control of landslide and tunnelling hazards / Bhawani Singh,
R.K. Goel.
      p. cm.
   Includes bibliographical references and index.
   ISBN 9058093603 -- ISBN 9058093611 (pbk.)
      1. Engineering--Computer programs. 2. Tunnelling--Safety measures--Computer
programs. 3. Landslides--Safety measures--Computer programs. I. Goel, R.K., 1960- II.
Title.

TA345 .S575 2002
624.1'5'0285--dc21

                                                                              2002026683

Cover design: Studio Jan de Boer, Amsterdam, The Netherlands
Typesetting: Charon Tec Pvt. Ltd, Chennai, India
Printed by: Krips, Meppel, The Netherlands

Published by: A.A. Balkema Publishers, a member of Swets & Zeitlinger
Publishers www. balkema.nl and www.szp.swets.nl

ISBN 90 5809 360 3 (hardback)

Contents

Preface

Until quite recently the basic approach in design has been an empirical one based on rock classification. This approach was the subject of an earlier book by the authors, *Rock Mass Classification: A Practical Approach in Civil Engineering* (1999), which has been enjoyed by experts all over the world. Lately however, a growing need for reliable software packages to aid engineering control of landslide and tunnelling hazards has inspired the writing of the present book.

Software for Engineering Control of Landslide and Tunnelling Hazards is based on the use of a rational approach to check the empirical predictions to be sure of the solution. The task of research and developing software packages on landslides for hill development was initially taken up at The Indian Institute of Technology, Roorkee, (Earlier known as The University of Roorkee, India) on the inspiration of Professor Jagdish Narain, a former Vice Chancellor. Subsequently, their reliability and utility were checked in a large number of consultancy projects in the seismic Himalaya, both in India and in Nepal as well as in other countries. The instant liking and success of these programs further boosted our morale. Latter research and software development on tunnelling hazards was also taken up. Finally, the intensive research work was systematically compiled into this book along with the source file, user's manual and input and output files to generate more confidence and interest among civil and mining engineers, geologists, geophysicists, planners, ecologists, researchers and students. The complementary set of books that we have produced has been possible due to God's grace, teamwork and worldwide acceptance, and moral support. In all, 18 programs on landslides and 6 programs on the tunnelling hazards are presented to tackle a variety of geohazards.

Emphasis is given to the practical analytical solution of geological hazard control rather than any rigorous analytical/numerical methods. The practical knowledge on engineering behaviour of rock masses, discontinuities, landslides and tunnelling hazards is also offered to add to the realistic input data. The real input data is the key to the success of any software package. Simple geological data is enough to run these programs. The advantage of software programs is that innovative solution and design is easier where it is not covered by empirical approaches.

Himalaya is a vast region, an amazingly beautiful creation which possesses extensive rejuvenating abilities. It is also one of the best field laboratories in which to study rock mechanics, geology and geohazards. The research experience gained in Himalaya is precious to the whole world.

The authors' foremost wish is to express their deep gratitude to Professor Charles Fairhurst, University of Minnesota; Professor J.A. Hudson, Imperial College of Science and Technology, London; Professor E. Hoek, International Consulting Engineer; Dr. N. Barton, Norway; Professor J.J.K. Daemen, University of Nevada; Dr. E. Grimstad, NGI; Professor G.N. Pandey, University of Swansea; Professor J. Nedoma, Academy of Sciences of the Czech Republic; Professor Zhao Jian, Nanyang Technological Univesity, Singapore; Professor V.D. Choubey; Professor T. Ramamurthy, IITD; Dr. R.K. Bhandari, CSIR; Mr. B.B. Deoja, Nepal; Mr. A. Wagner, Switzerland; Professor R.N. Chowdhary, Australia; Professor S. Sakurai, Japan; Dr. R. Anbalagan, IITR; Profosser M. Kwasniewski, Poland; Dr. B. Singh, Prof. B.B. Dhar, Dr. N.M. Raju; Dr. A.K. Dube, CRRI; Dr. J.L. Jethwa, CMRI; Dr. V.M. Sharma, ATES; Professor Gopal Ranjan, COER; Professor P.K. Jain, IITR; Dr. M.N. Viladkar, IITR; Dr. A.K. Dhawan, CSMRS; Dr. V.K. Mehrotra, Dr. H.S. Badrinath and Dr. Prabhat Kumar, CBRI; Dr. P.P. Bahuguna, ISM; Dr. Subhash Mitra, IITR; Dr. N.K. Samadhiya, IITR and Mr. H.S. Niranjan, HBTI for their constant moral support and vital suggestions and for freely sharing their precious field data with us. The authors are also grateful to the scientists of CMRI, CSMRS, UPIRI and IIT, Roorkee and to all project authorities for supporting field researches. We also thank Dr. V.V.R. Prasad, CMRI for helping in the preparation of these programs with graphics.

The authors also give thanks to Elsevier Science, U.K.; A.A. Balkema, The Netherlands; The American Society of Civil Engineers (ASCE); Reston, Ellis Horwood, U.K.; The Institution of Mining and Metallurgy, London; John Wiley & Sons, Inc., New York; Springer-Verlag, Germany; Trans Tech., Germany; Wilmington Publishing House, U.K.; Van Nostrand Reinhold, New York; and ICIMOD, Nepal for the kind permission, and also to all eminent professors, researchers and scientists whose work is referred to in the book.

All engineers and geologists and requested to kindly send their valuable suggestions for improving the book and software packages to the authors.

Bhawani Singh
Professor (Retired)
Department of Civil Engineering
Indian Institute of Technology
ROORKEE – 247 667, India

R. K. Goel
Scientist
Central Mining Research Institute
Regional Centre, CBRI Campus
ROORKEE – 247 667, India

PART 1: LANDSLIDE

CHAPTER 1

An introduction to landslide programs

Annual losses due to geological hazards are of the order of 100 billion dollars all over the world. Unfortunately, the frequency of landslides is increasing in many hilly regions especially in Himalayan region. It is now realized that the cost/benefit ratio of landslide control measures is quite high in over-populated hilly regions all over the world.

Thus, we are presenting a complete package of 18 Computer Programs on almost all kinds of Landslide Analysis and Control. You are welcome to use the following simple and field-tested Software Packages both for soil and rock slopes. These programs can be used reliably in the seismic hilly areas also.

1.1 A LIST OF SOFTWARE PACKAGES

Joint Analysis
STEREO: Plot of contours of joint sets on stereo net (Chapter 7).

Planar Slides
SASP: Stability analysis of slope with planar wedge failure and design of rock anchor system (Chapter 8).
SARP: Stability analysis of reservoir slope with planar wedge failure and toe cutting (Chapter 8).
BASP: Back analysis of slope with planar wedge failure and toe cutting (Chapter 8).
ASP: Optimum angle of cut slope with planar wedge failure (accounting for radius of curvature in plan) (Chapter 8).

3D Wedge Slide
SASW: Stability analysis of reservoir slope with 3D wedge failure and over-toppling failure (Chapter 9).
ASW: Optimum angle of cut slope with 3D wedge failure and over-toppling failure (Chapter 9).

WEDGE: Rigorous wedge analysis with tension crack (Chapter 10).

Toppling Failure
TOPPLE: Analysis of toppling failure and sliding failure in rock slopes (Chapter 11).

Circular/Rotational Slides
SARC: Stability analysis of reservoir slopes with circular wedge failure (Chapter 12).
BASC: Back analysis of slopes with circular wedge failure (Chapter 12).
ASC: Angle of cut slope with circular wedge failure (accounting for radius of curvature in plan) (Chapter 12).

Debris/Talus Slide
SAST: Stability analysis of slope with talus/debris deposit and design of remedial measures (Chapter 13).
BAST: Back analysis of slopes with talus/debris deposit on soil/rock slopes (Chapter 13).

Complex Slide
SANC: Stability analysis of slopes with non-circular wedge failure and non-vertical slices (Chapter 14).

Wave in Reservoir
WAVE: Height of wave in reservoir due to a landslide (Chapter 15).

Retaining Wall
RETAIN: Optimum design of rigid retaining wall and breast wall with inclined base and soil slopes (Chapter 16).

Footing on Hill Slope
QULT: Ultimate bearing capacity of rectangular footing on soil slopes (Chapter 17).

The computer programs on back analysis to calculate the strength parameters begin with B, on stability analysis begin with S and to obtain the factor of safety and angle of cut slope begin with A.

The commission on computer programs of the International Society for Rock Mechanics has compiled a list of other computer programs and published it in the International Journal of Rock Mechanics, Mining Sciences and Geomechanics Abstracts in 1988.

1.2   ASSUMPTIONS IN STABILITY ANALYSIS OF ROCK SLOPES

The following conventional assumptions are made in the stability analysis of slopes.
1. All rock joints are planar and not curved surfaces. The tension crack is vertical.
2. The rock wedge is a rigid body.

3. Pseudo-static analysis of stability of slope is adopted in the computer programs. However, dynamic settlement of slopes due to earthquakes is considered to be valid approximately.
4. Damage to the rock mass due to blasting is not considered. But, vibrations due to blasting may be considered in the computer programs.
5. Dynamic strength of rough rock joint is the same as its static strength.
6. Tensile strength across rock joints is zero.
7. The horizontal stresses in the slope mass are not considered. The horizontal in situ stress along the periphery of rock slope will try to pre-stress the rock wedge and thereby increase its stability. In concave slopes, this horizontal stress is quite high and exhibits existence of very steep rock and soil slopes. Thus, the proposed analysis in the computer programs leads to an estimate conservative value of the factor of safety of particularly deep landslides.

Section 4.4 lists assumptions in the analysis of soil slope.

The eye judges the angle of hill slopes on highly steeper side. Therefore, the slope angle should be measured by contour plan. Experience shows that surveying errors may be large in a hilly terrain.

1.3   SUGGESTED APPROACHES

Field experience suggests the following practical approaches:
1. Visual identification of a slope in distress and its mode of failure.
2. Back analysis of slope using appropriate program according to the mode of failure to obtain the strength parameters along failure surfaces.
3. Forward analysis of adjoining slopes using the proper programs according to the potential mode of failure.
4. Design of the remedial measures/bio-engineering controls wherever required.
5. Monitoring of landslide prone area near the important structures.

Back analysis of both models and its parameters should be done on the basis of observed behaviour of soil and rock masses.

1.4   FEATURES OF THE COMPUTER PROGRAM

A salient feature of the program is the estimation of dynamic settlement (from pseudo-static analysis) of rock/soil slopes which depends strongly upon the earthquake magnitude on Richter's scale. The emphasis is on proper design of remedial measures and not on rigorous analysis. Thus, simple data is enough. Programs will identify automatically critical failure surfaces both in soil and rock slopes. The strength parameters obtained from back analysis ensure that design of remedial measures is not over-conservative. The advantage of using a PC is the ability to access and perfect design. Feedback received from users in over 15 countries has aided the optimization of these programs.

Many civil and mining engineers and geologists have used these programs over the last 15 years for the design of dam reservoir slopes, rail and road side cut slopes, site development for building complexes in seismic hilly areas, landslide control, ropeway on steep hills, embankments, dam abutments, retaining walls, anchor design and planning of eco-development in seismic hilly regions. All the programs may be loaded on PC note book at the sites of landslides for an on-spot optimum design of remedial measures.

The uncertainties in geology, rock parameters and environmental conditions may be managed by a high factor of safety and redundancy (additional support systems, e.g. shotcreting of slopes, etc.). The experience with predictive tools and computer models has been discouraging. More important than prediction is the invention of a conservative solution of the problem in totality.

Happily, the software packages on LANDSLIDE have been found very useful in crisis management of major landslides in Himalaya. The International Centre for Integrated Mountain Development (ICIMOD) in their Mountain Risk Engineering Handbook recommended all of these software packages. The Commission on Education of International Society for Rock Mechanics (ISRM) has also encouraged the complete package in the report of its meeting on August 24, 1999.

These programs are not for taking away the authority or power of the decision-makers. In fact, the programs are an aid to the decision-makers. After all, we all are working for mutual benefits. Thus a team of young and inexperienced geologists and rock engineers may need to make several trials of computer models of geological and support systems until an experienced rock engineer gives approval.

1.5   UNITS USED

The suggested units in all the programs are (see also Appendix 1 for details),
   Tonne (T)
   Metre (M)
   Degree (°)
The SI units may also be used in the landslide programs which are discussed in Part 1.

1.6   USE OF COMPUTER PROGRAMS

All the computer programs are available in the directory SOFTWARE/SLOPE in the enclosed computer disc (CD). The Fortran source programs are also given in the directory SOFTWARE/SLOPE/SOURCE. These source programs may be written in C++ using program F2C. The programs are written in FORTRAN 77 and EXE files work in DOS environment. The attached CD also has user's manual file for almost all the programs. A user's manual is also presented in each chapter discussing a particular program.

For more clarifications of users, typical input data files are given beginning with I. Similarly, corresponding output files are also given, beginning with O. File details, for example, are as follows (Considering Program TOPPLE):

File Name	*Details*
Topple.txt	Users Manual
Itopple.dat	Typical input data file
Otopple.dat	Typical output data file

The typical computer commands to work in a DOS environment are (for example considering program TOPPLE):

Command	*Operation*
EDIT ITOPPLE.DAT	(TO PREPARE INPUT FILE)
TOPPLE	(TO RUN EXE FILE OF PROGRAM TOPPLE)
ITOPPLE.DAT	(TYPE INPUT FILE NAME)
OTOPPLE.DAT	(TYPE OUTPUT FILE NAME)
2	(TYPE 2 FOR EXECUTION)
EDIT OTOPPLE.DAT	(FOR VIEWING AND PRINTING OUTPUT FILE OTOPPLE.DAT)

Use of DOS Editor has been suggested to create an input data file. All the data files are in ASCII format. Users may use other file names for different sites. However, QULT (Chapter 17) is a user-friendly program with QULT.OUT as the output file.

A new file, say Ixyz.DAT, may be created by command EDIT Ixyz.DAT. Then the input data is typed per the sequence shown in the users manual. Similarly, a new output file Oxyz.DAT should be created and the empty file is closed. Finally, program XYZ is run with commands as shown above. After the execution of the program, a temporary empty output file is created which should be deleted.

Experience shows that an input file is a very good choice for frequent refinements in input data for optimizing design of support systems based on the experience, intuition and common sense. Appendix 4 may be used as a reference to run these softwares/programs in the WINDOWS environment. The next five chapters are devoted to support systems, types of landslides, approach of analysis, shear strength of discontinuities and rock masses. These chapters will provide the information on the selection of input parameters for various softwares discussed subsequently.

REFERENCE

List of computer programs in rock mechanics. *Int. J. Rock Mech. Min. Sci. Geomech. Abstr.* 25(4): 183–252, 1988.

CHAPTER 2

Inherently stable support systems in rock engineering

"Proper preparation prevents poor performance – five Ps"
Anonymous

2.1 DESIGN PHILOSOPHY IN CIVIL ENGINEERING

There is an urgent need for uncertainty management in rock and soil engineering. A great demand exists for new solutions to rock engineering problems so that design and construction techniques do not have to be changed frequently whenever departures in actual geology are seen to occur within reasonable limits. In reality, modern construction engineers prefer a rigid system of planning during construction. Consequently, any change in design may lead to a rise in the cost of construction from 30 percent to 200 percent, besides delays in the completion of hydro projects involving long tunnels. The only choice left is to invent a new type of inherently stable and robust support system which is stable structurally in almost all geological conditions, so that any error in the prediction of geology will not upset construction engineers who may take on-spot decisions to change the support density. For example, properly designed reinforced rock arch is a good design concept for most rock conditions in the underground openings. Similarly, reinforced breast wall is a good concept for stabilising slopes.

In civil engineering, unlike in mining, the prime need is for lifelong solutions. It must be ensured that the proposed solution needs the minimum maintenance in view of the utter dislike for maintenance in some Asian countries. Thus, philosophy of the support system is based on the following principles:
1. The support system should have adequate strength, and be robust and inherently stable,
2. The support system should be ductile (use ductile material for reinforcement) and,
3. In complex geological situations, there should be 2 or 3 lines of defence.
The limit state concept represents a logical design principle. However, this is not a radically new method compared to earlier design practice. Nonetheless, it does represent a clear formulation of some widely accepted principles in rock and soil engineering.

2.2   A CONCEPT OF LIMIT STATE DESIGN

In the method of design based on limit state concept, the structure is designed such that it shall (i) safely withstand all loads liable to act on it throughout its life, and (ii) satisfy the serviceability requirements, such as limitations on deflection and cracking. The acceptable limit for the safety and serviceability requirements before failure occurs is called a 'Limit State'. The limit states that are usually examined in design are:

1. *Limit state of collapse*: In this limit state, the resistance to bending, compression, shear and torsion at every section shall not be less than the value at that section produced by the probable most unfavorable combination of loads on the structure.
2. *Limit state of serviceability*: In this limit state, the deflection and the width of cracks shall not exceed the permissible values.
3. *In rock engineering*: It is desirable to ensure a safe mode of collapse or slow ductile failure, as rocks are brittle. As such, ductile materials (or reinforcement) should be preferred to brittle materials in support systems.

The aim of design is to achieve acceptable probabilities that the structure will not become unfit for the use for which it is intended, that is, it will not reach a limit state. This is accomplished by (i) decreasing the material strength and (ii) increasing the loads coming on the structure by certain factors known as partial safety factors. The condition (3) is suggested to ensure safety of life for miners, engineers and users.

2.3   THE MANAGEMENT OF UNCERTAINTIES

When one is talking about the behaviour of jointed rock masses or rock structures, one is really talking about the most probable behaviour of rock masses according to the Chaos theory. So one should expect some random mismatch between theory and observations. Some failures do not prove a theory wrong.

The ground displays a far greater range of both material properties and heterogeneity than manufactured materials such as steel and concrete. Even the best methods for obtaining the necessary geo-technical data and the most reliable calculation methods are therefore inadequate. In fact, the factors of safety act, to some an extent, as correction factors. For that reason, the best way of determining design criterion to be used in rock engineering practice is a combination of experience and back-calculation of successful structures.

During the years to come, the System of Partial Factors of Safety will find increasing use in rock engineering for the evaluation of the risk of failure and collapse. It may represent a useful tool for the design of traditional and routine rock structures, but it is not a universally applicable system which can readily be used with fixed numerical values for all geo-technical structures.

The best way to manage uncertainties in complex geo-environmental situations is continuous monitoring of the rock structures and to have a contingency plan if anything goes wrong. In fact, a second contingency plan should also be ready in case the first contingency plan fails. Furthermore, allowance should be made for uncertainties in the design of rock structures.

The most popular strategy is to follow a conservative approach (with a high factor of safety and 2 or 3 lines of defense). Normally, construction engineers suggest a higher factor of safety. The construction method should be invented so that future strengthening of rock structures/support system may be possible quickly and without many problems. NATM/NTM is one such flexible construction technology, hence its international popularity among designers.

A few centuries ago, geoscientists believed mankind to be a negligible force compared to the natural forces. With advent of the 21st century, however, geoscientists are now convinced that mankind has evolved into a mighty force comparable to the geological forces.

2.4 THE CHOICE OF ROCK BOLT

The choice of types of bolts, shotcrete and concept of rock reinforcement etc., are important factors in uncertainty management. Figure 2.1 shows the merits and demerits of grouted bolts compared to point anchored bolts. In the case of full-column grouted bolts or rock anchors, we can see the following advantages in the poor rock masses:

1. The rock mass in the roof is not fractured during pre-tensioning of rock bolts, as rock anchors need no pre-tensioning.
2. The rock mass is not fractured near the base plate due to no tension in the rock anchors, even if size of base plate is small.
3. Full-column grouted rock anchor makes the reinforced rock arch ductile, which fails slowly, giving enough time for strengthening (retro fitting) of the support system of an underground opening. (The law of averages is also applicable to fracture toughness on slightly lower bound side (Rao, 1983). Hence, rock arch reinforced by full-column grouted bolts has much higher fracture toughness than that of rock mass.)
4. The rock anchor is partially effective even if it is fractured, unlike point anchored bolt.
5. The required tension is induced in the rock anchors due to any slip along the rock joints which are usually rough and dilatant.
6. The stiffness of rock anchors is higher than that of the point anchored bolt.

According to Singh et al. (1995), the factor of safety mobilized by full-column grouted bolt in tunnels is

$$F_s = \frac{9.5}{(p_{roof})^{0.35}} \qquad (2.1)$$

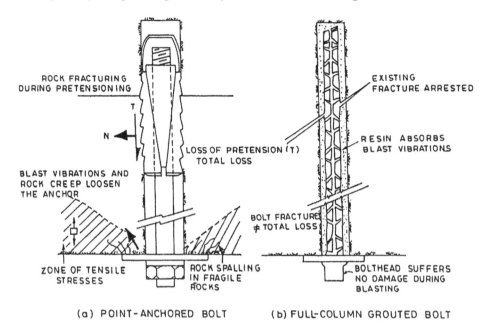

Figure 2.1. Comparison of point anchored and grouted bolts in poor rocks.

where

p_{roof} = ultimate support pressure in the roof of a tunnel/cavern (T/m^2), and

$$F_s = \frac{\text{Bolt capacity when fully tensioned}}{\text{Induced tension in the rock anchor}}$$

It may be noted that in the case of openings in the poor rock masses, the radial strains are higher than in the hard rock and p_{roof} is also more. So lesser F_s is mobilized and more tension is induced in the rock anchors in tunnels.

Furthermore, the factor of safety mobilized by point anchored bolts in tunnels is

$$F_s = 3.25(p_{roof})^{0.1} \tag{2.2}$$

Thus point-anchored bolts are slightly less effective as rock reinforcement in tunnels in the poor rock masses where p_{roof} is high.

2.5   THE CHOICE OF THE TYPE OF SHOTCRETE TO BE USED

The recent experience of tunnelling in complex and poor geological conditions in Himalaya has shown that the steel fibre reinforced shotcrete (SFRS) lining has performed well even in moderately squeezing rock conditions. So, the SFRS is ideally suited for uncertainty management, with the following advantages over the conventional shotcrete:

1. The shear strength of SFRS is almost twice of the conventional shotcrete due to steel fibre reinforcement. Hence, nearly half the thickness is all that is required when compared to the conventional shotcrete.
2. There is no need for welded mesh of steel bars which is a hindrance while shotcreting. In the case of conventional shotcrete, hollow gaps are left between shotcrete lining and the uneven rock surface. On the other hand, there is no such construction problem in spraying SFRS.
3. SFRS is ductile material due to steel fibres. So its failure will be slow, giving advance warning for the strengthening of the support system.

A good bond between shotcrete and rock surface helps to reduce the stresses in the shotcrete. Besides, such a good bond checks the extent of squeezing of weak rock mass inside the tunnels. This is indicated by non-linear finite element analysis (Kumar, 1999). It may be noted that roughness of the excavated surface prevents slip between the rock and the shotcrete.

It is heartening to note that the capacity of shotcrete lining is more than that of the rock bolts system. As such in poor rock masses, shotcrete lining is a primary support system. Of course, for hard rocks, a rock bolt system together with a welded mesh is a self-supporting system.

2.6　REINFORCED ROCK PORTAL

We all know that a portal frame is an inherently stable structure. This same concept is also an encouraging experience in the caverns. Figure 2.2 shows a typical reinforced rock portal frame. Even if unexpected rock loads exhibit from unknown geological conditions, this portal will swing and deform but is likely to regain its stability. To take care of uncertainties due to geology and geohydrology, the rock mass may be grouted all around this portal of reinforced rock mass.

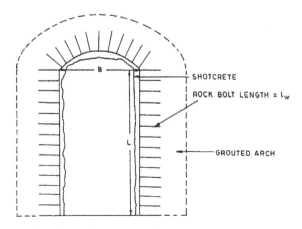

Figure 2.2. Design of wall reinforcement of caverns.

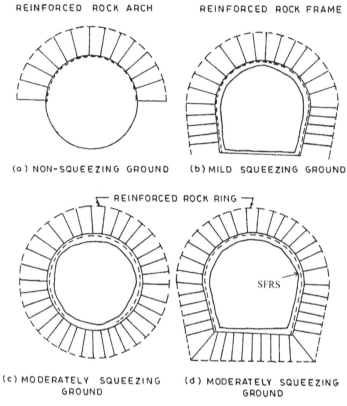

REINFORCED ROCK ARCH   REINFORCED ROCK FRAME

(a) NON-SQUEEZING GROUND   (b) MILD SQUEEZING GROUND

REINFORCED ROCK RING

SFRS

(c) MODERATELY SQUEEZING   (d) MODERATELY SQUEEZING
        GROUND   GROUND

Figure 2.3. Design philosophy of rock reinforcement in tunnelling (welded mesh not needed in SFRS).

2.7   REINFORCED ROCK ARCH/RING

Ancient experience proves beyond doubt that arch is an essentially very stable structure. A ring is similarly a stable structure. The same concept inspires tunnel engineers also. Figure 2.3 shows applications of this theory. In non-squeezing ground, reinforced rock arch may be sufficient. In the case of mild squeezing ground conditions (overburden $>350 \, Q^{1/3}$ m, $J_r/J_a < 0.5$), reinforced rock portal may be better. But reinforced rock ring may be ideal for moderately squeezing ground conditions. Steel supports embedded in shotcrete are needed in highly squeezing ground conditions.

Shear zones are often encountered in the tunnels and caverns. After, ground water is allowed to drain out. A thin shear zone (with a width greater than 0.5 m and less than 2 m) can be supported by special support system as shown in Figure 2.4 (Lang, 1961). First of all, the thin shear zone is excavated up to about 50 cm. Then it is backfilled by shotcrete. Simultaneously, long rock anchors are installed to stitch the shear zone as shown in Figure 2.4. The design of the usual support system is executed in the remaining part of the tunnel.

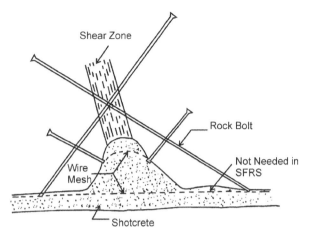

Figure 2.4. Typical treatment of narrow shear zone (Lang, 1961).

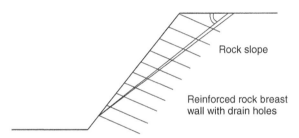

Figure 2.5. Reinforced rock breast wall.

2.8   REINFORCED ROCK BREAST WALL

The unstable tunnel portal and rock slope with an unfavourable orientation of joints may be stabilized effectively by rock anchoring which creates a reinforced rock breast wall (Fig. 2.5). The length of rock anchors is kept the same for easy construction and supervision. The horizontal drill holes are inclined slightly (say 10°) so that cement grout may enter by gravity flow. On the contrary, drainage holes should be inclined upwards at dip of about 5°–10° so that seepage water may flow downward easily.

2.9   REINFORCED SOIL BREAST WALL

A similar reinforced soil breast wall would stabilize an unstable soil slope as seen in Figure 2.6. Soil nailing of cut slopes in soil is becoming popular. The costs of soil nailing and using a cement stone breast wall are almost the same. The advantage lies in the huge saving of space at the bottom of the cut slope along hill roads.

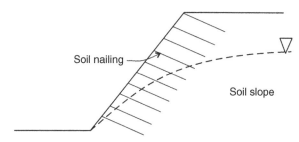

Figure 2.6. Reinforced soil breast wall.

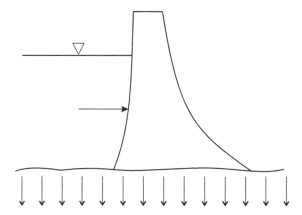

Figure 2.7. Concrete dam on undulating rock.

2.10   A CONCRETE DAM ON AN UNDULATING ROCK FOUNDATION

There is deep anxiety in minds of designers about stability of high gravity or arch dams in highly active seismic regions. It is difficult to attain desired factor of safety of 3 or 4 against sliding. Making the surface of dam foundation highly undulating as shown in Figure 2.7 solves this problem. This solution is better than shear keys. It must be ensured that no bedding plane, shear zone or joint is dipping upward below the dam foundation. In latter case dam foundation has to be anchored by cable anchors deep inside the rock mass in the foundation.

2.11   A RCC FRAME STRUCTURE IN SEISMIC HILLY AREAS

Hill development has become important nowadays, especially in Himalayan countries. The first step is the development of a safe terrace system, even in the landslide prone areas. The next step is selection of the type of structure. Experience in lower and upper Himalaya proves that a RCC frame structure with a wide base and a heavy plinth beam is the right choice for buildings (Fig. 2.8). This type of structure would be able to withstand slope movement and differential

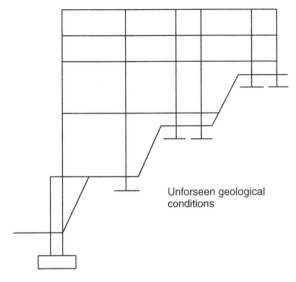

Figure 2.8. RCC frame structure of wide base (tilt of no consequence).

settlement due to a wide variation in the geological conditions. However, the joint between column and beams in a RCC structure is the weakest link which gives way during major earthquakes. Therefore, extra reinforcement should be provided in the joints.

The earthquake damage to life in well-built RCC buildings on slopes have been negligible, where chances of liquefaction are small, as compared to that in the plains.

A paradox is that a structure designed on the basis of least observed strength of materials may not be stable. The designer generally assumes that a foundation is homogeneous. The unknown inhomogeneity may cause differential settlement and then failure of structure. Everyone should take a calculated risk.

2.12 PRECAUTIONS

2.12.1 The importance of drainage

According to the anonymous quote below, there are three problems associated with hill development.
"First problem is drainage,
Second problem is drainage,
Third problem is also drainage."
Unfortunately, there is an intense dislike for maintenance in some Asian civilizations. Consequently, drainage systems are not cleaned regularly. A poorly maintained and choked-up drainage system is worse than no drainage system at all.

The level of negligence is such that polythene bags are littered on hill slopes and these polythene bags choke the inadequate drainage network. The result is that water seeps into the hill slope, which slides downward eventually. The life of even strong structures is finally reduced drastically in undrained hilly areas.

2.12.2 Water sensitive rocks

Some rocks disintegrate when placed under water. For example, schist may break up along planes of schistocity when water vapour seeps into the micro cracks and then propagates through fractures. (In micro-capillary, water cannot flow and only its vapour can travel under high pressure.) So dehumidification and humidification should be prevented by shotcreting the cut or the foundation pit immediately after excavation.

Other examples of water sensitive rocks are soft shales, sand rocks and conglomerate. Seepage of water may cause leaching of calcium-ions in limestone for several decades. Thus, the permeability of such rocks may keep on increasing with time. So seepage problems may be faced after several decades of excavation. Swelling of shales is a serious problem. Tunnels in swelling rocks should be shotcreted immediately after excavation and the tunnel closure should be monitored carefully. This phenomenon cannot be simulated in the computer modeling at present. Experts should be invited to share their past experiences of tackling these problems.

2.12.3 Coarse-grained rocks

The coarse-grained rock mass appears appealing to the eyes. Yet their strength is low because the size of cracks is of the order of the size of grains in a rock. Therefore, cores of adequate diameter should be tested to determine its UCS.

2.12.4 Safe town planning in hills

The modern trend is to plan towns along ridges or near top of the hills due to tourism trends. There are a number of benefits obtained, especially in the Himalaya (Deoja et al., 1991; Niranjan, 2000):
1. A ridge is generally a rocky area, which is more stable than its riverside area. Damage to houses on rocks is less than the houses on soil in the valley. The approach road will have a lesser number of bridges to cross natural drainage.
2. It is not prone to landslide generally.
3. Sunshine hours are more than that in valleys and the scenery is more beautiful.
4. A ridge is protected from the flashfloods due to the break down of landslides or cloudbursts.
5. There are more springs near to some ridges due to exposed rocks in hills made out of synclines. So, drinking water is available.

6. Drainage is excellent near the hilltop than in the valley. So, there are no mosquitoes. Thus, towns are more hygienic.
7. Apples grow near ridges only in a cold climate.
8. There is some snowfall near hilltops at some locations and this attracts the tourists.

It is costly to build towns near ridges but it is accepted due to safety. Life is very hard and risky for local people. Playgrounds are rare. However, hill towns are excellent for tourists and for short visits. Moreover, houses should not be built near faults, gullies and tilted trees. There should be a safe distance between the edge of the slope and the foundation of structures. Multi-storied buildings with more than four stories should not be planned in seismic hilly areas.

2.12.5 Hygienic planning

The old approach was to build hill roads and let the slopes come to a new equilibrium. During the last 30 years, there has been a trend towards eco-equilibrium. The new approach is investment in the safe and hygienic planning of hill infrastructures. In other words, cut slopes are stabilized by bioengineering methods. As such the attention (of civil and mining engineers) is focussed on a solution rather than on rigorous analysis of hill problems.

REFERENCES

Deoja, B., Dhital, M., Thapa, B. and Wagner, A. 1991. *Mountain Risk Engineering Handbook* (3 Vols). Kathmandu, International Centre of Integrated Mountain Development.

Kumar, Prabhat. 2000. Personal Communication.

Lang, T.A. 1971. Theory and practice of rock bolting. *AIME Trans* 220.

Niranjan, R.K. 2000. Personal Communication.

Rao, Kameshwar C.V.S. 1983. A note on fracture toughness of multiphase materials. *Eng. Fract. Mech.* 18(1): 35−38.

Singh, Bhawani,Viladkar, M.N., Samadhiya N.K. and Sandeep 1995. A semi-empirical method for the design of support systems in underground openings. *Tunnel. Underground Space Technol.* 10(3): 375−383.

CHAPTER 3

Types of failures of rock and soil slopes

"I render infinite thanks to God for being so kind as to make me first observer of marvels kept hidden in obscurity for all previous centuries"
Galileo Galilei (1609)

3.1 INTRODUCTION

The classification of rock and soil slopes is based on the mode of failure. In a majority of cases, the slope failures in rock masses are governed by joints and occur across surfaces formed by one or several joints. Some common modes of failure are described below which are frequently found in the field.

3.2 PLANAR (TRANSLATIONAL) FAILURE

Planar (Translational) failure takes place along prevalent and/or continuous joints dipping towards the slope, with strike nearly parallel ($\pm 5°$) to slope face (Fig. 3.1b). Stability condition occurs if:
1. critical joint dip is less than the slope angle, and
2. mobilized joint shear strength is not enough to assure stability.
Generally, a planar failure depends on joint continuity.

3.3 3D WEDGE FAILURE

Wedge failure occurs along two joints of different sets when these two discontinuities strike obliquely across the slope face and their line of intersection daylights in the slope face, as shown in Figure 3.1c (Hoek and Bray, 1981). The

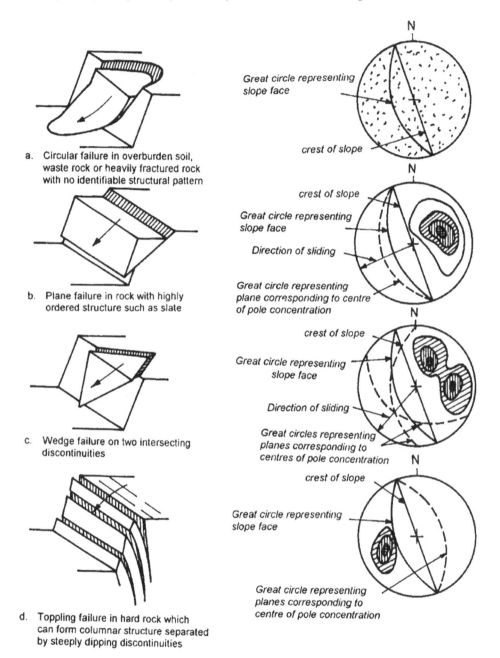

Figure 3.1. Main types of slope failure and stereo plots of structural conditions likely to give rise to these failures (Hoek and Bray, 1981).

wedge failure depends on joints' attitude and conditions and is more frequent than planar failure. The factor of safety of a rock wedge to slide increases significantly with the decreasing wedge angle for any given dip of the intersection of its two joint planes (Hoek and Bray, 1981).

3.4 CIRCULAR (ROTATIONAL) FAILURE

This occurs along a surface which develops only partially along joints, but mainly crosses them. These failures can only happen in heavily jointed rock masses with a very small block size and/or very weak or heavily weathered rock mass (Fig. 3.1a). It is essential that all the joints are oriented favourably so that planar and wedge failures are not possible.

The modes of failure which have been discussed so far involved the movement of a mass of material upon a failure surface. An analysis of failure or a calculation of the factor of safety for these slopes requires that the shear strength of the failure surface, defined by c and ϕ, be known. There are a few types of slope failures which cannot be analysed even if the strength of material is known, because failure does not involve simple sliding. These cases are discussed below.

3.5 TOPPLING FAILURE (TOPPLES)

Toppling failure with its stereo plot is shown in Figure 3.1d. This mode of rock slope failure is explained here. Consider a block of rock resting on an inclined plane as shown in Figure 3.2a. Here the dimensions of the block are defined by height 'h' and base length 'b' and it is assumed that the force resisting the downward movement of the block is friction only, i.e. the cohesion is almost zero.

When the vector representing weight of the block W falls within the base b, sliding of the block will occur if the inclination of the plane ψ is greater than the

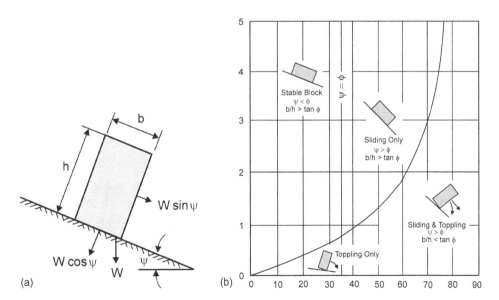

Figure 3.2. (a) Geometry of block on inclined plane. (b) Conditions for sliding and toppling of a block on an inclined plane (Hoek and Bray, 1981).

angle of friction ϕ. However, when the block is tall and slender (h > b), the weight vector W can fall outside the base b and, when this happens, the block will topple, i.e. it will rotate about its lowest contact edge (Hoek and Bray, 1981).

The conditions for sliding and or toppling for a rock block are defined in Figure 3.2b. The four regions in this diagram are defined as follows:

Region 1: $\psi < \phi$ and b/h > tan ϕ, the block is stable and will neither slide nor topple

Region 2: $\psi > \phi$ and b/h > tan ϕ, the block will slide but will not topple

Region 3: $\psi < \phi$ and b/h < tan ϕ, the block will topple but will not slide

Region 4: $\psi > \phi$ and b/h < tan ϕ, the block can slide and topple simultaneously.

Wedge toppling occurs along a rock wedge where a third joint set intersects the wedge and dips towards the hill side. Thus thin triangular rock wedges topple down successively. The process of toppling is slow.

3.6   RAVELLING SLOPES (FALLS)

Accumulation of screes or small pieces of rock detached from the rock mass at the base of steep slopes and cyclic expansion and contraction associated with freezing and thawing of water in cracks and fissures in the rock mass are the principal reasons of slope ravelling. A gradual deterioration of materials, which cement the individual rock blocks together, may also play a part in this type of slope failure.

Weathering or the deterioration of certain types of rock exposure will also give rise to the loosening of a rock mass and the gradual accumulation of materials on the surface, which falls at the base of the slope. It is important that the slope designer should recognize the influence of weathering on the nature of the materials with which he is concerned (see Section 2.12.5).

3.7   SLOPE HEIGHT AND GROUND WATER
EFFECT ON SLOPE ANGLE

Figure 3.3 illustrates the significant effect of slope height on stable slope angle for various modes of failure. The ground water condition also reduces the factor of safety. The University of Roorkee (now IIT Roorkee) has developed several software packages, SASP, SASW/WEDGE, SARC and SAST, for the analysis of planar, 3D wedge, circular and debris slides respectively (Singh and Anbalagan, 1997). A few deep-seated landslides such as planar and rotational are more catastrophic than millions of surfacial landslides along reservoir rims of dams. In the landslide hazard zonation, potential deep-seated landslides should therefore be identified.

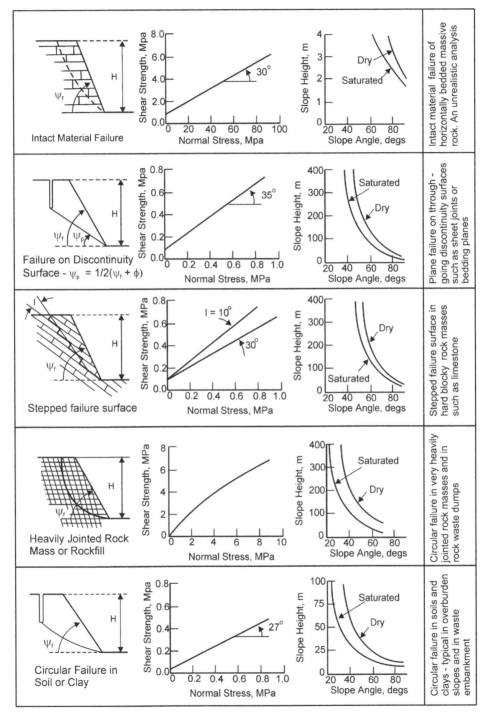

Figure 3.3. Slope angle versus height relationships for different materials (Hoek and Bray, 1981).

3.8   A BASIC LANDSLIDE CLASSIFICATION SYSTEM

The basic types of landslides/rockslides are summarized in Table 3.1.
   The landslides are defined as follows:

Debris slide	–	Sliding of debris or talus on rock slopes due to a temporary ground water table just after long rains.
Debris flow	–	Liquid flow of mixture of boulders, debris, clay and water along gully during rains or cloud burst.
Earth flow/mud flow	–	Liquid flow of mixture of soil, clay and water along a gully.

The landslide control measures may be selected from the last column in Table 3.1. Lien and Tsai (2000) have shown that the slit dams have been effective in trapping big boulders and retarding the debris flow in the Himalayas in China. The slit dam is like a check dam with many slits. They have given the design principles.

Table 3.1. Basic landslide classification system (Indian Standard Code).

Type of movement		Type of material			Recommended control measures
		Soils		Bedrock	
		Predominantly fine	Predominantly coarse		
Falls		Earth fall	Debris fall	Rock fall	Geotextile nailed on slope/spot bolting
Topples		Earth topple	Debris topple	Rock topple	Breast walls/soil nailing
Slides	Rotational	Earth slump	Debris slump	Rock slump	Alteration of slope profile and earth and rock fill buttress
	Translational	Earth block slide	Debris block slide	Rock block slide	Reinforced earth or rock reinforcement in rock slope
		Earth slide	Debris slide	Rock slide	Bio-technical measures
Lateral spreads		Earth spread	Debris spread	Rock spread	Check dams along gully
Flows		Earth flow	Debris flow	Rock flow	Series of check dams, slit dam
		(Soil creep)		(Deep creep)	Rows of deep piles
Complex		Combination of two or more principal types of movement			Combined system

3.9 CAUSATIVE CLASSIFICATION

Landslides may also be classified according to their causes as follows (Deoja et al., 1991).

1. Rainfall induced landslide – most of landslides and rock slides.
2. Earthquake induced landslides – generally rock falls and boulder jumping to long distances in hilly areas.
3. Cloudburst induced landslide – mostly mud flows, debris flows (and flash floods) along gullies in Himalayan region.
4. Landslide dam break – resulting in flash floods and large number of landslides due to the toe erosion along the hill rivers.
5. Glacial lake outburst flood (GLOF) is common in glaciated Himalayan ridges due to melting of nearby glaciers. Such a flood causes bank under-cutting, landslide and debris flows.
6. Freeze and thaw induced rock falls during sunny days in the snow bound steep Rocky Mountains.

Bhandari (1987) presented state of the art measures on landslides in the fragile Himalaya. He has also presented very economical landslide measures in the Himalayan region. Subsequently, Choubey (1998) has highlighted causes of rock slides in Himalaya and stressed upon the need of detailed field investigations at the sites of complex landslides.

3.10 A CLASSIFICATION SYSTEM OF LANDSLIDES

Hutchinson (1988) presented a detailed classification of landslides. It is a significant improvement over classification of Varnes (in Schuster and Krizek, 1978). It is surprising that there are so many types of different landslides. Table 3.2 lists a comprehensive classification system of landslides both for rocks and soils based on its slope movement. Figures 3.4–3.11 illustrate various modes of failure of rock and soil slopes. Recommended computer programs are also mentioned against various types of landslides. The authors' experience is that debris slides are most common along roads (Fig. 3.5). Engineers avoid generally landslide or landslide prone areas for hill development. Their interest lies in the development of a safe terrace system for at least 25 years. Therefore, site development is the real challenge. Experience of adjoining landslides gives the clue to the potential mode of failure.

3.11 LANDSLIDES IN OVER-CONSOLIDATED CLAYS

The expert advice is needed in tackling landslides in over consolidated clays. Progressive failure of slopes in clays and soft shales occur slowly. The slope failure may take place after say 30 years of temporary stability. It is recommended that

Table 3.2. Classification of sub-aerial slope movements (based principally on morphology with some account taken of mechanism, material and rate of movement; after Hutchinson, 1988).

A. Rebound (Fig. 3.4)
   Movements associated with:
   1. Man-made excavations
   2. Naturally eroded valleys
B. Creep
   1. Superficial, predominantly seasonal creeps; mantle creep:
      (a) Soil creep, talus creep (non-periglacial)
      (b) Frost creep and gelifluction of granular debris (periglacial)
   2. Deep-seated, continuous creep; mass creep
   3. Pre-failure creep; progressive creep
   4. Post-failure creep
C. Sagging of mountain slopes (Fig. 3.5)
   1. Single-aided sagging associated with the initial stages of landsliding:
      (a) Of rotational (essentially circular) type (R-sagging)
      (b) Of compound (markedly non-circular) type (C-sagging); (i) listric
          (CL), (ii) bi-planar (CB)
   2. Double-aided sagging associated with the initial stages of double
      landsliding, leading to ridge spreading:
      (a) Of rotational (essentially circular) type (DR-sagging)
      (b) Of compound (markedly non-circular) type (DC-sagging);
          (i) listric (DCL), (ii) bi-planner (DCB)
   3. Sagging associated with multiple toppling (T-sagging)
D. Landslides (Figs 3.6 and 3.7):
   1. Confined failures (Fig. 3.6)
      (a) In natural slopes
      (b) In man-made slopes
   2. Rotational slips:
      (a) Single rotational slips
      (b) Successive rotational slips
      (c) Multiple rotational slips
   3. Compound slides (markedly non-circular, with listric or bi-planar slip surfaces):
      (a) Released by internal shearing towards rear
          (i) In slide mass of low to moderate brittleness
          (ii) In slide mass of high brittleness
      (b) Progressive compound slides, involving rotational slip at rear and
          fronted by subsequent translational slide
   4. Translational slides (Fig. 3.7):
      (a) Sheet slides
      (b) Slab slides; flake slides
      (c) Peat slides
      (d) Rock slides:
          (i)   Planar slides; block slides
          (ii) Stepped slides
          (iii) Wedge failures
      (e) Slides of debris:
          (i) Debris-slides; debris avalanches (non-periglacial)
          (ii) Active layer slides (periglacial)
      (f) Sudden spreading failures

Table 3.2. (*continued*)

E. Debris movements of flow-like form (Fig. 3.8)
    1. Mudslides (non-periglacial):
       (a) Sheets
       (b) Lobes (lobate or elongate)
    2. Periglacial mudslides (gelifluction of very sensitive clays):
       (a) Sheets
       (b) Lobes (lobate or elongate, active and relict)
    3. Flow slides :
       (a) In loose, cohesionless materials
       (b) In lightly cemented, high porosity silts
       (c) In high porosity, weak rocks
    4. Debris flows, very rapid to extremely rapid flows of wet debris:
       (a) Involving weathered rock debris (except on volcanoes):
          (i) Hillslope debris flows
          (ii) Channelled debris flows; mud flows; mud-rock flows during heavy
              rains or cloud burst
       (b) Involving peat; bog flows, bog bursts
       (c) Associated with volcanoes; lahars:
          (i) Hot lahars
          (ii) Cold lahars
    5. Sturzstroms, extremely rapid flows of dry debris
F. Topples (Fig. 3.9)
    1. Topples bounded by pre-existing discontinuities:
       (a) Single topples
       (b) Multiple topples
    2. Topples released by tension failure at rear of mass
    3. Wedge toppling due to falling of thin triangular rock wedges slowly
G. Falls (Fig. 3.9)
    1. Primary, involving fresh detachment of material; rock and soil falls
    2. Secondary, involving loose material, detached earlier; stone falls
    3. Boulder jumping for long distances particularly just after earthquake
H. Complex slope movements (Figs 3.10 and 3.11)
    1. Cambering and valley bulging (Fig. 3.10)
    2. Block-type slope movements (Fig. 3.11)
    3. Abandoned clay cliffs
    4. Landslides breaking down into mudslides or flows at the toe:
       (a) Slump-earthflows
       (b) Multiple rotational quick-clay slides
       (c) Thaw slumps
    5. Slides caused by seepage erosion where ground water intersects a soil slope
    6. Multi-tiered slides
    7. Multi-storied slides

residual and drained shear strength parameters should be used in analysing the static stability of the clay slopes. In the dynamic analysis, peak undrained shear strength parameters should be used. It should be noted that orientation of platy clay particles takes place in a thin zone along the slip surface. As such, the

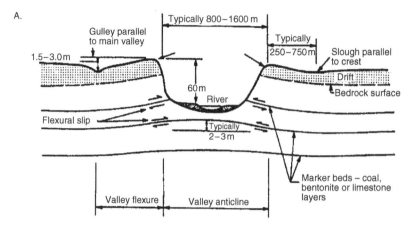

Figure 3.4. Valley rebound.

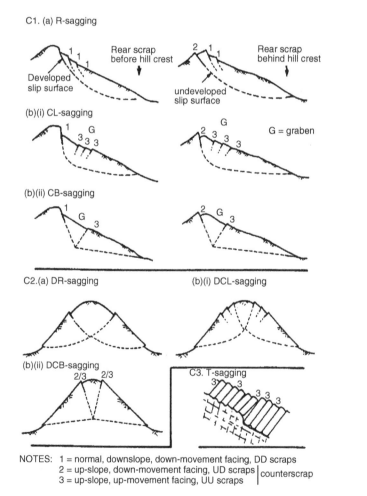

NOTES: 1 = normal, downslope, down-movement facing, DD scraps
2 = up-slope, down-movement facing, UD scraps | counterscrap
3 = up-slope, up-movement facing, UU scraps |

Figure 3.5. Main types of sagging (SANC is recommended for C1 and C2).

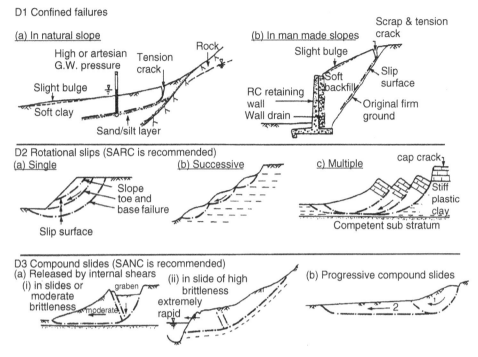

Figure 3.6. Main types of confined failure, rotational and compound slides.

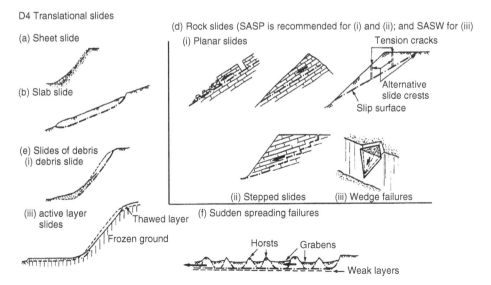

Figure 3.7. Main types of translational failure (SAST is recommended).

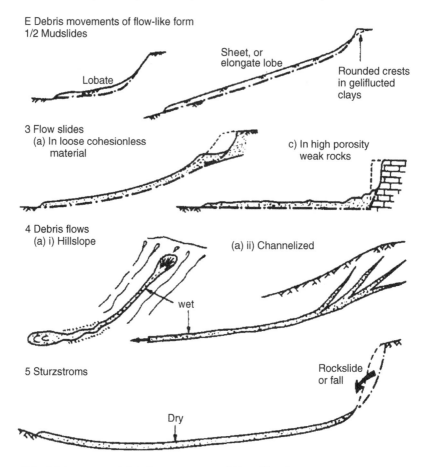

Figure 3.8. Main types of debris movement of flow-like form.

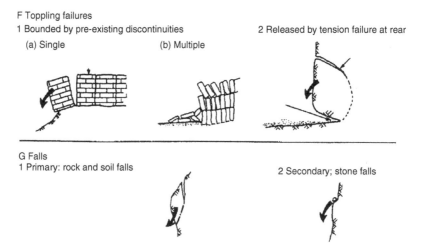

Figure 3.9. Main types of toppling failure and fall.

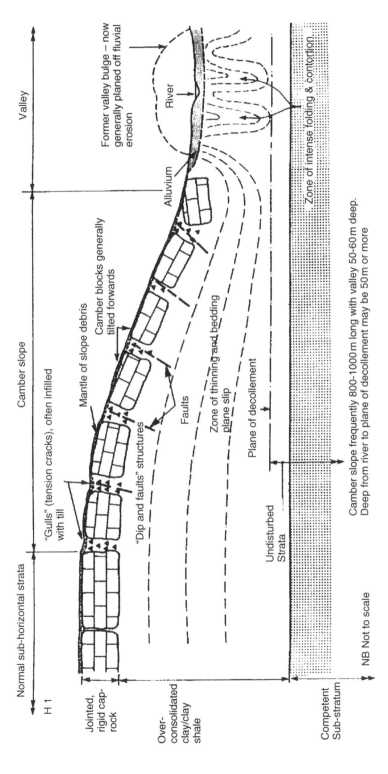

Figure 3.10. Schematic section of cambering and valley.

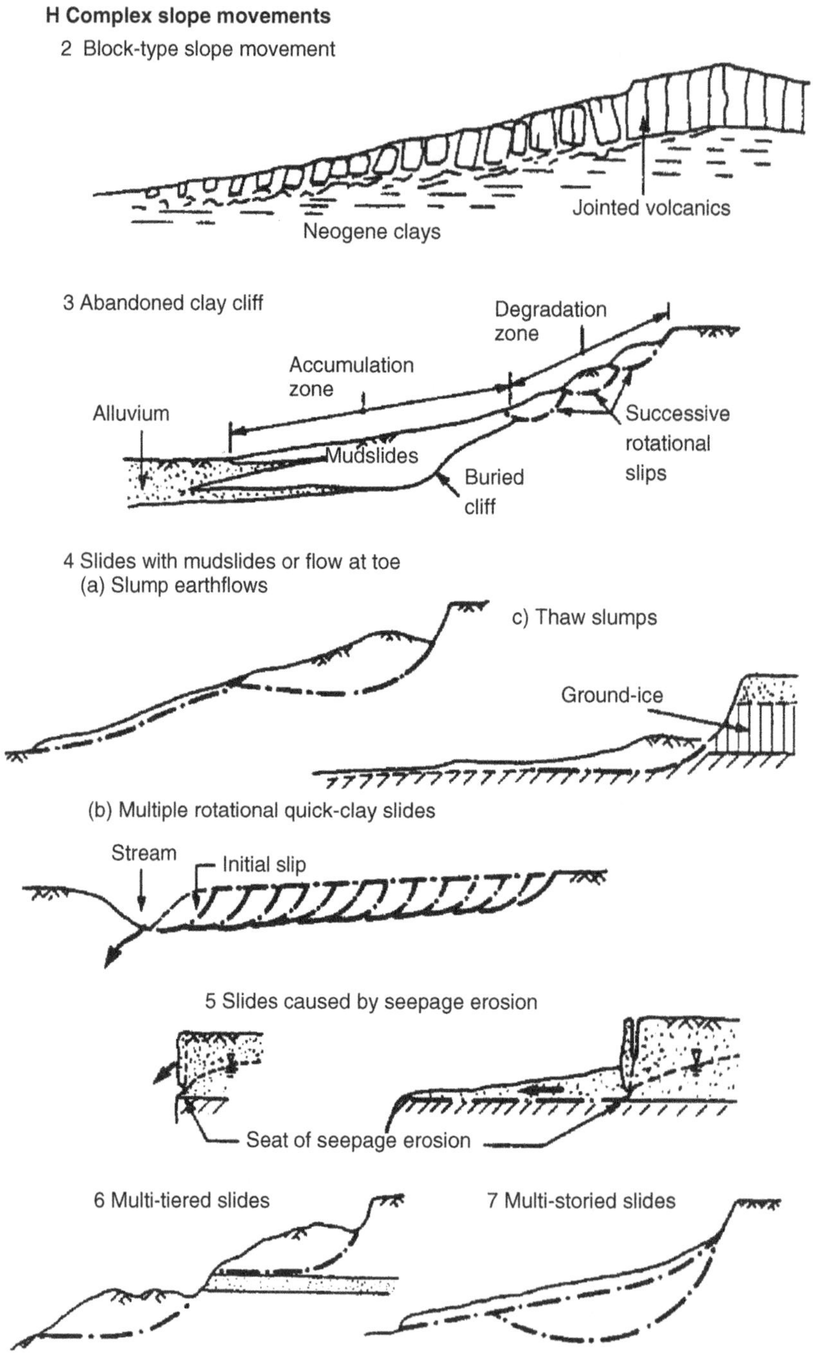

H Complex slope movements

2 Block-type slope movement

Jointed volcanics

Neogene clays

3 Abandoned clay cliff

Degradation zone

Accumulation zone

Alluvium

Mudslides

Buried cliff

Successive rotational slips

4 Slides with mudslides or flow at toe
(a) Slump earthflows

c) Thaw slumps

Ground-ice

(b) Multiple rotational quick-clay slides

Stream

Initial slip

5 Slides caused by seepage erosion

Seat of seepage erosion

6 Multi-tiered slides

7 Multi-storied slides

Figure 3.11. Some types of complex slope movements.

strength parameters along actual slip surface are lower significantly than those along any other assumed slip surface.

3.12 LANDSLIDE DAMS

Landslide dams are formed in steep valleys due to a deep-seated landslide in deforested hills. They are also created by a huge deposit of debris, which is brought by a network of gullies during cloud burst. The dam-river water quickly forms a reservoir submerging roads and houses (Bhandari, 1987; Choubey, 1998). The reservoir water may also enter in the tail race tunnels of nearby hydroelectric projects. This back-water has caused immense damage to the two underground powerhouses in Himalaya.

3.13 DO'S AND DON'TS IN LANDSLIDE AREAS

Minor landslide/rock falls are cleared soon by road and rail engineers. Major landslides need a cautious approach of clearance. Mud flows, debris flows and flash floods are disastrous to the people living downstream of gullies. Appendix 2 offers do's and don'ts for landslide clearance.

REFERENCES

Bhandari, R.K. 1987. Slope instability in the fragile Himalaya and strategy for development. *Indian Geotech. J.* 17(1): 1−78. Ninth IGS Annual Lecture.

Choubey, V.D. 1998. Landslide hazard assessment and management in Himalayas. *Int. Conf. Hydro Power Development in Himalayas*, Shimla, India, pp. 220−238.

Deoja, B., Dhital, M., Thapa, B. and Wagner A. 1991. *Mountain Risk Engineering Handbook*. Kathmandu, Nepal, International Centre of Integrated Mountain Development, Part I and Part II.

Hoek, E. and Bray, J.W. 1981. *Rock Slope Engineering*, Revised Third Edition. London, The Institution of Mining and Metallurgy, p. 358.

Hutchinson, J.N. 1988. Morphological and geotechnical parameters of landslides in relation to geology and hydrology, General Report. *Proc. 5th Int. Symp. Landslide*, Vol. 1, pp. 3−35.

Indian Standard Code on Landslide Control Guidelines. *Bureau of Indian Standards*, New Delhi, in Print.

Lien, hui-Pang and Tsai, Fang-Wu 2000. Debris flow control by using slit dams. *Int. J. Sedi. Res.* 15(4): 391−409.

Matheson, D.S. and Thomson, S. 1973. Geological implications of valley rebound. *Can. J. Earth Sci.* 10: 961−978.

Schuster, R.L. and Krizek, R.J. 1978. Landslides − Analysis and Control. National Academy of Sciences, Washington D.C., Special Report 176, 234 p.

Singh, Bhawani and Anbalagan, R. 1997. Evaluation of stability of dam and reservoir slopes – mechanics of landslide, seismic behaviour of ground and geotechnical structures. *Proc. Special Technical Session on Earthquake Geotechnical Engineering, XIV Int. Conf. on Soil Mech. and Foundation Engg.*, Hamburg, pp. 323–339.

Singh, B. and Goel, R.K. 1999. *Rock Mass Classification: A Practical Approach in Civil Engineering*, Chapter 16. Reproduced and revised, Permission from Elsevier Science Ltd., U.K. p. 268.

CHAPTER 4

A practical approach to stability analysis of slopes

"Everything should be made as simple as possible, but not simpler"
Albert Einstein

4.1 PHILOSOPHY

Slopes may be artificial, that is, man-made, as in cuttings and embankments for highways and railroads, earth dams, temporary excavations, spoil heaps and landscaping operations for development of sites, etc. Slopes can also be natural, as in hillside and valleys, coastal and river cliffs and so on. In all these cases, forces exist which tend to cause the soil/rock to move from high points to low points. In other words, there exists an inherent tendency in the slopes to assume a more stable configuration due to the law of potential energy. If there is only a tendency to move, it may be construed as instability. If actual movement of slope mass occurs, it is a slope failure.

The most important forces causing instability are the force of gravity and the force of seepage. In areas of seismic activity, earthquake forces may also be an important factor causing instability. The forces causing instability (called the actuating forces) induce shearing stresses throughout the slope mass. Unless the shearing resistance on every possible failure surface within the soil/rock mass is larger than the shearing stress, failure will occur in the form of mass movement of soil/rock along a slip/joint surface. The shearing resistance is derived mainly from the shear strength of the slope mass and from other natural factors such as roots of plants, lenses of ice, etc., which must be broken along the slip surface.

The stability analysis is made commonly by using a limit equilibrium approach. In this method, it is presumed that failure occurs along an assumed or known failure surface. The shearing resistance for maintaining a limiting equilibrium condition is compared with the available shearing strength of the slope mass. The ratio of the available shear strength to the mobilized shearing resistance is the average factor of safety along that particular slip surface.

The value of shear strength to be used in the stability analysis is, in many cases, not easy to determine. First, there is the problem of deciding whether to use the undrained shear strength or the drained shear strength in a given situation. The variation of shear strength with depth and with time has also to be considered. Also, the results of laboratory shear tests on soils and rock joints are likely to be vitiated if the test samples are not undisturbed and test conditions not properly controlled. Since the value of factor of safety is quite sensitive with respect to the shear strength, the necessity to exercise proper care and judgement in the estimation of shear strength cannot be overemphasized.

The analysis of stability requires the estimation of the actuating forces like the gravity forces, seepage forces and earthquake forces. In addition, a definition of the shape of the slope mass involved in failure is also a pre-requisite for the solution. It is relatively simple to define the shape of the failure surface, but several trials are needed to determine the critical failure surface which leads to the minimum value of the factor of safety.

The engineering practice is to assess the static factor of safety of a slope in the rainy season. It is also essential to estimate the dynamic settlement of slopes during an earthquake. It should be noted that the dynamic factor of safety less than 1 is allowed generally. The factor of safety (F) is defined as

$$F = \frac{\text{Shear strength along slip surface}}{\text{Shear stress along slip surface}} \tag{4.1}$$

In order to boost the confidence of project engineers and geologists, it is essential that economical and feasible control measures for arresting all the landslide are recommended. Thus, more attention is given to solution than to analysis of slopes.

4.2 A DEFINITION OF THE PROBLEM

In nature, a slope may fail in a complex manner or the slope failure looks complex. Landslides in general, however, show one dominant and characteristic mode of failure. A careful and detailed examination may indicate the dominant and characteristic mode of failure. Accordingly the landslide may be classified on the basis of observed mode of failure as follows.
1. Planar slide along a joint plane on a rock slope,
2. 3-D wedge slide along two joint planes on a rock slope,
3. Rotational slide along a cylindrical rupture in soil or rock slope, homogeneous earth dams and mine dumps,
4. Talus or debris slide along the contact of underlying rock bed, which is common in Himalaya and other fragile mountains,
5. Rock fall due to over-toppling of rock blocks, often seen on steep rock slopes having steep inward dipping joints. Rock falls damage nearby buildings during major earthquakes.
Varnes (Schuster and Krizek, 1978) classified landslides in more detail on the basis of type of the landslide and the nature of materials involved. At present, no realistic

mathematical model of such failures in seismic regions is available. This limitation should be kept in mind while applying a theory of landslide in nature.

4.3 AN APPROACH TO THE PROBLEM

There is another difficulty in applying analytical models in the field. It is related to collection of input data. Even in landslide prone areas, it is generally not possible to perform in situ tests for determining shear strength parameters of discontinuities as well as soil/rock mass. It is also hazardous to guess the future groundwater conditions within the slope during peak rainy season. It is also difficult to predict weathering in the future. Observations of deep pits show that density of soil increases with depth whereas the weathering of rock mass decreases rapidly with depth. That is why in situ direct-shear tests (near surface) give conservative shear strength parameters compared to those obtained by back analysis, except in rare cases of homogeneous slopes.

Thus, a new approach to the solution of this problem is required. Past experiences of many landslides show that shear strength parameters along slip surface may be obtained by back analysis of steep slopes in distress. Back analysis also provides a good check on other input parameters, e.g. ground water conditions and tension crack. If back analysis is not done correctly, the factor of safety of even a temporarily stable slope would be found to be less than unity.

The package of computer programs, therefore, also includes programs of back analysis of slopes for each mode of failure. Those with long experience advise the following approach to solve the problems of landslides (Deoja et al., 1991).
1. Back analyse slopes in distress or high and steep slopes,
2. Predict factors of safety of adjoining slopes in the same rock/soil mass (using strength parameters from the above back analysis),
3. Select and design the preventive measures for slope stability if factor of safety is not adequate.

Some idea of strength parameters may be found in published literature. Hoek and Bray (1981) have compiled peak and residual strength parameters along discontinuities. The typical strength parameters of rock mass may be picked up according to state of weathering and joint characteristics. Bieniawski (1984) also suggested strength parameters for rock mass according to their RMR (Rock Mass Rating). This is presented in Chapters 5 and 6.

4.4 LIMITATIONS OF ANALYTICAL MODELS

The following assumptions are generally made in limit state analysis of soil slopes:
1. Slope material is homogeneous and isotropic,
2. It obeys Coulomb's law of friction for both joints and rock/soil mass,

$$\tau = c + (\sigma - u)\tan \phi \tag{4.2}$$

where

τ = shear strength,

c = effective cohesion,

ϕ = effective angle of internal friction,

σ = normal stress across slip surface, and

u = pore water pressure.

3. Strength is uniformly mobilized along entire rupture surface. There is no progressive failure.
4. Tension crack is vertical.
5. Pseudo-static forces replace earthquake forces. (The horizontal component of earthquake acceleration in case of submerged slope and hill slopes in Himalaya is generally taken as 0.08 in rock and 0.25 in soil. It is taken as 0.10 to account for blasting vibrations. The earthquake acceleration is higher at the top of the slopes.)
6. In the case of submerged slopes, groundwater level is the same as river (reservoir) water level at least up to the slope surface. The earthquake forces act on the saturated soil and not on the submerged weight of soil.
7. In situ stresses do not affect stability of the slope.
8. Radius of curvature of slope in plan is infinite as compared to the height of slope.

Section 1.2 lists assumptions in the analysis of rock slopes.

4.5   THE MYTH OF LUBRICATION EFFECT DURING SATURATION

Terzaghi and Peck (1948) have shown beyond doubt that there is nothing like lubrication between rock/soil particles or joint surfaces during saturation in the rainy season. In reality, the water pressure (u) develops within soil pores or rock joint openings. As such, the shear strength of soil/rock mass decreases because of reduction in the effective stress $(\sigma - u)$ according to Equation 4.2. One can easily check that there is no significant difference in coefficient of friction of dry and wet joint surfaces. However, cohesion may be reduced drastically after saturation, particularly in clayey soils and rock masses with water sensitive minerals.

4.6   BACK ANALYSIS OF SLOPES

Back analysis of steep slopes in distress as discussed earlier will give the much needed strength parameters of a slope material. For that purpose, the mode of failure has to be identified initially. In case of slope failure, F is taken equal to 1.0 in Equation 4.1. The cohesion values are calculated for different values of angle of sliding friction (ϕ_j). Computer program BASP for planar slide is available for this purpose. A judicious choice of c_j and ϕ_j is then made on the basis of past experience.

Similarly, Equation 4.1 may be used for rotational slide to back-calculate the strength parameters by substituting F equal to 1.0. Computer program BASC is used for this purpose. However, one should select realistic values of strength parameters of soil/rock mass. It is very surprising to note that the back analysis of natural rotational slides gives a linear relationship between c and ϕ (Singh and Ramaswamy, 1979). Future research may provide its proof.

In the case of talus or debris slide, computer program BAST is also available for back-calculating strength parameters. Generally cohesion is small or negligible in talus deposit. So, the choice of angle of internal friction is not a difficult problem.

Back analysis is a very useful experience. It saves us from being too conservative as factor of safety cannot be less than 1.0. Many times nature provides indications on slope movement. Tilted trees and tension cracks are often seen at sites of distress (Fig. 4.1).

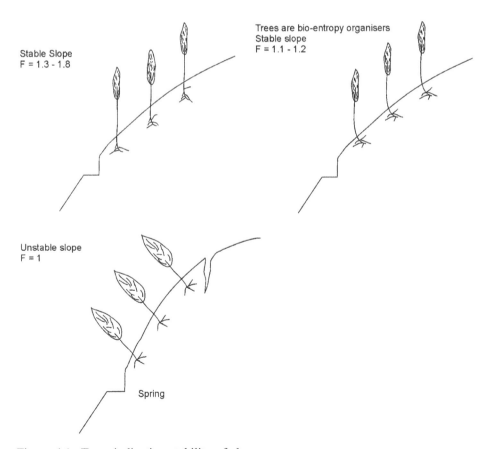

Figure 4.1. Trees indicating stability of slopes.

4.7   DYNAMIC SETTLEMENT DURING EARTHQUAKES

Experience of computations by various computer programs suggest surprisingly a good and simple correlation for the dynamic factor of safety F_{dyn} (Singh and Anbalagan, 1997),

$$F_{dyn} = F_{static}/(1 + 3.3\alpha_h) \tag{4.3}$$

where

F_{static} = static factor of safety, and

α_h = horizontal component of peak ground acceleration due to earth-quake.

The above correlation has been found for nearly dry natural soil and rock slopes. However, the constant 3.3 in Equation 4.3 may vary slightly for different slopes. The critical acceleration (α_{cr}) may be obtained approximately from the above correlation for a unit dynamic factor of safety. But, in the dry cohesionless soil, $\alpha_{cr} = \tan(\phi - \psi_f)$, where ψ_f is the dip of slope and ϕ is the angle of internal friction of the soil. In the planar slide of rock slope, $\alpha_{cr} = \tan(\phi_j - \psi_p)$, where ψ_p is the dip of continuous joint plane and ϕ_j is the sliding angle of friction of the joint plane.

It is important to understand that a slope may not fail even if the dynamic factor of safety is less than 1.0. It is because dynamic displacement of the wedge may not be significant. Slopes are subjected to much severe vibration and fracturing during blasting. But significant dynamic displacement does not occur because waves due to blasting contain only one peak, whereas earthquake waves contain many peaks.

Jansen (1990) has developed a simple semi-empirical equation for determining total dynamic settlement in meters during major earthquakes:

$$S_{dyn} = 5.8(0.1M)^8(\alpha_h - \alpha_{cr})/\sqrt{\alpha_{cr}} \tag{4.4}$$

< 1 m or 1 percent of height of slope for soil slopes

where

M = magnitude of design earthquake on Richter's scale,

α_{cr} = average coefficient of critical acceleration at factor of safety of unity, and

α_h = coefficient of peak ground acceleration in horizontal direction at the crest or near the crest of slope ($\alpha_h > \alpha_{cr}$ and $\alpha_h < 1$, $\alpha_v < 1$; 2m $>$ $S_{dyn} > 0$).

Further research is needed to develop appropriate co-relations for various types of landslides/rock slides. The magnitude of earthquake on Richter's scale (M) is determined from the amplitude of body waves which are recorded by seismographs within 100 km of the epicenter. The M is indirectly correlated to the magnitude of earthquake (M_s) on the basis of surface waves beyond 100 km but $M \ll M_s$.

According to Pederson et al. (1994), the observed coefficient of peak ground acceleration (PGA) due to earthquake is amplified at the rock slope crest between 2 and 2.5 times of that at the base of a rock slope. The amplification of PGA is more for thin debris on the rock slope, i.e., up to 3.5 times of that at the base of rock slope. Jansen (1990) also reported similar amplification for embankments. The natural frequency of the debris of thickness Z is $V_p/4Z$, where V_p is the P wave velocity. So the natural frequency is quite high for thin debris and chances of the resonance of debris is low during major earthquakes.

According to Krishna (1992), the observed PGA increases with earthquake magnitude up to 7M (on Richter's scale) and becomes constant at 0.70g up to 8M. The reason is that an epicentre is not a centre, but a long fault of the order of 500 km which could be generated by an earthquake of magnitude 8M. Hence, such a long fault may not add to PGA near the epicentre region.

All computer programs (SASP, SARC, SASW and SAST) calculate dynamic settlement, assuming the validity of Equation 4.4 in the natural soil and rock slopes approximately. It is an amazing experience that high earth dams have not failed during the period 1995–2000, when many major earthquakes occurred world-wide.

A simple experiment may be conducted to convince ourselves that a dynamic pulse causes only small dynamic displacement and not a failure of slope. Kindly place a paperweight on a tilted notebook and apply a dynamic pulse by hitting the notebook with your palm. You will see that the paperweight slides only a little due to instantaneous reduction in the factor of safety. Fear of dynamic settlement may be removed by demonstrating this experiment to all concerned.

Example
The static factor of safety of a 200 m-high natural soil slope is estimated to be 1.30 by computer program SARC. The design coefficient of horizontal component of an earthquake in this area is 0.4 and the design magnitude of an earthquake is 8M on Richter's scale. Estimate the dynamic settlement of slope approximately.

Solution
The dynamic factor of safety of slope is given by Equation 4.3,

$$F_{dyn} = \frac{1.30}{1 + 3.30 \times 0.40} = 0.56$$

Equation 4.3 also gives the following critical acceleration α_{cr} for a dynamic factor of safety of 1.0

$$\alpha_{cr} = \frac{1}{3.3} \left[F_{static} - 1 \right]$$
$$= \frac{1}{3.3} \left[1.30 - 1 \right] = 0.09$$

The approximate value of the dynamic settlement (S_{dyn}) of slope is found from Jansen's Equation 4.4 as follows

$$S_{dyn} = 5.8 \frac{(0.1 \times 8)^8 (0.40 - 0.09)}{\sqrt{0.09}}$$

$$= 1\,m\,(Safe)$$

It should be noted that this natural slope is predicted to be stable even under shaking by earthquake of magnitude 8M which reduces the dynamic factor of safety to as low as 0.56. Hence the importance of concept of dynamic settlement of slopes is more prominent.

4.8 THE VALUE OF JUDGEMENT

Harr (1966) pointed out that trees were better indicators of the stability of slopes than many computer programs of that time. Further experiences of the authors in Himalaya proved the same and these are summarized in Figure 4.1. A soil or rock slope is highly stable if tall trees aged more than 30 years are standing vertical. Its factor of safety is likely to be adequate and in between 1.3 and 1.8. In the case of creeping slope, trees are likely to be bent initially but become vertical after root reinforcement makes the slope stable. Its factor of safety may be taken to lie between 1.1 and 1.2. A slope is definitely in distress if many trees are inclined or a tension crack is seen at the top of a slope. If the seepage flow through cut slopes in rock or soil increases slowly with time, it is an indicator of landslide. Failure of slope is a fatigue failure generally, except in case of rock falls. Thus, engineers get an advance warning of slope failure for their stabilization.

REFERENCES

Bieniawski, Z.T. 1984. *Rock Mechanics Design in Mining and Tunnelling*. Rotterdam, A.A. Balkema, pp. 97–133.

Deoja, B., Dhital, M., Thapa, B. and Wagner, A. 1991. *Mountain Risk Engineering Handbook*, Part I, II and III. Kathmandu, International Centre of Integrated Mountain Development.

Harr, M.E. 1966. Personal Communication.

Hoek, E. and Bray, J.W. 1981. *Rock Slope Engineering*. London, Institution of Mining and Metallurgy, Revised Third Edition.

Jansen, R.B. 1990. Estimation of embankment dam settlement caused by earthquake. *Water Power and Dam Construction*. U.K., Wilmington Publishing House, pp. 35–40.

Krishna, Jai 1992. Seismic zoning maps in India. *Current Science*, India, 62(1&2): 17–23.

Pederson, H., Brun, L.B., Hatzfeld, D., Campillo, M. and Bard, P.V. 1994. Ground motion amplitude across ridges. *Bull. Seismol. Soc. Am.* 84(6): 1786–1800.

Schuster, R.L. and Krizek, R.J. 1978. Landslides – Analysis and Control. National Academy of Sciences, Washington D.C., Special Report 176, 234 p.

Singh, Bhawani and Anbalagan, R. 1997. Evaluation of stability of dam and reservoir slopes mechanics of landslide. 14th *Int. Conf. of Soil Mech. and Found. Eng.* Technical Session on Earthquake Geo-technical Engineering, pp. 323–339.

Singh, Bhawani and Ramaswamy, G. 1979. Back analysis of natural slopes for evaluation of strength parameters. *Proc. Int. Conf. Comp. Applic. Civil Eng.*, vol. 1, Roorkee, India.

Terzaghi, K. and Peck, R.B. 1948. *Soil Mechanics in Engineering Practice.* John Wiley & Sons, Second Edition, Section 17, p. 729.

CHAPTER 5

The strength of discontinuities

5.1 INTRODUCTION

Rock mass is a heterogeneous, anisotropic and discontinuous mass. When civil engineering structures like dams are founded on rock, they transmit normal and shear stresses on discontinuities in rock mass. Failure may be initiated by sliding along a joint plane near or along the foundation or along the abutments of dam. For a realistic assessment of the stability of structure, estimation of the shear resistance of a rock mass along any desired plane of potential shear or along the weakest discontinuity becomes essential. The shear strength of discontinuities depends upon the alteration of joints or the discontinuities, the roughness, the thickness of infillings or the gouge material or the moisture content, etc.

The mechanical difference between contacting and non-contacting joint walls will usually result in widely different shear strengths and deformation characteristics. In the case of unfilled joints, the roughness and compressive strength of the joint walls are important, whereas in the case of filled joints the physical and mineralogical properties of the gouge material separating the joint walls are of primary concern.

To quantify the effect of these on the strength of discontinuities, various researchers have proposed different parameters and correlations for obtaining strength parameters. Barton et al. (1974), probably for the first time, have considered joint roughness (J_r) and joint alteration (J_a) in their Q-system to take care of the strength of clay coated discontinuities in the rock mass classification. Later, Barton and Choubey (1977) defined two parameters, joint wall roughness coefficient (JRC) and joint wall compressive strength (JCS), and proposed an empirical correlation for friction of rock joints without fillings which can be used both for extrapolating and predicting shear strength data accurately.

5.2   JOINT WALL ROUGHNESS COEFFICIENT (JRC)

The wall roughness of a joint or discontinuity is a potentially very important component of its shear strength, especially in the case of undisplaced and interlocked features (e.g. unfilled joints). The importance of wall roughness declines as thickness of aperture filling or the degree of any previous displacement increases.

JRC$_o$ (JRC at laboratory scale) can be obtained by visual matching of actual roughness profiles with the set of standard profiles proposed by Barton and Choubey (1977). As such, the joint roughness coefficients are suggested for ten types of roughness profiles of joints (Fig. 5.1). The core sample will be intersected by joints at angles varying from 0 to 90° to the axis. Joint samples will therefore vary in some cases from a metre or more in length (depending upon the core length) to 100 mm (core diameter). Most samples are expected to be in the range of 100 to 300 mm in length.

The recommended approximate sampling frequency for the above profile matching procedure is 100 samples per joint set per 1000 m of core. The two most adverse prominent sets should be selected which must include the adverse joint set selected for J_r and J_a characterization.

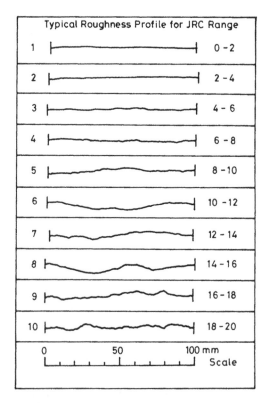

Figure 5.1. Standard profiles for visual estimation of JRC (Barton and Choubey, 1977).

Roughness amplitude per length, i.e., a and L measurements will be made in the field for estimating JRC_n (JRC at large scale). The maximum amplitude of roughness (in mm) should be usually estimated or measured on profiles of at least two lengths along the joint plane, for example 100 mm length and 1 m length.

It has been observed that the JRC_n can also be obtained from JRC_o using the following equation,

$$JRC_n = JRC_o (L_n/L_o)^{-0.02 JRC_o} \qquad (5.1)$$

where L_o is the laboratory scale length, i.e., 100 mm and L_n represents the larger scale length. The chart of Barton (1982) presented in Figure 5.2 is easier for evaluating JRC_n according to the amplitude of asperities and the length of joint profile studied in the field.

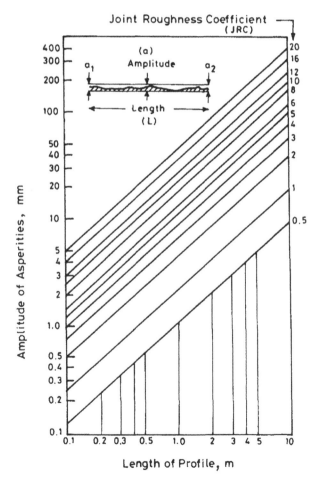

Figure 5.2. Assessment of JRC from amplitude of asperities and length of joint profile (Barton, 1982).

5.2.1   The relationship between J_r and JRC roughness descriptions

The description of roughness given in the Q-system by the parameter J_r, and the JRC are related. Figure 5.3 has been prepared by Barton (1993) for the benefit of users of these rock mass descriptions. The ISRM (1978) methods suggested for a visual description of joint roughness profiles have been combined with profiles given by Barton et al. (1980) and with Equation 5.1, to produce some examples of the quantitative description of joint roughness that these parameters provide.

The roughness profiles shown in Figure 5.3 are assumed to be at least 1 m in length. The column of J_r values would be used in Q-system, while the JRC values for 20 cm and 100 cm block size could be used to generate appropriate shear stress displacement and dilation − displacement curves.

5.3   JOINT WALL COMPRESSIVE STRENGTH (JCS)

The joint wall compressive strength (JCS) of a joint or discontinuity is a potentially very important component of its shear strength, especially in the case of undisplaced and interlocked discontinuities, e.g., unfilled joints (Barton and Choubey, 1977). As in the case of JRC, the wall strength JCS decreases as

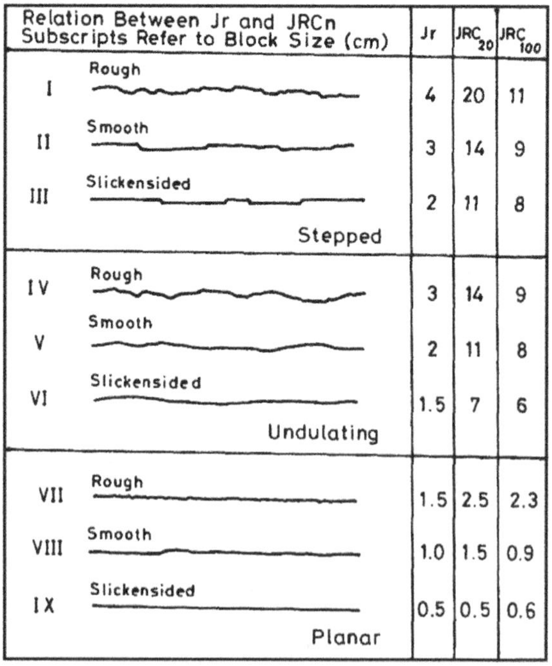

Relation Between Jr and JRCn Subscripts Refer to Block Size (cm)		Jr	JRC_{20}	JRC_{100}
I	Rough	4	20	11
II	Smooth	3	14	9
III	Slickensided	2	11	8
	Stepped			
IV	Rough	3	14	9
V	Smooth	2	11	8
VI	Slickensided	1.5	7	6
	Undulating			
VII	Rough	1.5	2.5	2.3
VIII	Smooth	1.0	1.5	0.9
IX	Slickensided	0.5	0.5	0.6
	Planar			

Figure 5.3. Suggested methods for the quantitative description of different classes of joints using J_r and JRC. Subscript refer to block size in cms.

aperture or filling thickness or the degree of any previous displacement increases. JCS, therefore, need not be evaluated for thickly (>10 mm) filled joints.

In the field, JCS is measured by performing Schmidt hammer (L-type) tests on the two most prominent joint surfaces where it is smooth and averaging the highest 10 rebound values. JCS_o, the small scale value of wall strength relative to a nominal joint length (L_o) of 100 mm, may be obtained from the Schmidt hammer rebound value (r) as follows, or by using Figure 5.4.

$$JCS_o = 10^{(0.00088r\gamma+1.01)} \text{ MPa} \tag{5.2}$$

where

    r = rebound number, and
    γ = dry density of rocks.

In case the Schmidt hammer is not used vertically downward, the rebound values need correction as given in Table 5.1.

The joint wall compressive strength may be equal to the uniaxial compressive strength of rock material for unweathered joints, otherwise it should be estimated indirectly from the Schmidt hammer index test. Experience has shown that the Schmidt hammer is found to give entirely wrong results on rough joints. Therefore, it is advisable not to use Schmidt hammer rebound for JCS in case of rough joints. Lump tests on saturated small lumps of asperities will give better UCS or JCS_o.

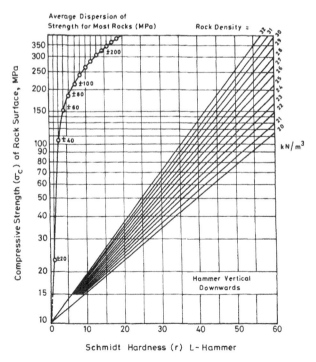

Figure 5.4. Correlation chart for compressive strength with rock density and Schmidt hammer rebound number on smooth surfaces (Miller, 1965).

Table 5.1. Corrections for the orientation of Schmidt hammer (Barton and Choubey, 1977).

Rebound r	Downward		Upward		Horizontal $\alpha = 0°$
	$\alpha = -90°$	$\alpha = -45°$	$\alpha = +90°$	$\alpha = +45°$	
10	0	-0.8	$-$	$-$	-3.2
20	0	-0.9	-8.8	-6.9	-3.4
30	0	-0.8	-7.8	-6.2	-3.1
40	0	-0.7	-6.6	-5.3	-2.7
50	0	-0.6	-5.3	-4.3	-2.2
60	0	-0.4	-4.0	-3.3	-1.7

For larger blocks or joint lengths (L_n), the value of JCS reduces to JCS_n, where the two are related by the following empirical equation:

$$JCS_n = JCS_o(L_n/L_o)^{-0.03JRC_o} \text{ MPa} \tag{5.3}$$

where JCS_n is the joint wall compressive strength at a larger scale.

5.4 JOINT MATCHING COEFFICIENT (JMC)

Zhao (1997) suggested a new parameter, joint matching coefficient (JMC), in addition to JRC and JCS for obtaining shear strength of joints. JMC may be obtained by observing the approximate percentage area in contact between the upper and the lower walls of a joint. Thus, JMC has a value between 0 and 1.0. A JMC value of 1.0 represents a perfectly matched joint, i.e., with 100 percent surface contact. On the other hand, a JMC value close to 0 (zero) indicates a totally mismatched joint with no or minimum surface contact.

5.5 THE ANGLE OF INTERNAL FRICTION

The residual friction angle ϕ_r of a joint is a very important component of its total shear strength, whether the joint is rock-to-rock interlocked or clay filled. The importance of ϕ_r increases as the clay coating or filling thickness increases of course up to a certain limit.

An experienced field observer may make a preliminary estimate of ϕ_r. The quartz rich rocks and many igneous rocks have ϕ_r between 28° and 32°, whereas, mica-rich rock masses and rocks having considerable effect of weathering have somewhat lower values of ϕ_r than mentioned above.

In the Barton–Bandis joint model it is proposed to add an angle of primary roughness for obtaining the field value of a peak friction angle for a natural joint (ϕ_j) without fillings

$$\phi_j = \phi_r + i + JRC \log_{10}(JCS/\sigma) < 70°; \quad \text{for } \sigma/JCS < 0.3 \tag{5.4}$$

where JRC accounts for secondary roughness in laboratory tests, 'i' represents angle of primary roughness (undulations) of natural joint surface and is generally $\leq 6°$ and σ is the effective normal stress across joint.

The expression [JRC $\log_{10}(\text{JCS}/\sigma)$] in the above equation represents approximately the dilation angle of a joint. It may be noted that at high pressures ($\sigma = \text{JCS}$), no dilatation will take place as all the asperities will get sheared. It may be noted here that the value of ϕ_r is important as roughness (JRC) and wall strength (JCS) reduce through weathering.

Residual frictional angle ϕ_r may also be estimated by the equation

$$\phi_r = (\phi_b - 20°) + 20(r/R) \tag{5.5}$$

where ϕ_b is the basic frictional angle obtained by sliding or tilt tests on dry, planar (but not polished) or cored surface of the rock (Barton and Choubey, 1977), R is the Schmidt rebound on fresh, dry unweathered smooth surfaces of the rock and r is the rebound on the smooth natural, perhaps weathered and water-saturated joints.

According to Jaeger and Cook (1969), enhancement in the dynamic angle of sliding friction ϕ_r of smooth rock joints may be about 2° only. For clay-coated joints, the sliding angle of friction (ϕ_j) is found to be

$$\phi_j = \tan^{-1}(J_r/J_a) \geq 14° \tag{5.6}$$

5.6 THE SHEAR STRENGTH OF JOINTS

Barton and Choubey (1977) have proposed the following non-linear correlation for shear strength of natural joints which is found surprisingly accurate.

$$\tau = \sigma \cdot \tan[\phi_r + \text{JRC}_n \log_{10}(\text{JCS}_n/\sigma)] \tag{5.7}$$

where τ is the shear strength of joints, JRC_n may be obtained easily from Figure 5.2, JCS_n from Equation 5.3 and the rest of the parameters are defined above.

The effect of mismatching of joint surface on its shear strength has been proposed by Zhao (1997) in his JRC-JCS shear strength model as

$$\tau = c_j + \sigma \cdot \tan[\phi_r + \text{JMC} \cdot \text{JRC}_n \log_{10}(\text{JCS}_n/\sigma)] \tag{5.8}$$

The minimum value of JMC in the above equation should be taken as 0.3. The cohesion along discontinuity is c_j.

Field experience shows that natural joints are not continuous as assumed in theory and laboratory tests. There are rock bridges in between them. The shear strength of these rock bridges add to the cohesion of overall rock joint (0−0.1MPa). The real discontinuous joint should be simulated in the theory.

In the case of highly jointed rock masses, failure takes place along the shear band (kink band) and not along the critical discontinuity. Thus, the value of JCS in a rock mass is suggested to be its uniaxial compressive strength q_{cmass}. More attention should be given to the strength of discontinuity in the jointed rock masses.

For joints filled with gouge, the following correlation of shear strength is used for low normal stresses (Barton and Bandis, 1990)

$$\tau = \sigma \cdot (J_r/J_a) \qquad (5.9)$$

Sinha and Singh (2000) have proposed an empirical criterion for shear strength of filled joints.

The angle of internal friction is correlated to the plasticity index (PI) of normally consolidated clays (Lamb and Whitman, 1979). The same may be adopted for thick and normally consolidated clayey gouge in the rock joints as follows:

$$\sin \phi_j = 0.81 - 0.23 \log_{10} PI \qquad (5.10)$$

Choubey (1998) suggested that the peak strength parameters should be used in the case of designing rock bolt system and retaining walls, where control measures do not permit large deformations along joints. For long-term stability of unsupported rock and soil slopes, residual strength parameters of rock joint and soil should be chosen in the analyses respectively as large displacement may reduce the shear strength of rock joint to its residual strength eventually.

Misra (1997) recommended the use of advanced model of rock joints in the finite element analysis of rock structures. His model is based on micro-mechanics of interaction between asperities.

5.7   THE DYNAMIC SHEAR STRENGTH OF ROUGH ROCK JOINTS

Jain (2000) performed a large number of dynamic shear tests on dry rock joints at Nanyang Technological University (NTU), Singapore. He observed that significant dynamic normal stress (σ_{dyn}) is developed across the rough rock joints. Hence there is high rise in the dynamic shear strength. Thus, the effective normal stress ($\bar{\sigma}$) in Equation 5.8 should be as follows:

$$\begin{aligned} \bar{\sigma}_{dyn} &= \sigma_{static} - u_{static} + \sigma_{dyn} - u_{dyn} \\ &\geq \bar{\sigma}_{static} \end{aligned} \qquad (5.11)$$

It is also imagined that negative dynamic pore water pressure (u_{dyn}) will develop in the water charged joints due to dilatancy. This phenomenon is likely to be similar to undrained shearing of dilatant and dense sand/ over consolidated clay. Further research is needed to develop correlations for σ_{dyn} and u_{dyn} from dynamic shear tests on rock joints. There is likely to be significant increase in the dynamic shear strength of rock joints due to shearing of more asperities.

5.8 THE NORMAL AND SHEAR STIFFNESS OF ROCK JOINTS

The values of static normal and shear stiffness are used in finite element method and distinct element method of analysis of rock structures. Appendix 3 lists their suggested values on the basis of experiences of back analysis of uniaxial jacking tests in USA and India.

Barton and Bandis (1990) have also found correlation for shear stiffness. The shear stiffness of joint is defined as the ratio between shear strength τ in Equation 5.7 above and the peak slip. The latter may be taken equal to $(S/500) (JRC/S)^{0.33}$, where S is equal to the length of a joint or simply the spacing of joints. The normal stiffness of a joint may be 10 to 30 times its shear stiffness. This is the reason why the shear modulus of jointed rock masses is considered to be very low as compared to that for an isotropic elastic medium (Singh, 1973). Of course the dynamic stiffness is likely to be significantly more than their static values.

REFERENCES

Barton, N. 1982. Shear strength investigations for surface mining, Ch. 7. *3rd Int. Conf. on Surface Mining*, Vancouver, SME 1982, pp. 171–196.

Barton, N. 1993. Predicting the behaviour of underground openings in rock, *Proc. Workshop on Norwegian Method of Tunnelling, CSMRS-NGI Institutional Co-operation Programme*, September, New Delhi, India, pp. 85–105.

Barton, N. and Bandis, S. 1990. Review of predictive capabilities of JRC-JCS model in engineering practice, Reprinted from: Barton, N.R. & O. Stephansson (eds), *Rock Joints Proc. of a Regional conference of the International Society for Rock Mechanics*, Leon, 4–6.6.1990. Rotterdam, A.A. Balkema, 820pp.

Barton, N. and Choubey, V.D. 1977. The shear strength of rock joints in theory and practice. *Rock Mechanics*. Springer-Verlag, No. 1/2, pp. 1–54. Also NGI-Publ. 119, 1978.

Barton, N., Lien, R. and Lunde, J. 1974. Engineering classification of rock masses for the design of tunnel support. *Rock Mechanics*, Vol. 6, No. 4., Springer-Verlag, pp. 189–236.

Barton, N., Loset, F., Lien, R. and Lunde, J. 1980. Application of Q-system in design decisions concerning dimensions and appropriate support for underground installations. *Int. Conf. on Sub-surface Space, Rock Store*, Stockholm, Sub-Surface Space, Vol. 2, pp. 553–561.

Choubey, V.D. 1998. Landslide hazard assessment and management in Himalayas. *Int. Conf. Hydro Power Development in Himalayas*, Shimla, India, pp. 220–238.

ISRM 1978. Suggested methods for the quantitative description of discontinuities in rock masses (Co-ordinator N. Barton). *Int. J. Rock Mech. Min. Sci. Geomech. Abstr.* 15: 319–368.

Jaeger, J.C. and Cook, N.G.W. 1969. *Fundamentals of Rock Mechanics*. Mathew and Co. Ltd., Section 3.4

Jain, Mukesh 2000. Personal communication.

Lamb, T.W. and Whitman, R.V. 1979. *Soil Mechanics* SI Version. Wiley Eastern Ltd., Section 21.1, p. 553.

Miller, R.P. 1965. *Engineering classification and index properties for intact rock*. Ph.D. Thesis, University of Illinois, USA, pp. 1–282.

Misra, Anil 1997. Mechanistic model for contact between rough surfaces. *J. Eng. Mech.*, ASCE 123(5): 475–484.

Sinha, U.N. and Singh Bhawani 2000. Testing of rock joints filled with gouge using a triaxial apparatus. *Int. J. rock Mech. Mining Sci.* 37: 963–981.

Singh, Bhawani 1973. Continuum characterization of jointed rock mass: Part II – significance of low shear modulus. *Int. J. Rock Mech. Min. Sci. Geomech. Abstr.* 10: 337–349.

Singh, B. and Goel, R.K. 1999. *Rock Mass Classification: A Practical Approach in Civil Engineering*. Chapter 14 reproduced and revised, Permission from Elsevier Science Ltd., U.K., p. 268.

Tse, R. and Cruden, D.M. 1979. Estimating joint roughness coefficients. *Int. J. Rock Mech. Min. Sci. Geomech. Abstr.* 16: 303–307.

Zhao, J. 1997. Joint surface matching and shear strength, Part B: JRC-JMC shear strength criterion. *Int. J. Rock Mech. Min. Sci. Geomech. Abstr.* 34(2): 179–185.

CHAPTER 6

Shear strength of rock masses in slopes

"Failure does not take place homogeneously in a material, but failure occurs by strain localization along shear bands, tension cracks in soils, rocks, concrete, masonry and necking in ductile material"
Prof. G.N. Pandey (1997)

6.1 MOHR-COULOMB STRENGTH PARAMETERS

Stability analysis of a rock slope requires assessment of shear strength parameters, i.e., cohesion c and angle of internal friction ϕ of the rock mass. Estimates of these parameters are usually not based on extensive field tests. Mehrotra (1993) has carried out extensive block shear tests to study the shear strength parameters of the rock masses. The following inferences may be drawn from the study of Mehrotra (1993):

1. RMR system may be used to estimate the shear strength parameters c and ϕ of the weathered and saturated rock masses. It was observed that the cohesion c and the angle of internal friction ϕ increase with the increase in RMR (Fig. 6.1).
2. The effect of saturation on shear strength parameters has been found to be significant. For poor saturated (wet) rock masses, a maximum reduction of 70 percent has been observed in cohesion c, while the reduction in angle of internal friction ϕ is of the order of 35 percent when compared to those for dry the rock masses.
3. Figure 6.1 shows that there is a non-linear variation of the angle of internal friction with RMR for dry rock masses. The study also shows that ϕ values of Bieniawski (1989) are somewhat conservative.

6.2 NON-LINEAR FAILURE ENVELOPES FOR ROCK MASSES

Dilatancy in a rock mass is unconstrained near slopes as normal stress on joints is fixed by weight of the wedge. Therefore, the failure of a rock mass occurs partially along joints and partially in non-jointed portions, i.e., solid rocks. But in

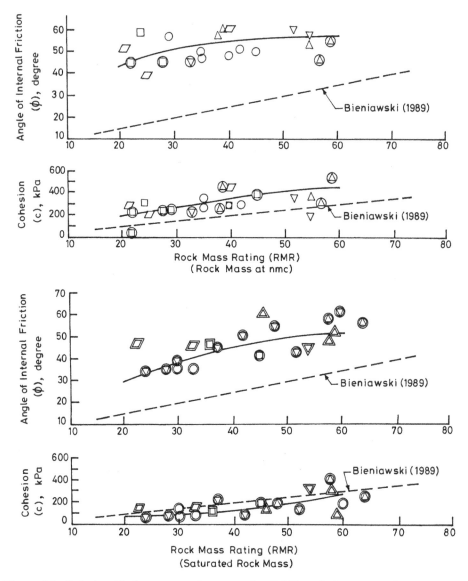

Figure 6.1. Relationship between rock mass rating RMR and shear strength parameters, cohesion c and angle of internal friction φ (Mehrotra, 1993) [nmc: natural moisture content].

massive rocks, it may occur entirely in solid rocks. Therefore, the failure of a rock mass lie within the area bounded by the failure envelope for a solid rock and that of a joint. The mode of failure thus depends on the quality and the type of the rock mass under investigation.

In the case of poor rock masses, the magnitude of normal stress σ influences the shear strength significantly. A straight-line envelope is therefore not a proper fit for such data and is likely to lead to over-estimation of angle of internal friction ϕ at higher normal stresses.

The failure envelopes for the rock masses generally show a non-linear trend. A straight line criterion may be valid only when loads are small ($\sigma \ll q_c$) which is generally not the case in civil engineering (hydroelectric) projects where the intensity of stresses is comparatively high. The failure envelopes based on generalized empirical power law may be expressed as follows (Hoek and Brown, 1980):

$$\tau = A(\sigma + T)^B \qquad (6.1)$$

where

  τ = tensile strength of rock mass,
  A, B and T = rock mass constants.

For known values of power factor B, constants A and T have been worked out from a series of block shear test data. Consequently, empirical equations for the rock masses, both at natural moisture content and at saturation, have been calculated for defining failure envelopes. The values of the power factor B have been assumed to be the same as in the equations proposed by Hoek and Brown (1980) for heavily jointed rock masses.

Mehrotra (1993) has plotted the Mohr envelopes for four different categories of rock masses namely: (i) limestones, (ii) slates, xenoliths, phyllites, (iii) metabasics, traps and (iv) sandstones and quartzites. One such typical plot is shown in Figure 6.2. The constants A and T have been estimated using the results obtained from the in situ block shear tests carried out on the lesser Himalayan rocks. Recommended non-linear strength envelopes (Table 6.1) may be used only for preliminary designs of dam abutments and rock slopes. There is scope of refinement if the present data are supplemented with in situ triaxial test data. For RMR > 60, shear strength will be governed by strength of rock material because failure plane will partly pass through solid rock.

Results of the study of Mehrotra (1993) for poor and fair rock masses are presented below.

Poor Rock Masses (RMR = 23 to 37)
1. It is possible to estimate the approximate shear strength from the data obtained from in situ block shear tests.
2. Shear strength of the rock mass is stress-dependent. The cohesion of the rock mass varies from 0.13 MPa to 0.16 MPa for saturated and about 0.22 MPa for naturally moist rock masses.
3. Beyond the normal stress σ value of 2 MPa, there is no significant change in the values of tan ϕ. It is observed that the angle of internal friction ϕ of rock mass is asymptotic at 20°.

Bieniawski (1989) has suggested that ϕ may decrease to zero if RMR reduces to zero. This is not borne out by field experience. Even sand has much higher angle of internal friction. Limited direct shear tests by The University of Roorkee (now

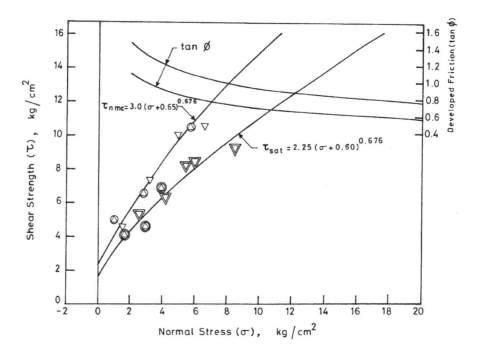

Figure 6.2. Failure envelopes for jointed trap and metabasic rocks at natural moisture content (nmc) and under saturated conditions.

I.I.T, Roorkee), India suggest that ϕ is above 15° for very poor rock masses (RMR = 0–20).

Fair Rock Masses (RMR = 41 to 58)
1. It is possible to estimate approximate shear strength from in situ block shear test data.
2. Shear strength of a rock mass is stress dependent. At natural moisture content the cohesion intercept of the rock mass is about 0.3 MPa. At saturation, the cohesion intercept varies from 0.23 to 0.24 MPa.
3. Beyond a normal stress (σ) value of 2 MPa, there is no significant change in the values of tan ϕ. It is observed that the angle of internal friction of a rock mass is asymptotic at 27°.
4. The effect of saturation on the shear strength is found to be significant. When saturated, the reduction in the shear strength is about 25 percent at the normal stress (σ) of 2 MPa.

6.3   THE STRENGTH OF ROCK MASSES IN SLOPES

1. E_d and q_{cmass} are significantly higher in deep tunnels than those near the ground surface and rock slopes for the same value of rock mass quality except near faults and thrusts.

Table 6.1. Recommended Mohr envelopes for slopes in jointed rock masses (Mehrotra, 1993).

S.No.	Rock type/quality	Limestone	Slate, Xenolith, Phyllite	Sandstone, Quartzite	Trap, Metabasics
1	Good rock mass RMR = 61–80 Q = 10–40	$\tau_n(nmc) = 0.38(\sigma_n + 0.005)^{0.669}$ $\tau_n(sat) = 0.35(\sigma_n + 0.004)^{0.669}$ [s = 1]	$\tau_n(nmc) = 0.42(\sigma_n + 0.004)^{0.683}$ $\tau_n(sat) = 0.38(\sigma_n + 0.003)^{0.683}$ [s = 1]	$\tau_n(nmc) = 0.44(\sigma_n + 0.003)^{0.695}$ $\tau_n(sat) = 0.43(\sigma_n + 0.002)^{0.695}$ [s = 1]	$\tau_n(nmc) = 0.50(\sigma_n + 0.003)^{0.698}$ $\tau_n(sat) = 0.49(\sigma_n + 0.002)^{0.698}$ [s = 1]
2	Fair rock mass RMR = 41–60 Q = 2–10	$\tau_{nmc} = 2.60(\sigma + 1.25)^{0.662}$ $\tau_{sat} = 1.95(\sigma + 1.20)^{0.662}$ [s = 1]	$\tau_{nmc} = 2.75(\sigma + 1.15)^{0.675}$ [$S_{av} = 0.25$] $\tau_{sat} = 2.15(\sigma + 1.10)^{0.675}$ [s = 1]	$\tau_{nmc} = 2.85(\sigma + 1.10)^{0.688}$ [$S_{av} = 0.15$] $\tau_{sat} = 2.25(\sigma + 1.05)^{0.688}$ [s = 1]	$\tau_{nmc} = 3.05(\sigma + 1.00)^{0.691}$ [$S_{av} = 0.35$] $\tau_{sat} = 2.45(\sigma + 0.95)^{0.691}$ [s = 1]
3	Poor rock mass RMR = 21–40 Q = 0.5–2	$\tau_{nmc} = 2.50(\sigma + 0.80)^{0.646}$ [$S_{av} = 0.20$] $\tau_{sat} = 1.50(\sigma + 0.75)^{0.646}$ [s = 1]	$\tau_{nmc} = 2.65(\sigma + 0.75)^{0.655}$ [$S_{av} = 0.40$] $\tau_{sat} = 1.75(\sigma + 0.70)^{0.655}$ [s = 1]	$\tau_{nmc} = 2.80(\sigma + 0.70)^{0.672}$ [$S_{av} = 0.25$] $\tau_{sat} = 2.00(\sigma + 0.65)^{0.672}$ [s = 1]	$\tau_{nmc} = 3.00(\sigma + 0.65)^{0.676}$ [$S_{av} = 0.15$] $\tau_{sat} = 2.25(\sigma + 0.60)^{0.676}$ [s = 1]
4	Very poor rock mass RMR < 21 Q < 0.5	$\tau_{nmc} = 2.25(\sigma + 0.65)^{0.534}$ $\tau_{sat} = 0.80(\sigma)^{0.534}$ [s = 1]	$\tau_{nmc} = 2.45(\sigma + 0.60)^{0.539}$ $\tau_{sat} = 0.95(\sigma)^{0.539}$ [s = 1]	$\tau_{nmc} = 2.65(\sigma + 0.55)^{0.546}$ $\tau_{sat} = 1.05(\sigma)^{0.546}$ [s = 1]	$\tau_{nmc} = 2.90(\sigma + 0.50)^{0.548}$ $\tau_{sat} = 1.25(\sigma)^{0.548}$ [s = 1]

$\tau_n = \tau/q_c$; $\sigma_n = \sigma/q_c$; σ in kg/cm²; $\tau = 0$ if $\sigma < 0$: S = degree of saturation [average value of degree of saturation is shown by S_{av}] = 1 for completely saturated rock mass.

2. The Hoek et al. (1992) criterion is applicable to rock slopes and open-cast mines with weathered and saturated rock masses. Block shear tests suggest q_{cmass} to be $0.38\gamma Q^{1/3}$ MPa ($Q < 10$), as joint orientation becomes a very important factor due to unconstrained dilatancy and negligible intermediate principal stress unlike in tunnels. So, block shear tests are recommended only for slopes and not for supported deep underground openings.
3. The angle of internal friction of rock masses with mineral coated joint walls may be assumed as $\tan^{-1}(J_r/J_a)$ approximately for low normal stresses.
4. In case of rock slopes both σ_2 and σ_3 are negligible, there is insignificant or no strength enhancement. As such, block shear tests on rock masses give realistic results for rock slopes and dam abutments only; because σ_2 is zero in these tests. It is most important that the blocks of rock masses are prepared with extreme care to represent the undisturbed rock mass.
5. In rock slopes, E_d is found lower due to complete relaxation of in situ stress, low confining pressures σ_2 and σ_3, excessive weathering and longer length of joints. For the same Q, q_{cmass} will also be low near rock slopes.

6.4   BACK ANALYSIS OF DISTRESSED SLOPES

The most reliable method of estimating strength parameters along discontinuities of rock masses is by appropriate back analysis of distressed rock slopes. Software packages BASP, BASC and BAST have been developed at The University of Roorkee to back-calculate strength parameters for planar, circular and debris slides respectively. The experience of careful back analysis of rock slopes also supports Bieniawski's values of strength parameters.

REFERENCES

Bieniawski, Z.T. 1989. *Engineering Rock Mass Classification*. John Wiley & Sons, p. 251.
Hoek, E. and Brown E.T. 1980. *Underground Excavations in Rock*. London, Institution of Mining and Metallurgy, Revised Edition.
Hoek, E. 1994. Strength of rock and rock masses. *ISRM News J.* 2: 4–16.
Hoek, E., Wood, D. and Shah, S. 1992. A modified Hoek-Brown failure criterion for jointed rock masses. *ISRM Symposium, Eurock '92 on Rock Characterization*, ed. J.A. Hudson, London, Thomas Telford.
Mehrotra, V.K. 1993. *Estimation of engineering parameters of rock mass*. Ph.D Thesis, University of Roorkee, Roorkee, India, 267 p.
Singh, B. and Goel, R.K. 1999. *Rock Mass Classification: A Practical Approach in Civil Engineering*. Chapter 15. Reproduced and Revised, Permission from Elsevier Science Ltd., U.K. p. 268.
Singh, Bhawani, Goel, R.K., Mehrotra, V.K., Garg, S.K. and Allu, M.R. 1998. Effect of intermediate principal stress on strength of anisotropic rock mass. *J. Tunnel. Underground Space Technol.* 13(1): 71–79.

CHAPTER 7

Stereographic analysis of joint data – STEREO

7.1 GENERAL

Joint is the general term for any mechanical discontinuity in a rock mass, along which the tensile strength is zero or negligible. It is also the general term for most type of discontinuities, bedding planes, weak schistocity planes, zones of weaknesses, shear zones and faults for engineering purposes.

The orientation of joints with respect to an engineering structure controls significantly the degree of instability of rock wedges or excessive deformations and movements. The importance of orientation of joints increases where other unfavourable conditions for deformation are present, such as low shear strength. The orientation of a joint is defined by the dip direction (α) measured clockwise from true north and by the dip (ψ_j or ψ_p), of the line of the steepest declination measured from the horizontal. The clinometer compass or the Brunton is used to measure true dip and its direction of dip (α). Figure 7.1 explains measurement of strike, true dip and dip direction with respect to three different directions of north.

The identification of total number of joint sets and their dips and dip directions is very important in the analysis of stability of wedges both in rock slopes and underground openings. As such, true dip and dip direction of a large number of exposed joints are measured at the site. Then the poles of these joints are plotted on the stereonet. Finally contours of pole concentration are drawn manually on the stereonet and true dip and dip direction of pole with maximum pole concentration is read from the stereonet. This exercise is done for all the joint sets. Help of a geologist should be taken in the survey and analysis of joint data.

The manual method of joint analysis is time-consuming and prone to mistakes. Zhang and Tong (1988) developed a computer program, STEREO, for automatic plotting of pole concentration. By full utilization of the property of the Wuff Stereonet projection, a large hemisphere is considered to reduce the errors in counting of poles in a circle.

In pole plot, it will print a dot up to 10 plotting, when more than 10 plotting points coincide, then it will give a symbol. In the concentration plot, numerical characters directly indicate percentage, and beyond 9 percent, alpha characters from A to Q represent 10 percent to 25 percent of the concentration, and above 25 percent at one point it will be shown by character V.

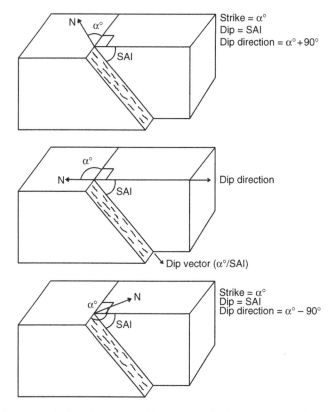

Figure 7.1. Diagram indicating the strike, dip and dip direction of three differently oriented planes.

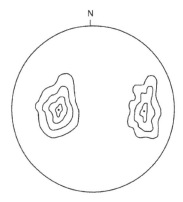

Figure 7.2. Output results of STEREO.

Sometimes there is difficulty in 3D imagination of orientation of joints. For fun, one may take a potato and cut it along two joints in the nature. The chip of the cutout potato may show the shape of actual rock wedge crudely. This may also give some idea of stability of a rock wedge.

7.2   USER'S MANUAL – STEREO

THIS PROGRAM DEALS WITH POLES PLOTTING ON AN EQUAL AREA MERIDINAL NET AND POLES CONCENTRATION PLOT (FIG. 7.2)

NAME OF PROGRAM–   STEREO.FOR
UNITS USED –   DEGREE EITHER FROM NORTH OR FROM HORIZONTAL

GIVE INPUT DATA IN FOLLOWING SEQUENCE

TITLE OF PROBLEM IN ONE LINE (< 80 CHARACTERS)
TOTAL SETS OF DISCONTINUITIES (NUMBER OF SITES)
NUMBER OF DISCONTINUITIES IN Ist SET (SITE)
DIP DIRECTIONS (DD) AND DIP AMOUNTS (DA) OF ALL JOINTS IN A SITE
PLEASE REPEAT ABOVE TWO LINES FOR OTHER SITES
ENTER   0 FOR TERMINATION
             1 FOR FURTHER HELP REGARDING OUTPUT
             2 FOR EXECUTION

OUTPUT DATA INCLUDE

  I: TITLE
 II: SET NUMBER OF DATA
III: TOTAL DATA OF DD AND DA
IV: POLE PLOT
 V: CONCENTRATION PLOT

CLUES REGARDING OUTPUT PLOTS
IN POLE PLOT
UPTO 10 PLOTTING IT WILL PRINT A SYMBOL •
WHEN MORE THAN 10 PLOTTING POINTS COINCIDE
THEN IT WILL GIVE A SYMBOL *
IN CONCENTRATION PLOT
NUMERICAL CHARACTER DIRECTLY INDICATES PERCENT
AND BEYOND 9 PERCENT ALPHA CHARACTERS FROM A TO Q
REPRESENT 10 PERCENT TO 25 PERCENT OF CONCENTRATION
AND ABOVE 25 PERCENT AT ONE POINT IT WILL PRINT V

(a) Input file name: STEREO.DAT
GRAPHICAL STEREO PLOTTING OF JOINT DATA

1		230	85	290	44	40	48
61		296	56	324	53	41	45
280	60	298	55	55	61	36	46
350	48	304	56	44	45	310	74
352	45	302	54	238	62	315	70
295	65	300	52	240	65	310	75
300	60	302	58	37	48	312	70
298	49	304	52	310	72	310	72
310	52	306	58	220	56	314	71
305	58	298	48	56	60	37	50
308	54	230	55	36	45	38	45
310	55	36	32	54	62	309	71
306	62	36	32	52	60		
305	64	310	73	51	64		

300	27	220	70	38	44
200	65	280	30	39	45
286	70	218	60	40	45

(b) Output file name: STEREO.OUT
GRAPHICAL STEREO PLOTTING OF JOINT DATA

INPUT FILE NAME –	STEREO.DAT
OUTPUT FILE NAME –	STEREO.OUT
UNITS USED –	DEGREE EITHER FROM NORTH OR FROM HORIZONTAL

DATA SET # 1

{DD#&DD#}		{DD#&DD#}		{DD#&DD#}		{DD#&DD#}	
280	60	230	85	290	44	40	48
350	48	296	56	324	53	41	45
352	45	298	55	55	61	36	46
295	65	304	56	44	45	310	74
300	60	302	54	238	62	315	70
298	49	300	52	240	65	310	75
310	52	302	58	37	48	312	70
305	58	304	52	310	72	310	72
308	54	306	58	220	56	314	71
310	55	298	48	56	60	37	50
306	62	230	55	36	45		
305	64	36	32	54	62		
300	27	36	32	52	60		
200	65	310	73	51	64		
286	70	220	70	38	44		
38	45	280	30	39	45		
309	71	218	60	40	45		

Please see the output data file STEREO.OUT for plots and contours. These plots are obtained on a printer of 132 columns in the natural scale. These are not reproduced here due to the large size of computer output.

REFERENCES

Deoja, B., Dhital, M., Thapa, B. and Wagner, A. 1991. *Mountain Risk Engineering Handbook*. Kathmandu, International Centre of Integrated Mountain Development, Part I, Chap. 7.

Hoek, E. and Bray, J.W. 1981. *Rock Slope Engineering*. London, Institution of Mining and Metallurgy, Revised Third Edition, Chapters 3 and 4.

Zhang, S. and Tong, G. 1988. Computerized pole concentration graphs using the wuff stereographic projection. *Int. J. Rock Mech. Min. Sci. Geomech. Abstr.* 25(1): 45–51.

CHAPTER 8

Planar slide

"Let us learn to live with landslide danger"
E. Hoek (1998)

8.1 GENERAL

The plane wedge slide occurs under gravity alone where rock mass rests on an inclined geological discontinuity, such as bedding/weakness plane that day-lights into the free slope ($\Psi_f > \Psi_p > \phi_j$). A caution should be exercised that a rock slope may fail by toppling if dip of cross joint is more than $90° - \Psi_f + \phi_j$, where Ψ_f is the slope angle and ϕ_j is the sliding angle of friction along joint plane.

8.2 DESIGN OF THE ROCK ANCHOR SYSTEM

The rock bolt or cable reinforcement system should be designed to stabilise the slip of the rock wedge (natural/artificial), which occurs due to the geological, climatic, hydrological, tectonic movements and human activities in the immediate and/or adjacent area of the structures.

For stabilising the rock slopes, cement grouted rock-anchors/bolts/cables should be used. The time interval between drilling the hole and its grouting should always be kept minimum and as a policy, one should always drill and grout the anchor on the same day. The shrinkage in the grout separating the intimate contact between the grout and rock provides a weaker surface. Hence, it should be ensured that the grout includes a small amount of a suitable expanding element. Unfortunately expanding admixture is forgotten and the consequence is a drastic reduction in the pull-out capacity of rock anchors.

Sometimes anchor bars are pulled out during a landslide because of insufficient fixed anchor length (FAL), which is the length of anchor bars beyond the critical discontinuity. The fixed anchor length should not be less than 60 and 100 times

Table 8.1. Fixed anchor length (FAL) for different rock conditions.

S.No.	Rock condition	Fixed anchor length (m)
1	Very good (RMR = 81 to 100)	2
2	Good (RMR = 61 to 80)	3
3	Fair/poor (RMR = 21 to 60)	4
4	Very poor (RMR = 0 to 20)	6

the diameter of deformed anchor bars and plain anchor bars respectively. The fixed anchor length should not be less than 2, 3, 4 and 6 m for very good, good, fair/poor and very poor rock conditions respectively (Table 8.1).

The diameter of the drill-hole should be about 30 mm more than the diameter of the anchor so that sufficient grout-cover prevents the corrosion. Suitable preventive measures should be adopted to prevent corrosion depending upon the aggressive nature of rock mass.

For the analysis of the stability of slope, partial factors of safety for each parameter should be used instead of using the single factor of safety. A higher factor of safety should be applied to the ill-defined parameters such as water pressure and cohesive strength. A low factor of safety should be applied to those quantities (weight of wedge), which are known to a greater degree of precision. The following partial factors of safety should be used for rock slopes (Hoek and Bray, 1981):

F_c = 1.5 (for cohesive strength),

$F\phi$ = 1.2 (for frictional strength),

F_w = 1.0 (for weight of wedge),

F_s = 1.5 (for steel and grout anchor), and

F_v = F_u = 2 (for water pressure).

These partial factors of safety are nearly the same as recommended by Eurocode 7 for soils (Ovesen, 2000).

A drainage system should be provided if the drained slope has a static factor of safety more than 1.2, and a dynamic factor of safety more than 1.0. In the case of smaller factors of safety, both rock anchors and drainage systems should be provided. A drainage system may consist of drain holes of 38 mm diameter dipping 5° towards the valley, at a spacing of 3 m × 3 m. The rolled tube of (geo) wire net should be inserted into the drainage holes to prevent their choking.

There should be a catch drain at the toe of the cut slope to drain off water. If seepage is occurring even in normal season and increasing with time, it is an indication that slope will give way soon. Drainage is then immediately required.

Rock reinforcement is needed if the static factor of safety and the dynamic factor of safety are less than 1.2 and 1.0 respectively even after complete drainage. An unreinforced rock slope or cut may be considered stable if the static factor of safety is more than 1.2 and dynamic displacement is less than say 0.2 percent of height of slope or 20 cm, whichever is less.

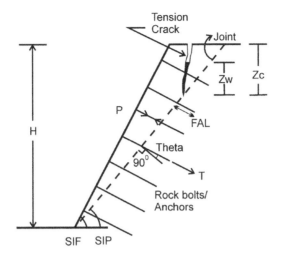

Figure 8.1. Input variable for SASP.

8.3 ANALYSIS OF A SLOPE WITH PLANAR SLIDE – SASP

Hoek and Bray (1981) have suggested a theory of design of the rock anchor system. The software SASP (stability analysis of slope with planar slide) has been developed on their theory and the above requirements. It is suggested that all the anchor bars should be of the same length so that there is no problem of supervision during construction. Figure 8.1 shows the definition of input parameters for the SASP program.

The SASP software is based on Barton and Bandis' (1990) theory of the shear strength of joints. It is assumed here, however, that $\phi_j \leq 45°$ in weathered joints in nature. The weakness of Hoek and Bray's (1981) theory is that the depth of tension crack (Zc) is predicted to be equal to height of slope (H) where the angle of slope is vertical. In nature $Zc < 2H/3$. Both of the above checks have been included in SASP.

The computer program also designs the rock anchor system. In some cases the program gives too often long anchors which are not available in the market (say 12 m). Then, the spacing of anchors may be reduced proportionately. Thus, the effective number of anchor bars are the same.

8.4 USER'S MANUAL – SASP

THIS PROGRAM IS FOR THE STABILITY ANALYSIS OF ROCK SLOPE WITH PLANE WEDGE FAILURE AND ROCK REINFORCEMENT (FIG. 8.1).

NAME OF PROGRAM –	SASP.FOR
UNITS USED –	TONNE – METER – DEGREE

GIVE INPUT DATA IN THE FOLLOWING SEQUENCE

NO
TITLE OF PROBLEM IN ONE LINE (< 80 CHARACTERS)
H,SIP,ZWR,AH,GAMA,GAMAW,ZC,SIF,ANCH,EQM
CJ,PHIR,JRC,JCS
FU,FW,FPHI,FC,THETA,P,FAL,QA,DS
REPEAT ABOVE FOUR LINES NO TIMES

DO YOU WANT HELP REGARDING DEFINITIONS OF VARIABLES USED
ENTER   0 FOR TERMINATION
        1 FOR FURTHER HELP
        2 FOR EXECUTION

NO	=	NUMBER OF SLOPES
H	=	HEIGHT OF SLOPE
SIP	=	DIP OF JOINT PLANE
ZWR	=	ZW/ZC
ZW	=	DEPTH OF WATER IN TENSION CRACK
AH	=	COEFF. OF HORIZONTAL ACCELERATION OF EARTHQUAKE NEAR CREST OF SLOPE
EQM	=	CORRESPONDING EARTHQUAKE MAGNITUDE (RICHTER SCALE)
GAMA	=	UNIT WEIGHT OF ROCK
GAMAW	=	UNIT WEIGHT OF WATER
SIF	=	SLOPE ANGLE
ZC	=	DEPTH OF TENSION CRACK (IF 0, PROGRAM WILL CALCULATE IT)
ANCH	=	0.0 MEANS THAT ANCHOR SYSTEM IS NOT TO BE DESIGNED
	=	1.0 MEANS RATIONAL DESIGN OF ANCHORS
	=	2.0 RECOMMENDATION FOR CONSTRUCTION
CJ	=	COHESION ALONG JOINT PLANE
PHIR	=	RESIDUAL SLIDING ANGLE OF FRICTION ALONG JOINT + AVERAGE ANGLE OF LARGE SCALE ROUGHNESS (0°−6°)
JRC	=	JOINT ROUGHNESS COEFF. (SMALLER SCALE ROUGHNESS)
JCS	=	JOINT-WALL COMPRESSIVE STRENGTH
FU	=	FACTOR OF SAFETY FOR HYDRAULIC FORCES
FW	=	FACTOR OF SAFETY FOR WEIGHT
FPHI	=	FACTOR OF SAFETY FOR FRICTION
FC	=	FACTOR OF SAFETY FOR COHESION
THETA	=	ANGLE OF ANCHOR W.R.T. NORMAL OF JOINT PLANE
P	=	SAFE ANCHOR CAPACITY
FAL	=	FIXED ANCHOR LENGTH
QA	=	ALLOWABLE BEARING PRESSURE
DS	=	DIAMETER OF ANCHOR

(a) Input file name: ISASP.DAT

2
STABILITY ANALYSIS OF ROCK SLOPE WITH PLANAR FAILURE
30.,35.,1.,0.1,2.5,1.,0.0,70.,2.,7.
10.,25.,15.,150.
1.,1.,1.2,1.2,55.,30.,3.,50.,.04
STABILITY ANALYSIS OF ROCK SLOPE WITH PLANAR FAILURE AND ANCHORS

30.,35.,1.,0.,2.5,1.,0.0,45.,2.,0.
0.,25.,5.,150.
1.,1.,1.2,1.2,55.,30.,3.,50.,.04

(b) Output file name: OSASP.DAT

STABILITY ANALYSIS OF ROCK SLOPE WITH PLANAR FAILURE

UNITS USED –   TONNE – METER – DEGREE
INPUT FILE NAME –   ISASP.DAT
OUTPUT FILE NAME –   OSASP.DAT

Case No. 1

COHESION	=	10.0000
RESIDUAL ANGLE OF FRICTION	=	25.0000
JOINT ROUGHNESS COEFFICIENT	=	15.0000
JOINT WALL COMP. STRENGTH	=	150.0000
HEIGHT	=	30.0000
DIP OF JOINT PLANE	=	35.0000
DEPTH OF WATER IN TENSION CRACK	=	14.8551
COEFF. OF HORIZONTAL ACCELERATION	=	0.1000
FOR EARTHQUAKE MAGNITUDE (RICHTER SCALE)	=	7.0000
UNIT WEIGHT OF ROCK	=	2.5000
UNIT WEIGHT OF WATER	=	1.0000
DEPTH OF TENSION CRACK	=	14.8551
SLOPE ANGLE	=	70.0000

UNREINFORCED SLOPE MAY FAIL BY OVERTOPPLING IF CONTINUOUS CROSS
JOINT DIPS MORE THAN 54 DEGREES

STATIC FACTOR OF SAFETY	=	1.0851
DYNAMIC FACTOR OF SAFETY	=	0.9068
DYNAMIC SETTLEMENT IN METRE	=	0.2604
CRITICAL ACCELERATION	=	0.0126
FACTOR OF SAFETY–DRAINED SLOPE	=	1.7697
DYNAMIC FACTOR OF SAFETY-DRAINED SLOPE	=	1.4752
SLIDING ANGLE OF FRICTION	=	39.9587

PROVIDE 38 MM DIAMETER DRAIN HOLES (DIP = 5 DEGREES) AT SPACING OF
3M × 3M WITH ROLLED GEOTEXTILE FABRIC/PERFORATED PIPES TO ENSURE
GOOD DRAINAGE

STABILITY ANALYSIS OF ROCK SLOPE WITH PLANAR FAILURE AND ANCHORS

UNITS USED –   TONNE – METER – DEGREE
INPUT FILE NAME –   ISASP.DAT
OUTPUT FILE NAME –   OSASP.DAT

Case No. 2

COHESION	=	0.0000
RESIDUAL ANGLE OF FRICTION	=	25.0000
JOINT ROUGHNESS COEFFICIENT	=	5.0000
JOINT WALL COMP. STRENGTH	=	150.0000
HEIGHT	=	30.0000

DIP OF JOINT PLANE	=	35.0000
DEPTH OF WATER IN TENSION CRACK	=	4.8965
COEFF. OF HORIZONTAL ACCELERATION	=	0.0000
FOR EARTHQUAKE MAGNITUDE (RICHTER SCALE)	=	0.0000
UNIT WEIGHT OF ROCK	=	2.5000
UNIT WEIGHT OF WATER	=	1.0000
DEPTH OF TENSION CRACK	=	4.8965
SLOPE ANGLE	=	45.0000

UNREINFORCED SLOPE MAY FAIL BY OVER-TOPPLING IF CONTINUOUS CROSS JOINT DIPS MORE THAN 77 DEGREES

STATIC FACTOR OF SAFETY	=	0.5896
DYNAMIC FACTOR OF SAFETY	=	0.5896
DYNAMIC SETTLEMENT IN METRE	=	0.0000
CRITICAL ACCELERATION	=	−0.2986
FACTOR OF SAFETY − DRAINED SLOPE	=	0.8971
DYNAMIC FACTOR OF SAFETY-DRAINED SLOPE	=	0.8971
SLIDING ANGLE OF FRICTION	=	32.1362

CASE OF REINFORCED SLOPE:		
ALLOWABLE BEARING PRESSURE	=	50.0000
DIAMETER OF ANCHOR	=	0.0400
FACTOR OF SAFETY FOR HYDRAULIC FORCES	=	1.0000
FACTOR OF SAFETY FOR WEIGHT	=	1.0000
FACTOR OF SAFETY FOR FRICTION	=	1.2000
FACTOR OF SAFETY FOR COHESION	=	1.2000
ANGLE OF ANCHOR W.R.T. NORMAL OF JOINT	=	36.7801
SAFE ANCHOR CAPACITY	=	30.0000
FIXED ANCHOR LENGTH	=	4.0000

UNIFORM LENGTH OF ANCHORS CREATE A REINFORCED ROCK BREAST WALL WHICH CAN STABILISE UNSTABLE ROCK MASS

LENGTH FOR ANCHOR NO. 1	=	13.1128
LENGTH FOR ANCHOR NO. 2	=	13.1128
LENGTH FOR ANCHOR NO. 3	=	13.1128
LENGTH FOR ANCHOR NO. 4	=	13.1128
LENGTH FOR ANCHOR NO. 5	=	13.1128
LENGTH FOR ANCHOR NO. 6	=	13.1128
LENGTH FOR ANCHOR NO. 7	=	13.1128
LENGTH FOR ANCHOR NO. 8	=	13.1128
LENGTH FOR ANCHOR NO. 9	=	13.1128
LENGTH FOR ANCHOR NO. 10	=	13.1128
LENGTH FOR ANCHOR NO. 11	=	13.1128
LENGTH FOR ANCHOR NO. 12	=	13.1128
LENGTH FOR ANCHOR NO. 13	=	13.1128

DYNAMIC FACTOR OF SAFETY	=	1.1402
DYNAMIC FACTOR OF SAFETY − DRY SLOPE	=	1.4692
SPACING OF ANCHORS (SQ. PATTERN)	=	3.0023
SIZE OF BASE PLATE/SLAB	=	0.7746
TOTAL ANCHOR PULL (T)	=	130.6878

8.5 RESERVOIR SLOPE WITH PLANAR SLIDE – SARP

Planar slides occur due to toe-cutting or the erosion by the river flow along its concave curve (outer bank). This happens particularly during a high flood or a flash flood. Seepage erosion has also been observed to be the cause of the toe erosion in rock slopes in soluble rocks like limestone. The toe cutting reduces the factor of safety of a planar slide significantly. (Apparently, toe-cutting may also undermine seriously the stability of a 3D-wedge failure.) Toe-cutting has resulted in the washing out of several kilometres of riverside roads in Himalaya within a short period of five years.

Thus, the computer program SASP has been extended to analyse a planar slide with toe-cutting, and subsequently, program SARP has been developed. Figure 8.2 shows the definition of input parameters for SARP. It is prudent to assume a toe cut of 1−2 m along riverside slopes for probable instability in future. Observations along a river can also reveal the possible extent of toe-cutting.

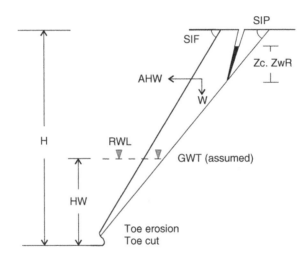

Figure 8.2. Input variable for SARP.

8.6 USER'S MANUAL – SARP

THIS PROGRAM IS FOR THE STABILITY ANALYSIS OF RESERVOIR SLOPE WITH PLANE WEDGE FAILURE BECAUSE OF TOE-CUTTING (FIG. 8.2).

NAME OF PROGRAM –	SARP.FOR
UNITS USED –	TONNE – METER – DEGREE

GIVE INPUT DATA IN THE FOLLOWING SEQUENCE

NO
TITLE OF PROBLEM IN ONE LINE (< 80 CHARACTERS)

H,SIP,ZWR,AH,GAMA,GAMAW,ZC,SIF,HW,TOECUT,EQM
CJ,PHIR,JRC,JCS
PLEASE REPEAT ABOVE THREE LINES NO TIMES

DO YOU WANT HELP REGARDING DEFINITION OF VARIABLES USED
ENTER   0 FOR TERMINATION
        1 FOR HELP
        2 FOR EXECUTION

NO	=	NUMBER OF SLOPES
H	=	HEIGHT OF SLOPE
SIP	=	DIP OF JOINT PLANE
ZWR	=	ZW/ZC
ZW	=	DEPTH OF WATER IN TENSION CRACK
AH	=	COEFFICIENT OF HORIZONTAL ACCELERATION OF EARTHQUAKE NEAR CREST OF SLOPE
EQM	=	CORRESPONDING EARTHQUAKE MAGNITUDE (RICHTER SCALE)
GAMA	=	UNIT WEIGHT OF ROCK
GAMAW	=	UNIT WEIGHT OF WATER
ZC	=	DEPTH OF TENSION CRACK (IF 0, PROGRAM WILL CALCULATE IT)
SIF	=	SLOPE ANGLE
HW	=	HEIGHT OF RESERVOIR WATER LEVEL ABOVE TOE
CJ	=	COHESION ALONG JOINT PLANE
PHIR	=	RESIDUAL ANGLE OF FRICTION ALONG JOINT + AVERAGE ANGLE OF WAVINESS OF LARGE SCALE ROUGHNESS ($0°-6°$)
JRC	=	JOINT ROUGHNESS COEFF. (SMALLER SCALE ROUGHNESS)
JCS	=	JOINT WALL COMPRESSIVE STRENGTH
TOECUT	=	HORIZONTAL DEPTH OF TOE EROSION IN SOLUBLE ROCK DUE TO SEEPAGE/TOE CUTTING BY RIVER IN FLOOD

(a) Input file name: ISARP.DAT

1
STABILITY ANALYSIS OF RESERVOIR SLOPE WITH PLANAR FAILURE &
TOE CUTTING
20.,65.,1.,0.05,2.35,1.,2.5,75.,5.0,2.,7.
8.5,25.,0.,100.

(b) Output file name: OSARP.DAT

STABILITY ANALYSIS OF RESERVOIR SLOPE WITH PLANAR FAILURE &
TOE CUTTING

UNITS USED –	TONNE – METER – DEGREE
INPUT FILE NAME –	ISARP.DAT
OUTPUT FILE NAME –	OSARP.DAT

Case No. 1

COHESION   =   8.5000

RESIDUAL ANGLE OF FRICTION	=	25.000
JOINT ROUGHNESS COEFFICIENT	=	0.0000
JOINT WALL COMP. STRENGTH	=	100.00
HORIZONTAL DEPTH OF TOE EROSION	=	2.0000
HEIGHT	=	20.000
DIP OF JOINT PLANE	=	65.000
DEPTH OF WATER IN TENSION CRACK	=	2.5000
COEFF. OF HORIZONTAL ACCELERATION	=	0.0500
FOR EARTHQUAKE MAGNITUDE ON RICHTER SCALE	=	7.0000
UNIT WEIGHT OF ROCK	=	2.3500
UNIT WEIGHT OF WATER	=	1.0000
DEPTH OF TENSION CRACK	=	2.5000
SLOPE ANGLE	=	75.000
HEIGHT OF RESERVOIR WATER ABOVE TOE	=	5.0000

UNREINFORCED SLOPE WILL FAIL BY OVER-TOPPLING IF CONTINUOUS
CROSS JOINT DIPS MORE THAN 37 DEGREES

CASE BEFORE SUBMERGENCE:		
STATIC FACTOR OF SAFETY	=	1.1183
DYNAMIC FACTOR OF SAFETY	=	1.0704
DYNAMIC SETTLEMENT OF SLOPE IN METER	=	0.0000
CRITICAL ACCELERATION	=	0.1279
WEIGHT OF WEDGE	=	183.8041
SLIDING ANGLE OF FRICTION	=	25.0000
CASE AFTER SUBMERGENCE:		
STATIC FACTOR OF SAFETY	=	1.2039
DYNAMIC FACTOR OF SAFETY	=	1.1506
DYNAMIC SETTLEMENT OF SLOPE IN METER	=	0.0000
CRITICAL ACCELERATION	=	0.2056
WEIGHT OF WEDGE	=	171.3246
SLIDING ANGLE OF FRICTION	=	25.0000

8.7 BACK ANALYSIS OF A SLOPE WITH PLANAR SLIDE – BASP

Experience of back analysis suggests that there is some long-term cohesion along natural discontinuous joints. No empirical theory is available to the best knowledge of the authors to estimate the cohesion along joints. Along continuous joints like bedding planes, however, cohesion is zero. The empirical theory of Barton and Bandis (1990) is accepted internationally to be valid in field also. Zhao (1997) has suggested to adopt the modified JRC which is the product of JRC and JMC for natural joints. Difficulty has been experienced in finding out the values of JRC, JMC and JCS particularly for weathered rock joints (Chapter 5).

The JCS may be determined approximately by conducting lump strength index tests on small lumps of asperities of a rock joint. The approach of back analysis is therefore suggested to assess the cohesion and modified JRC for a known JCS. In fact judgement has to be exercised to select realistic values of long-term cohesion and modified JRC from many possible values obtained from the back analysis, as unfortunately single back analysis does not give unique strength parameters. Nevertheless, back analysis of a distressed rock slope whose factor of safety tends to be unity saves us from being over-conservative.

Thus, computer program SARP has been rewritten to develop program BASP to back analyse planar slides due to toe-cutting along a river. Normal planar slide is a special case with no toe cutting or toe-erosion. The output of the program BASP is a set of values for cohesion (c_j) back analysed for different sliding angles of friction (ϕ_j) along the critical joint. The plot between c_j and ϕ_j is a convex curve as expected (Fig. 8.3). In case of continuous joint or bedding plane, $c_j = 0$. For a discontinuous and clean joint, a low value of cohesion ($1-10 \text{ T/m}^2$) may be assumed to read the corresponding value of ϕ_j. For a filled up joint, ϕ_j may be assumed approximately equal to $\tan^{-1}(J_r/J_a)$ and corresponding c_j may be read from Figure 8.3.

The success of the back analysis depends on the judgement on probable causes of distress or failure of a slope. A discussion with site engineers and geologists may indicate if distress has occurred a few years back or a long time back. In recent landslides the cause of distress (e.g. tension crack at top and increase in seepage in rains) might be rains. On the other hand, a long past history of distress is indicative of cause of distress probably due to an earthquake in seismic hilly areas. Both these forces may also contribute to instability of slope.

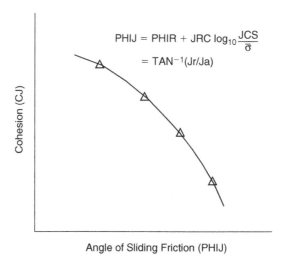

Figure 8.3. Output results of BASP.

8.8   USER'S MANUAL – BASP

THIS PROGRAM IS FOR THE BACK-ANALYSIS OF ROCK SLOPE WITH PLANE WEDGE FAILURE (FIG. 8.3).

NAME OF THE PROGRAM –   BASP.FOR

UNITS –   TONNE – METER – DEGREE

GIVE INPUT DATA IN THE FOLLOWING SEQUENCE

TITLE OF PROBLEM IN ONE LINE (< 80 CHARACTERS)
H,SIP,ZWR,AH,GAMA,GAMAW,ZC,SIF,TOECUT
JCS,F
NPHI
(PHIR(I),JRC(I),I = 1,NPHI)

DO YOU WANT HELP REGARDING DEFINITION OF VARIABLES USED
ENTER   0 FOR TERMINATION
         1 FOR HELP
         2 FOR EXECUTION

H	=	HEIGHT OF SLOPE
SIP	=	DIP OF JOINT PLANE
ZWR	=	ZW/ZC
AH	=	COEFF. OF HORIZONTAL ACCELERATION OF EQ.
GAMA	=	UNIT WEIGHT OF ROCK
GAMAW	=	UNIT WEIGHT OF WATER
ZC	=	DEPTH OF TENSION CRACK
		(IF 0, PROGRAM WILL CALCULATE IT)
ZW	=	DEPTH OF WATER IN TENSION CRACK
SIF	=	SLOPE ANGLE
JCS	=	JOINT-WALL COMPRESSIVE STRENGTH
F	=	FACTOR OF SAFETY OF EXISTING SLOPE
NPHI	=	NO OF SETS OF POSSIBLE VALUES OF PHIR AND JRC
PHIR	=	RESIDUAL ANGLE OF FRICTION ALONG JOINT
		+AVERAGE ANGLE OF WAVINESS OF LARGE SCALE ROUGHNESS (0°–6°)
JRC	=	JOINT ROUGHNESS COEFF. (SMALLER SCALE ROUGHNESS)
TOECUT	=	HORIZONTAL DEPTH OF TOE EROSION IN SOLUBLE ROCK DUE TO SEEPAGE OR TOE CUTTING BY RIVER DURING FLOOD

(a) Input file name: IBASP.DAT

BACK ANALYSIS OF ROCK SLOPE WITH PLANAR FAILURE AND TOE CUTTING
30.,35.,1.,0.,2.5,1.,0.,70.,2.
150.,1.
5
25.,5.,30.,6.,35.,7.,40.,8.,45.,9.

(b) Output file name: OBASP.DAT

BACK ANALYSIS OF ROCK SLOPE WITH PLANAR FAILURE AND TOE CUTTING

UNITS USED –   TONNE – METER – DEGREE

INPUT FILE NAME –   IBASP.DAT
OUTPUT FILE NAME –   OBASP.DAT

HEIGHT			=	30.0000
DIP OF JOINT PLANE			=	35.0000
DEPTH OF WATER IN TENSION CRACK			=	14.8551
COEFF. OF HORIZONTAL ACCELERATION			=	0.0000
UNIT WEIGHT OF ROCK			=	2.5000
UNIT WEIGHT OF WATER			=	1.0000
DEPTH OF TENSION CRACK			=	14.8551
SLOPE ANGLE			=	70.0000
JOINT WALL COMPRESSIVE STRENGTH			=	150.0000
CJ	PHIJ	PHIR	JRC	
13.0013	29.40	25.00	5.00	
10.1561	35.28	30.00	6.00	
6.8636	41.16	35.00	7.00	
2.9194	47.04	40.00	8.00	
−2.0059	52.92	45.00	9.00	

UNREINFORCED SLOPE WILL FAIL BY OVER-TOPPLING IF CONTINUOUS CROSS
JOINT DIPS MORE THAN 73 DEGREES

8.9   SAFE ANGLE OF A CUT SLOPE – ASP

In nature, there are many dip slopes. Dip slope means the dip of slope which
is the same as a dip of joint or bedding plane. However, it may be too costly to suggest an angle of cut slope equal to the dip angle of joint which is flatter than 45°.

The computer program ASP is a modified version of SASP used to calculate
the safe angle of cut slope in the area of potential planar slide for a desired factor
of safety. Further stability may be achieved by draining the potentially unstable
rock slopes. As such, long drainage holes of 38 mm diameter are made and
wrapped geomembrane is inserted in them to prevent their choking. In bi-annual
maintenance, these geomembranes are pulled out, cleaned, and again inserted in
the drainage holes.

Hoek and Bray (1981) reported their experience on the effect of curvature (in
plan) of rock slope on the safe angle of a cut slope in the open-cast mines. This
experience is confirmed in lower Himalaya also. Their recommendation has been
incorporated in the computer program ASP. Figure 8.4 shows the curved surface
of a planar slide.

Hence, program ASP may be used in designing slopes in the open-cast mines
also. The vibrations due to heavy rock blasting may be taken into account by
assigning the coefficient of horizontal component of acceleration between 0.1
and 0.3.

Small concrete gravity dams have been built in area of potential planar slide.
Not only abutment had been excavated to a safe slope angle, but cable anchors

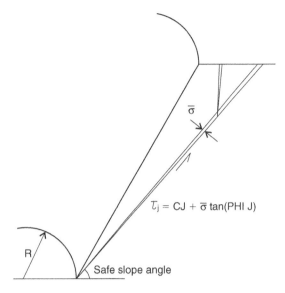

$$T_j = CJ + \overline{\sigma} \tan(PHI\ J)$$

Figure 8.4. Input variables for ASP.

have been installed across the unfavourably oriented bedding planes to prevent movement of risky abutment. The choice of earth and rockfill dam may be better in the highly seismic hilly areas. Obviously concrete arch dams are not recommended in potentially planar slide area with the present state of art of Rock Mechanics.

8.10 USER'S MANUAL – ASP

THIS PROGRAM CALCULATES THE OPTIMUM ANGLE OF ROCK CUT SLOPE WITH PLANE WEDGE FAILURE (FIG. 8.4).

NAME OF PROGRAM –   ASP.FOR
UNITS USED –   TONNE – METER – DEGREE

GIVE INPUT DATA IN THE FOLLOWING SEQUENCE

TITLE OF PROBLEM IN ONE LINE (<80 CHARACTERS)
H,SIP,ZWR,AH,GAMA,GAMAW,ZC
CJ,PHIR,JRC,JCS
FA,FADYN,R

DO YOU WANT HELP REGARDING DEFINITION OF VARIABLES USED
ENTER   0 FOR TERMINATION
            1 FOR FURTHER HELP
            2 FOR EXECUTION

H	=	HEIGHT OF SLOPE
SIP	=	DIP OF JOINT PLANE
ZWR	=	ZW/ZC

ZW	=	DEPTH OF WATER IN TENSION CRACK
AH	=	COEFF. OF HORIZONTAL ACCELERATION OF EARTH-QUAKE.
GAMA	=	UNIT WEIGHT OF ROCK
GAMAW	=	UNIT WEIGHT OF WATER
ZC	=	DEPTH OF TENSION CRACK
		(IF 0, PROGRAM CALCULATES IT)
CJ	=	COHESION
PHIR	=	RESIDUAL ANGLE OF FRICTION ALONG JOINT
		+AVERAGE ANGLE OF WAVINESS OF LARGE SCALE ROUGHNESS (0°–6°)
JRC	=	JOINT ROUGHNESS COEFF. (SMALLER SCALE ROUGHNESS)
JCS	=	JOINT WALL COMPRESSIVE STRENGTH
FA	=	ALLOWABLE STATIC FACTOR OF SAFETY
FADYN	=	ALLOWABLE DYNAMIC FACTOR OF SAFETY FOR DRY SLOPE
R	=	RADIUS OF CURVATURE OF SLOPE IN PLAN

(a) Input file name: IASP.DAT

OPTIMUM CUT SLOPE ANGLE OF ROCK SLOPE WITH PLANAR FAILURE
80., 35., 1., 0., 2.5, 1., 0.0
6., 25., 5., 150.
1.2, 1.0, 150.

(b) Output file name: OASP.DAT

OPTIMUM CUT SLOPE ANGLE OF ROCK SLOPE WITH PLANAR FAILURE

UNITS USED –   TONNE – METER – DEGREE
INPUT FILE NAME –   IASP.DAT
OUTPUT FILE NAME –

COHESION	=	6.0000
RESIDUAL ANGLE OF FRICTION	=	25.0000
JOINT ROUGHNESS COEFF.	=	5.0000
JOINT WALL COMP. STRENGTH	=	150.0000
HEIGHT	=	80.0000
DIP OF JOINT PLANE	=	35.0000
DEPTH OF WATER IN TENSION CRACK	=	1.4633
COEFF. OF HORIZONTAL ACCELERATION	=	0.0000
UNIT WEIGHT OF ROCK	=	2.5000
UNIT WEIGHT OF WATER	=	1.0000
DEPTH OF TENSION CRACK	=	1.4633
SLIDING ANGLE OF FRICTION	=	29.4046
OPTIMUM SLOPE ANGLE	=	47.5000
STATIC FACTOR OF SAFETY	=	1.1940
DYNAMIC FACTOR OF SAFETY (DRY SLOPE)	=	1.2235
WEIGHT OF WEDGE	=	3421.3570
WATER THRUST IN CRACK	=	1.0706
UPLIFT IN JOINT	=	100.1783

DYNAMIC NORMAL LOAD ON JOINT	=	2701.8160
STATIC NORMAL LOAD ON JOINT	=	2802.6080
RADIUS OF CURVATURE OF SLOPE	=	150.0000

REFERENCES

Barton, N. and Bandis, S.C. 1990. Review of predictive capabilities of JRC-JCS model in engineering practice. *Proc. Int. Symp. Rock Joints*, Norway, pp. 603–610.

Deoja, B., Dhital, M., Thapa, B. and Wagner, A. 1991. *Mountain Risk Engineering Handbook*. Kathmandu, International Centre of Integrated Mountain Development, Part I, Chaps 10 and 13 and Part II, Section 24.9.

Jansen, R.B. 1990. Estimation of embankment dam settlement caused by earthquake. *Water Power and Dam Construction*, December, pp. 35–40.

Hoek, E. and Bray, J.W. 1981. *Rock Slope Engineering*. London, Institution of Mining and Metallurgy, Revised Third Edition, Chaps 5 and 7.

Schuster, R.L. and Krizek, R.J. 1978. *Landslides – Analysis and Control*. Washington, D.C., National Academy of Sciences, Special Report 176, 234 p.

Ovesen, N.K. 2000. *Workshop on limit analysis in design codes of geotechnical engineering*. University of Roorkee, February 24, p. 70.

Zhao, J. 1997. Joint surface matching and shear strength, Part B: JRC-JMC shear strength criterion. *Int. J. Rock Mech. Mining Sci.* 34(2): 179–185.

CHAPTER 9

A simple analysis of 3D rock wedge

9.1 GENERAL

The failure of (three-dimensional) rock wedge occurs when two discontinuities strike obliquely across the slope face and their line of intersection daylights in the slope face. The rock wedge resting on these discontinuities may slide down along the line of intersection, provided that the inclination of this line is significantly greater than the sliding angle of friction. Alternatively the wedge may slide on any plane of discontinuity. Further, the wedge may fail by floating due to high seepage pressures on the joints.

9.2 A SIMPLE ANALYSIS OF HOEK AND BRAY (1981)

A wedge is considered with a horizontal slope crest and with no tension crack. Each plane may have a different sliding angle of friction and cohesion. The influence of water pressure on each plane is included in the solution. The influence of an external force is not considered in this solution. The solution presented is for the computation of the factor of safety for translational slip of a tetrahedral wedge which is formed in a rock slope by two intersecting discontinuities, the slope face and the upper ground surface. It does not take into account of rotational slip or toppling, nor does it include a consideration of those cases in which more than two intersecting discontinuities isolate tetrahedral or tapered wedges of rock. In other words, the influence of a tension crack is not considered in this solution. The rigorous solution is presented in Chapter 10.

When a pair of discontinuities are selected at random from a set of field data, it is not known whether,
1. The planes could form a wedge (the line of intersection may plunge too steeply to daylight the slope face or it may be too flat to intersect the upper ground surface).
2. One of the planes overlies the other.
3. One of the planes lies to the right or the left of the plane when viewed from the bottom of the slope.

In order to resolve these uncertainties, the simple solution (Hoek and Bray, 1981) has been derived in such a way that either of the planes may be labeled 1 (or 2) and allowance has been made for one plane overlying the other. In addition, a check on whether the two planes do form a wedge is included in the solution at an early stage. Depending upon the geometry of the wedge and the magnitude of the water pressure acting on each plane, contact may be lost on either plane and this contingency is provided in the solution.

The geometry of the wedge problem is shown in Figure 9.1. The discontinuities are denoted by 1 and 2, the upper ground surface is shown by 3 and slope face by 4. A computer program SASW is developed on the basis of simple

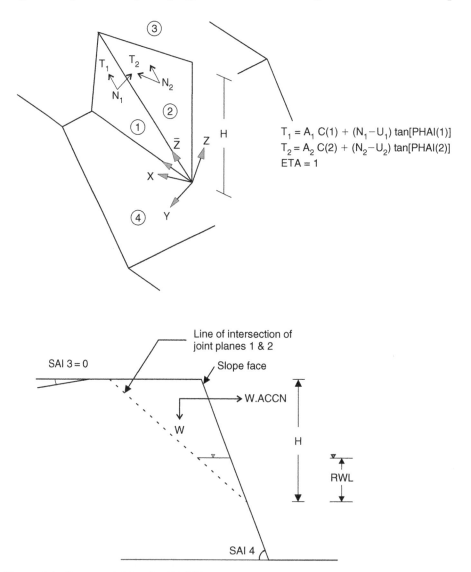

$$T_1 = A_1\, C(1) + (N_1 - U_1)\, \tan[PHAI(1)]$$
$$T_2 = A_2\, C(2) + (N_2 - U_2)\, \tan[PHAI(2)]$$
$$ETA = 1$$

Figure 9.1. Input variables for SASW.

solution of Hoek and Bray (1981). It is found that the factor of safety increases rapidly with decreasing angle of the rock wedge.

9.3   ASSUMPTIONS

The discontinuities are completely filled with water and that the water pressure varies from zero at the free faces to a maximum at some point on the line of intersection. Further, there is no joint set which intersects the rock wedge and initiates toppling of thin triangular wedges.

9.4   FIELD VERIFICATION

Deoja et al. (1991) checked SASW for rock slopes in Kathmandu. Tabatabei (1993) evaluated the program SASW in Garhwal Himalaya. The program predicted that rock wedges under study will not fail during earthquake (Magnitude < 7 on the Richter Scale) in the dry season. The Uttarkashi earthquake confirmed this prediction. In one case SASW predicted the wedge to fail by over-floating due to high seepage pressures in the rainy season. In fact the wedge was seen in an intact condition at the bottom of the rock slope. This program has been used for a stability analysis of a 125 m-high steep rock slope of the Ganwi Mini Hydel Project, H.P. It is heartening to note that SASW has been used in many countries with great satisfaction.

9.5   DYNAMIC SETTLEMENT

The simple solution of Hoek and Bray (1981) has been extended for a dynamic stability analysis of rock wedge. The rock wedge is likely to slide and experience dynamic settlement during an earthquake of high intensity. The dynamic settlement of rock wedge is estimated approximately using Jansen's (1990) correlation. This value of dynamic settlement may be conservative as he has developed correlation on the basis of observed dynamic settlement of earth dams during earthquakes of magnitude M on Richter's scale causing peak ground acceleration $\alpha_h \cdot g$. This step is necessitated, as no other simple correlation for dynamic settlement of rock slope is available to the best knowledge of the authors.

9.6   USER'S MANUAL – SASW

THIS PROGRAM CALCULATES THE FACTOR OF SAFETY OF TETRAHEDRAL WEDGE WITH HORIZONTAL SLOPE CREST AND WITH NO TENSION CRACK (FIG. 9.1)

NAME OF PROGRAM –	SASW.FOR
UNITS USED–	TONNE – METER – DEGREE

GIVE INPUT DATA IN THE FOLLOWING SEQUENCE

NO

(PLEASE REPEAT ALL FOLLOWING LINES NO TIMES FOR NO SLOPES)

TITLE OF PROBLEM IN ONE LINE (<80 CHARACTERS)

NJT,NCASE

SAI(I),ALPHA(I),C(I),PHAI(I)

(PLEASE REPEAT ABOVE LINE NJT TIMES)

SAI3,ALPHA3,SAI4,ALPHA4

H,GAMA,GAMAW,ETA,ACCN,RWL,PORE,EQM

(PLEASE REPEAT ABOVE LINE NCASE TIMES)

DO YOU WANT HELP REGARDING DEFINITION OF VARIABLES USED

ENTER   0 FOR TERMINATION

1 FOR FURTHER HELP

2 FOR EXECUTION

THE DISCONTINUITIES ARE DENOTED BY 1 AND 2, THE UPPER GROUND SURFACE
BY 3 AND THE SLOPE FACE BY 4

NO	=	NUMBER OF SLOPES
NJT	=	NUMBER OF JOINT SETS
NCASE	=	NO OF CASES
SAI(I)	=	DIP OF I TH JOINT PLANE (DEG.)
ALPHA(I)	=	DIP DIRECTION OF I TH JOINT PLANE (DEG.)
C(I)	=	COHESION OF I TH JOINT PLANE (T/SQ.M)
PHAI(I)	=	FRICTION ANGLE OF I TH JOINT PLANE (DEG.)
	=	ARC TAN(Jr/Ja) FOR CLAY COATED JOINTS
SAI3	=	ANGLE OF SLOPE OF UPPER GROUND SURFACE
ALPHA3	=	DIP DIRECTION OF THE UPPER GROUND SURFACE
SAI4	=	ANGLE OF ROCK SLOPE
ALPHA4	=	DIP DIRECTION OF THE ROCK SLOPE
H	=	HEIGHT OF THE CREST OF THE SLOPE ABOVE TOE OF INTERSECTION
GAMA	=	UNIT WEIGHT OF ROCK(T/CU.M)
GAMAW	=	UNIT WEIGHT OF WATER(T/CU.M)
ETA	=	− 1 MEANS SLOPE FACE OVERHANGS TOE OF THE SLOPE
	=	+ 1 MEANS SLOPE FACE DOES NOT OVERHANG
ACCN	=	COEFFICIENT OF HORIZONTAL ACCELERATION OF EARTHQUAKE NEAR CREST OF SLOPE
EQM	=	CORRESPONDING EARTHQUAKE MAGNITUDE (RICHTER SCALE)
RWL	=	WATER LEVEL ABOVE TOE OF INTERSECTION
PORE	=	PORE WATER PRESSURE FACTOR
	=	0 (FOR DRY SLOPE) OR 1 (FOR WET SLOPE)

(a) Input file name: ISASW.DAT

1

STABILITY ANALYSIS OF ROCK SLOPE WITH WEDGE FAILURE/OVER TOPPLING

3,2

66.,298.,0.,30.

58.,76.,0.,30.

75.,120.,0.,30.
0.,0.,70.,300.
120.,2.5,1.0,1.0,0.3,0.,0.,0.
120.,2.5,1.0,1.0,0.3,20.,0.,6.

(b) Output file name: OSASW.DAT

STABILITY ANALYSIS OF ROCK SLOPE WITH WEDGE FAILURE/OVER TOP-
PLING

UNITS USED –	TONNE – METER – DEGREE
INPUT FILE NAME –	ISASW.DAT
OUTPUT FILE NAME –	OSASW.DAT

Slope No. 1, Case No. 1
SLOPE IS LIKELY TO FAIL BY OVER-TOPPLING DUE TO JOINT NO. 3

PLANE	1	2	3	4		
SAI(DEGREE)	66.0	58.0	0.0	70.0		
ALPHA(DEGREE)	298.0	76.0	0.0	300.0		
H					=	120.000 M
GAMA					=	2.500 T/CU.M
C1					=	0.000 T/SQ.M
C2					=	0.000 T/SQ.M
PHAI1					=	30.000 DEG.
PHAI2					=	30.000 DEG.
U1					=	0.000 T/SQ.M
U2					=	0.000 T/SQ.M
EQ.ACCN					=	0.300
EQM					=	0.000
RWL					=	0.000 M

*   SLOPE DOES NOT OVERHANG
*   THERE IS CONTACT ON BOTH PLANES 1 AND 2

DYNAMIC SETTLEMENT		=	0.0000 M
CRITICAL ACCELERATION		=	0.2100
FACTOR OF SAFETY		=	0.8156

*   HENCE THE SLOPE IS UNSTABLE IN SLIDING

PLANE	1	2	3	4		
SAI(DEGREE)	66.0	75.0	0.0	70.0		
ALPHA(DEGREE)	298.0	120.0	0.0	300.0		
H					=	120.000 M
GAMA					=	2.500 T/CU.M
C1					=	0.000 T/SQ.M
C2					=	0.000 T/SQ.M
PHAI1					=	30.000 DEG.
PHAI2					=	30.000 DEG.
U1					=	0.000 T/SQ.M
U2					=	0.000 T/SQ.M
EQ.ACCN					=	0.300
RWL					=	0.000 M

```
*    SLOPE DOES NOT OVERHANG
*    NO WEDGE IS FORMED BETWEEN PLANES 1 AND 3
```

PLANE	1	2	3	4
SAI(DEGREE)	58.0	75.0	0.0	70.0
ALPHA(DEGREE)	76.0	120.0	0.0	300.0

H	=	120.000 M
GAMA	=	2.500 T/CU.M
C1	=	0.000 T/SQ.M
C2	=	0.000 T/SQ.M
PHAI1	=	30.000 DEG.
PHAI2	=	30.000 DEG.
U1	=	0.000 T/SQ.M
U2	=	0.000 T/SQ.M
EQ.ACCN	=	0.300
EQM	=	0.000
RWL	=	0.000 M

```
*    SLOPE DOES NOT OVERHANG
*    NO WEDGE IS FORMED BETWEEN PLANES 2 AND 3
```

Slope No. 1, Case No. 2

SLOPE IS LIKELY TO FAIL BY OVER-TOPPLING DUE TO JOINT NO. 3

PLANE	1	2	3	4
SAI(DEGREE)	66.0	58.0	0.0	70.0
ALPHA(DEGREE)	298.0	76.0	0.0	300.0

H	=	20.000 M
GAMA	=	2.500 T/CU.M
C1	=	0.000 T/SQ.M
C2	=	0.000 T/SQ.M
PHAI1	=	30.000 DEG.
PHAI2	=	30.000 DEG.
U1	=	0.000 T/SQ.M
U2	=	0.000 T/SQ.M
EQ.ACCN	=	0.300
EQM	=	6.000
RWL	=	20.000 M

```
*    SLOPE DOES NOT OVERHANG
*    THERE IS CONTACT ON BOTH PLANES 1 AND 2
```

DYNAMIC SETTLEMENT	=	0.0191 M
CRITICAL ACCELERATION	=	0.2100
FACTOR OF SAFETY	=	0.8147

```
*    HENCE THE SLOPE IS UNSTABLE IN SLIDING
```

PLANE	1	2	3	4
SAI(DEGREE)	66.0	75.0	0.0	70.0
ALPHA(DEGREE)	298.0	120.0	0.0	300.0

H	=	120.000 M
GAMA	=	2.500 T/CU.M

C1					=	0.000 T/SQ.M
C2					=	0.000 T/SQ.M
PHAI1					=	30.000 DEG.
PHAI2					=	30.000 DEG.
U1					=	0.000 T/SQ.M
U2					=	0.000 T/SQ.M
EQ.ACCN					=	0.300
EQM					=	6.000
RWL					=	20.000 M

*   SLOPE DOES NOT OVERHANG
*   NO WEDGE IS FORMED BETWEEN PLANES 1 AND 3

PLANE	1	2	3	4		
SAI(DEGREE)	58.0	75.0	0.0	70.0		
ALPHA(DEGREE)	76.0	120.0	0.0	300.0		
H					=	120.000 M
GAMA					=	2.500 T/CU.M
C1					=	0.000 T/SQ.M
C2					=	0.000 T/SQ.M
PHAI1					=	30.000 DEG.
PHAI2					=	30.000 DEG.
U1					=	0.000 T/SQ.M
U2					=	0.000 T/SQ.M
EQ.ACCN					=	0.300
EQM					=	6.000
RWL					=	20.000 M

*   SLOPE DOES NOT OVERHANG
*   NO WEDGE IS FORMED BETWEEN PLANES 2 AND 3

9.7   THE ANGLE OF CUT SLOPE WITH WEDGE FAILURE – ASW

In nature, slopes are safe with a factor of safety lying mainly between 0.8 and 1.8. However, the safest angle of cut slope is the apparent dip of line of intersection of any two discontinuities. This principle may be used to plan safe cut slopes along roads and rail lines in hilly areas (H < 20 m and $\psi_f < 45°$). However, the angle of dry and drained cut slope should not be less than the sliding angle of friction as the chances of wedge slide are much less in this case.

The computer program for simple wedge analysis (SASW) has been modified slightly as program ASW to determine the optimum angle of rock cut slope with desired factor of safety (say 1.2 to 1.3 in static and 1.0 in dynamic conditions). The user should also prescribe a minimum value of cut slope angle (SAI 4M) in the input data based on past experience for the site-specific geological conditions. The criterion of erosion of soft and erodible rock slopes should be kept in mind, as ASW does not take into account erosion. For example, vertical cut slopes (H < 10 m) have been used successfully in conglomerate to minimize erosion of

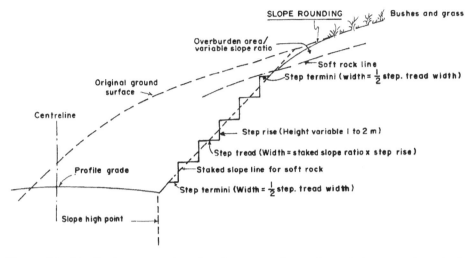

Figure 9.2. Idealised cross-section showing stepped cut slope design in soft rock (in Deoja et al., 1991).

the surface of rock slope. Schuster and Krizek (1978) recommended the stepped surface of the cut slope in soft rocks (e.g. soft sandstones, sand rock, mudstone and soft shale etc.). Thus, it is suggested that the average angle of cut slope should be obtained by ASW, but the excavation of cut slope should be as shown in Figure 9.2 in the soft and erodible rocks to reduce the velocity of flow of the surface run off and hence minimize the erosion of rocks. A major advantage of terrace planting in hilly areas is that local bushes with deep roots are helpful in reducing further erosion. Thus, the cut slopes appear pleasing to our eyes and in addition plantation improves the micro ecosystem along the road or rail lines.

The deep pits for foundation of piers on the rock masses are excavated to the designed depth. Although pits are temporary, but this stability needs to be ensured. Again, a safe angle of cut slopes of the pit may be determined by computer program ASW using a lower static factor of safety of, say, 1.1. It should be noted that the angles of cut slopes would be different for all the four walls of the pit for given sets of joints. However, one should recommend the minimum value of cut slope angle on all sides in all the pits to be on safe side and for easier construction.

The program ASW may also be used to recommend safe abutment slopes of dams (H < 75 m) as per specified factor of safety. For higher dams, computer programs of the distinct element method (e.g. UDEC and 3DEC, etc.) are suggested (Sharma et al., 1999).

Lemos and Cundall (Sharma et al., 1999) have reported that increased water flow through dam foundations has been sometimes observed immediately after an earthquake. One of the possible causes for the temporary increase in seepage is the change in the permeability of rock mass due to shearing along dilatant joints during an earthquake. This phenomenon should be considered in the detailed stability analysis of concrete dams.

In case of open cast mines also, the program ASW may be ideal. The correction for radius of plane of curvature of rock slopes (Hoek and Bray, 1981) has not been applied in 3D wedge failures, as the effect of curvature will be more pronounced in planar slides and circular slides than in 3D wedge slides.

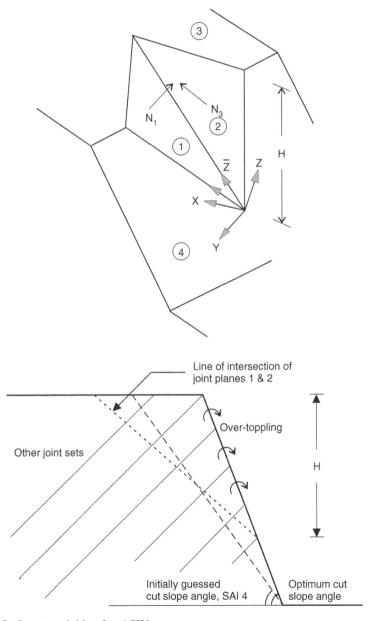

Figure 9.3. Input variables for ASW.

9.7.1 Precautions

Some material such as decomposed granite will ravel continuously due to weathering, filling ditch lines. Flattening the slope usually does not control this problem unless vegetation can be established. A successful solution is to step the slope in small benches, say 50 cm wide, and with height determined by the slope. The material will then ravel, filling the benches and at the same time protecting the underlying material, aiding plant growth to start and preventing further raveling. This procedure can be applied to many types of material.

In badly jointed or weathered rock, some means of reducing the hazard must be provided. This can be done by:
- rock bolting,
- grouting of fissures,
- horizontal berms or a wide ditch to catch falling rock,
- fences, wire mesh, or walls to catch falling rock, and
- pneumatically applied mortar to prevent ravelling.

In addition, the prevention of water pressure building in seams must be prevented by the proper drainage or sealing of the entrances. Any design must start with acomplete picture of the joints, fissures, and bedding planes of the area of concern.

9.8 USER'S MANUAL – ASW

THIS PROGRAM CALCULATES THE OPTIMUM ANGLE OF ROCK CUT SLOPE WITH TETRAHEDRAL WEDGE FAILURE AND OVER-TOPPLING FAILURE (FIG.9.3)

INPUT FILE NAME –	IASW.DAT
OUTPUT FILE NAME –	OASW.DAT
NAME OF PROGRAM –	ASW.FOR
UNITS USED –	TONNE – METER – DEGREE

GIVE INPUT DATA IN THE FOLLOWING SEQUENCE

NO
(PLEASE REPEAT ALL FOLLOWING LINES NO TIMES FOR NO SLOPES)
TITLE OF PROBLEM IN ONE LINE (<80 CHARACTERS)
NJT,NCASE
SAI(I),ALPHA(I),C(I),PHAI(I)
(PLEASE REPEAT ABOVE LINE NJT TIMES)
SAI3,ALPHA3,SAI4M,ALPHA4
H,GAMA,GAMAW,ETA,ACCN,RWL,PORE,FOS
(PLEASE REPEAT ABOVE LINE NCASE TIMES)

DO YOU WANT HELP REGARDING DEFINITION OF VARIABLES USED
ENTER 0 FOR TERMINATION
       1 FOR FURTHER HELP
       2 FOR EXECUTION

THE DISCONTINUITIES ARE DENOTED BY 1 AND 2, THE UPPER GROUND
SURFACE BY 3 AND THE SLOPE FACE BY 4

NO	=	NUMBER OF SLOPES
NJT	=	NUMBER OF JOINT SETS
NCASE	=	NO OF CASES
SAI(I)	=	DIP OF I TH JOINT PLANE (DEG.)
ALPHA(I)	=	DIP DIRECTION OF I TH JOINT PLANE (DEG.)
C(I)	=	COHESION OF I TH JOINT PLANE (T/SQ.M)
PHAI(I)	=	FRICTION ANGLE OF I TH JOINT PLANE (DEG.)
SAI3	=	ANGLE OF SLOPE OF UPPER GROUND SURFACE
ALPHA3	=	DIP DIRECTION OF THE UPPER GROUND SURFACE
SAI4M	=	MINIMUM CUT SLOPE ANGLE FROM FIELD EXPERIENCE
ALPHA4	=	DIP DIRECTION OF THE ROCK CUT SLOPE
H	=	HEIGHT OF THE CREST OF THE SLOPE ABOVE TOE OF INTERSECTION
GAMA	=	UNIT WEIGHT OF ROCK (T/CU.M)
GAMAW	=	UNIT WEIGHT OF WATER (T/CU.M)
ETA	=	− 1 MEANS SLOPE FACE OVERHANGS TOE OF THE SLOPE
	=	+ 1 MEANS SLOPE FACE DOES NOT OVERHANG
ACCN	=	COEFFICIENT OF EARTHQUAKE ACCELERATION
RWL	=	WATER LEVEL ABOVE TOE OF INTERSECTION
PORE	=	PORE WATER PRESSURE
	=	0 (FOR DRY SLOPE) OR 1 (FOR WET SLOPE)
FOS	=	ALLOWABLE FACTOR OF SAFETY

(a) Input file name: IASW.DAT

1
OPTIMUM CUT SLOPE ANGLE OF ROCK SLOPE WITH WEDGE
FAILURE/OVER TOPPLING
5,1
66.,298.,0.,30.
68.,320.,0.,30.
60.,358.,0.,30.
58.,76.,0.,30.
61.,118.,0.,30.
0.,0.,65.,300.
120.,2.5,1.0,1.0,0.,0.,0.,1.2

(b) Output file name: OASW.DAT

OPTIMUM CUT SLOPE ANGLE OF ROCK SLOPE WITH WEDGE
FAILURE/OVER TOPPLING

UNITS USED –   TONNE – METER – DEGREE
INPUT FILE NAME –   IASW.DAT
OUTPUT FILE NAME – OASW.DAT

Slope No.1, Case No. 1

PLANE	1	2	3	4
SAI(DEGREE)	66.0	58.0	0.0	65.0
ALPHA(DEGREE)	298.0	76.0	0.0	300.0

H	=	120.000 M
GAMA	=	2.500 T/CU.M
C1	=	0.000 T/SQ.M
C2	=	0.000 T/SQ.M
PHAI1	=	30.000 DEG.
PHAI2	=	30.000 DEG.
U1	=	0.000 T/SQ.M
U2	=	0.000 T/SQ.M
EQ.ACCN	=	0.000
RWL	=	0.000 M

*   SLOPE DOES NOT OVERHANG

DYNAMIC DISPLACEMENT	=	0.0000 M
CRITICAL ACCLERATION	=	0.2100
FACTOR OF SAFETY	=	1.5210
OPTIMUM CUT SLOPE ANGLE	=	65.0000 DEG.

INPUT CUT SLOPE ANGLE (SAI4M) APPEARS UNSAFE

REFERENCES

Deoja, B., Dhital, M., Thapa, B. and Wagner, A. 1991. *Mountain Risk Engineering Handbook.* Kathmandu, International Centre of Integrated Mountain Development, Part I, Chaps 10 and 13, and Part II, Section 24.9

Jansen, R.B. 1990. Estimation of embankment dam settlement caused by earthquake. *Water Power and Dam Construction*, December, pp. 35−40.

Hoek, E. and Bray, J.W. 1981. *Rock Slope Engineering.* London, Institution of Mining and Metallurgy, Revised Third Edition, Chaps 5 and 8 and Appendix II − Short Solution.

Schuster, R.L. and Krizek, R.J. 1978. Landslides − Analysis and Control. Washington D.C., National Academy of Sciences, Special Report 176, p. 234.

Sharma, V.M., Saxena, K.R. and Woods, R.D. 1999. *Distinct Element Modelling in Geomechanics.* New Delhi, India, Oxford & IBH Publishing Co. Pvt. Ltd., (Chapter 2 by C. Fairhurst and L. Long and Chapter 5 by J.V. Lemos and P.A. Cundall), p. 222.

Tabatabei, S.A. 1993. *Terrain studies for slope instability in parts of Tehri District, Garhwal Himalaya, U.P., India.* Ph.D. Thesis. Department of Earth Sciences, University of Roorkee, India.

CHAPTER 10

A rigorous analysis of 3D rock wedge – WEDGE

> *"The word impossible in itself says, I am possible!"*
> Anonymous

10.1 WEDGE FAILURE

When two discontinuities strike obliquely across the slope face and their line of intersection daylights in the slope face, the wedge of rock resting on these discontinuities will slide down along the line of intersection, provided that the inclination of this line is significantly greater than the angle of sliding friction. This type of failure is a much more general case as illustrated in Figure 10.1. The wedge may also slide downwards on any one discontinuity.

As in the case of planar failure, a condition of sliding is defined by $\psi_{fi} > \psi_i > \phi$ where ψ_{fi} is the inclination of the slope face, measured in the view at right angles to the line of the intersection and ψ_i is the dip of the line of intersection as in Figure 10.1d. It may be noted that ψ_{fi} would be the same as ψ_f, the true dip of the slope face, if the dip direction of the line of intersection was the same as the dip direction of the slope face.

10.2 AN ANALYSIS OF TETRAHEDRAL WEDGE

Two solutions designed for maximum speed and efficiency of calculation are given by Hoek and Bray (1981). These solutions are:
- A simple solution for a wedge with a horizontal slope crest and with no tension crack. The influence of an external force is not included in the solution. The computer program SASW has already been discussed in Chapter 9.
- A rigorous solution which includes the effects of a superimposed load, a tension crack and an external force such as that applied by a tensioned cable. In this chapter, the rigorous solution by Hoek and Bray (1981) has been used to develop computer program WEDGE.

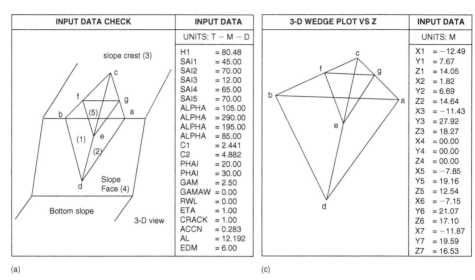

(a)

(c)

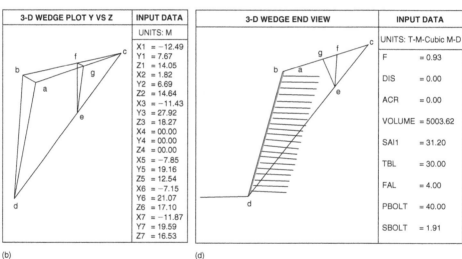

(b)

(d)

Figure 10.1. Design of reinforcement for stabilising wedge of a rock slope.

10.3   THE RIGOROUS SOLUTION

The rigorous solution is for computation of the factor of safety for translational slip of a tetrahedral wedge formed in a rock slope by two intersecting discontinuities, the slope face and the upper ground surface. The influence of a tension crack is included in the solution. The solution does not take into account rotational slip and toppling.

The solution allows for different strength parameters and water pressures on the two planes of weakness and for water pressures in the tension crack. There is

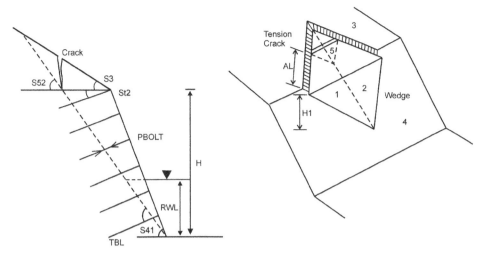

Figure 10.2. Input variables for WEDGE (ETA = 1 SE2 = 0 AE2 = 0).

no restriction on the inclination of the crest of slope. The influence of an external load 'E' and a cable tension 'T' are included in the analysis. By trial and error, one can compute the minimum factor of safety for a given external load (e.g. a blast acceleration acting in a known direction) and minimize the cable force required for a given factor of safety.

The solution considers the water pressures on joint planes (u_1 and u_2) and on tension crack (u_5), it is assumed that extreme conditions of very heavy rainfall occur and that in consequence the fissures are completely full of water. Again, it is assumed that the pressure varies from zero at the free faces to a maximum value at some point on the line of intersection of the two failure planes.

In the cases where no tension crack exists, $u_1 = u_2 = \gamma_w.Hw/6$, where Hw is the total height of the wedge. In the cases where tension crack exists, $u_1 = u_2 = u_5 = \gamma_w.H5w/3$, where H5w is the depth of the bottom vertex of the tension crack below the upper ground surface.

Allowance is made for the following:
1. Interchange of planes 1 and 2,
2. The possibility of one of the planes overlying the other,
3. The situation where the crest overhangs the base of the slope (in which case $\eta = -1$),
4. The possibility of contact being lost on either planes, and
5. Computer analysis of all possible wedges in case of several joint sets.

In addition to detecting whether or not a wedge can form, the solution also examines how the tension crack intersects the other planes and only accepts those cases where the tension crack truncates the wedge in the manner shown in Figure 10.1. The input parameters of the WEDGE program are defined in Figure 10.2.

10.4   DYNAMIC SETTLEMENT

An earthquake of magnitude M on Richter's scale will cause only a dynamic settlement of the rock wedge, as the dynamic force is instantaneous. The program WEDGE uses Jansen's correlation for computing approximately the dynamic settlement of a rock wedge. In fact, one can permit dynamic factor of safety less than 1.0 if dynamic settlement is less than 20 cm or H/500, where H is the height of rock slope.

10.5   THE STABILITY OF FOUNDATION ON ROCK WEDGE

The program WEDGE has been used for analysing stability of abutments on a rock wedge for several bridges in Garhwal and Kashmir Himalaya. The stability of a foundation of abutment or pier of a bridge may be analysed easily. In the first computer run, the weight of the critical rock wedge (W1) is computed so that this rock wedge gives a minimum factor of safety among all other possible wedges which are formed by multiple joint sets. Then, the weight of foundation (W2) of abutment or pier is added to W1. The unit weight of rock mass is increased by the factor (W1 + W2)/W1. In the second run, modified value of unit weight of rock mass is given in the input in place of the actual unit weight. The minimum computed factor of safety corresponds to the total weight (W1 + W2) of the foundation and the rock wedge.

10.6   REMEDIAL MEASURES

Stability may be improved by adding to the shearing resistance of the rock mass and thus an acceptable level of the factor of safety is achieved. This is achieved either by:
- *Grouting* the rock mass with cement grout chiefly to improve the overall shear strength parameters, i.e. c and φ values. Grouting however, should not be done in wet slopes. It may also endanger stability temporarily due to fluid pressure of the grout.
- *Anchoring*, i.e. providing anchors mainly to add to the effective normal pressure and thus locking together the wedge with the rock below the probable shear surfaces or critical discontinuities. This computer program also helps in the design of system of rock anchors.
- *De-pressurization of slope*, i.e. slope drainage which increases stability of the slope by preventing surface water from entering the slope through open tension cracks and fissures and thus reducing the water pressure in the vicinity of the potential failure surface by selective surface and sub-surface drainage.
- *Sealing of tension cracks* by filling and tamping silty soil inside cracks and fractures to prevent ingress of rain water inside the wedge.

Depending upon the conditions of individual site, any of the above or a combination of two or more may be adopted.

10.6.1 Grouting

The purpose of grout used in the bolt hole is to transfer the tensile forces from the anchor to ground and also to increase shear strength of the rock mass. The grout can be of (i) good quality of portland cement; and (ii) synthetic resin. Synthetic resin is preferred to portland cement grout, if early strength development is required in anchor. Resin is quick setting and it exhibits excellent resistance to corrosion. The best resin is non-saturated polyester resin.

10.6.2 Anchoring

The objective of anchoring is to stitch rock wedges with each other and also with stable adjoining rock masses. The sliding of the wedge pulls out rock anchors and induces tensile stresses in grouted rock anchors. The normal component of induced pretension in the anchors develops the shearing resistance along joint planes. This is in addition to the resistance offered by parallel component of tension in the anchors. This computer program helps in the design of system of rock anchors, i.e. length, spacing, anchor pull out capacity and direction of anchoring. The rock anchor system also forms a breast wall of reinforced rock mass which stabilizes a rock slope of even highly fractured rock mass.

Theoretically, one strong rock anchor is sufficient to stabilise a rock wedge. However, a rock slope may contain joint set with many joints. In such case large numbers of rock wedges are formed. All the rock wedges may be stabilised by installing a set of rock anchors at regular spacing such that a reinforced rock breast wall is formed (Fig. 10.1d). Cable anchors have to be used in high rock slopes where length of anchors is more than 12 m.

Progressive failure of rock wedges is observed if more than two joint sets are prevalent. Initially, main rock wedge having the lowest factor of safety slides down. Then it creates two new faces of side slopes which free the adjoining wedges to slip down. This progressive failure continues in successive rainy seasons. The area of rockslide increases exponentially with the number of joint sets (Deoja et al., 1991). However, the rock reinforcement designed by program WEDGE will stitch all the rock wedges and thereby stabilise them adequately.

10.6.3 Slope de-pressurization

Water pressures along failure surface reduces the stability of slope. Hence, to reduce the risk of failure, water pressure should be reduced by draining away surface water from tension cracks and fissures and not allowing water to seep along the potential failure surface. Drainage holes are to be made across the wedge. In dam abutments, drainage galleries are to be made. Lined surface drains are also helpful but unmaintained blocked drains activate landslide due to seepage of water into tension cracks and critical discontinuities during rainy season. Further, it is recommended to drill holes of 38 mm diameter dipping about 5° upward to act as drainage holes. Wrapped up geo-membranes (or wire net) should be inserted in these drainage holes to prevent their choking. These drainage holes are made all along the rock wedge to drain water from the rock joints.

10.6.4   Sealing of tension cracks

Water should be prevented from entering into the tension cracks by sealing them with a flexible impermeable material such as silt or clay. When the crack is more than a few centimeters wide, it should be filled with gravel or waste rock before flexible seal is placed. The purpose of this fill is to allow any water which finds its way into the crack to flow out again as freely as possible. Under no circumstances should the crack be filled with concrete or grout if this would result in the creation of an impermeable dam which could cause the build up of high water pressures in the slope.

10.7   USER'S MANUAL – WEDGE

THIS PROGRAM CALCULATES THE FACTOR OF SAFETY OF SUBMERGED TETRAHEDRAL WEDGE WITH INCLINED SLOPE CREST AND WITH TENSION CRACK AND ROCK ANCHORS BY RIGOROUS DYNAMIC ANALYSIS OF HOEK AND BRAY, 1981 (FIG. 10.2)

NAME OF PROGRAM: WEDGE.FOR
INPUT FILE NAME: IWEDGE.DAT
OUTPUT FILE NAME: OWEDGE.DAT
UNITS USED: TONNE – METER – DEGREE

GIVE INPUT DATA IN THE FOLLOWING SEQUENCE

NO
TITLE OF PROBLEM IN ONE LINE (<80 CHARACTERS)
NJT,NCASE
SI(I),ALPH(I),C(I),PHI(I)
(PLEASE REPEAT ABOVE LINE NJT TIMES)
S31,A31,S41,A41,S52,A52,ST2,AT2,SE2,AE2
H1,GAMA,GAMAW,ETA,AH,RWL,CRACK,AL,FS,EQM
TBL,FAL,PBOLT
(PLEASE REPEAT ABOVE TWO LINES NCASE TIMES)

{KINDLY REPEAT ENTIRE BLOCK [FROM NJT TO PBOLT] NO TIMES FOR EACH SECTION OF ROCK SLOPE}

DO YOU WANT HELP REGARDING DEFINITION OF VARIABLES USED
ENTER   0 FOR TERMINATION
             1 FOR FURTHER HELP
             2 FOR EXECUTION

THE DISCONTINUITIES ARE DENOTED BY 1 AND 2, THE UPPER GROUND SURFACE
BY 3 AND THE SLOPE FACE BY 4 AND PLANE OF CRACK BY 5

NO	=	NUMBER OF SLOPES
NJT	=	NUMBER OF JOINT SETS
NCASE	=	NO OF CASES
SI(I)	=	DIP OF ITH JOINT PLANE (DEG.)
ALPH(I)	=	DIP DIRECTION OF ITH JOINT PLANE (DEG.)
C(I)	=	COHESION OF I TH JOINT PLANE (T/SQ.M)
PHI(I)	=	FRICTION ANGLE OF I TH JOINT PLANE (DEG.)
S31	=	ANGLE OF SLOPE OF UPPER GROUND SURFACE
A31	=	DIP DIRECTION OF THE UPPER GROUND SURFACE
S41	=	ANGLE OF ROCK SLOPE
A41	=	DIP DIRECTION OF THE ROCK SLOPE
S52	=	DIP OF CRACK PLANE
A52	=	DIP DIRECTION OF CRACK PLANE
ST2	=	PLUNGE OR DIP OF ANCHOR BAR (NEGATIVE FOR UPWARD)
AT2	=	TREND OF ANCHOR BAR
SE2	=	PLUNGE OR DIP OF EARTHQUAKE FORCE (IF 0, PROGRAM CALCULATES IT)
AE2	=	TREND OF EARTHQUAKE FORCE (IF 0, PROGRAM ASSUMES IT SAME AS THAT OF INTERSECTION OF JOINTS
H1	=	HEIGHT OF SLOPE (PLANE 4) WITH RESPECT TO POINT OF INTERSECTION OF PLANE 1,3,4 ABOVE BOTTOM OF WEDGE
GAMA	=	UNIT WEIGHT OF ROCK (T/CU.M)
GAMAW	=	UNIT WEIGHT OF WATER (T/CU.M.). 0. MEANS DRY WEDGE
ETA	=	− 1 MEANS SLOPE FACE OVERHANGS TOE OF THE SLOPE + 1 MEANS SLOPE FACE DOES NOT OVERHANG TOE OF SLOPE
AH	=	COEFFICIENT OF EARTHQUAKE ACCELERATION
EQM	=	CORRESPONDING EQ. MAGNITUDE ON RICHTER SCALE
RWL	=	RESERVOIR WATER LEVEL ABOVE TOE OF WEDGE
CRACK	=	−1.,0.,1.,2. REFERS TO NO CRACK AND DRY CONDITION OF SLOPE; WET SLOPE WITH ZERO CRACK DEPTH; AND PRESENCE OF CRACK WITH WATER SLOPE WITH DRY CRACK
AL	=	DISTANCE OF CRACK POSITION ON TOP PLANE FROM TOP EDGE OF SLOPE ALONG PLANE 1
FS	=	RECOMMENDED FACTOR OF SAFETY FOR REINFORCED SLOPE
TBL	=	TOTAL BOLT/ANCHOR LENGTH
FAL	=	FIXED ANCHOR LENGTH
PBOLT	=	CAPACITY OF EACH BOLT/ANCHOR

(a) Input file name: IWEDGE.DAT

1
DESIGN OF REINFORCEMENT FOR STABILIZING WEDGE OF A ROCK SLOPE
2 1
45. 105. 2.441 20.
70. 235. 4.882 30.
30. 250. 65. 185. 80. 175. 30. 5. 0. 0.
30. 2.5 1. 1. 0. 0. 1. 12. 1.2 5.
12. 4. 50.

(b) Output file name: OWEDGE.DAT

DESIGN OF REINFORCEMENT FOR STABILIZING WEDGE OF A ROCK SLOPE

Slope No. 1, Case No. 1

HEIGHT OF WEDGE ON PLANE 1	=	30.0000
UNIT WEIGHT OF ROCK	=	2.5000
UNIT WEIGHT OF WATER	=	1.0000
DISTANCE OF CRACK FROM FACE	=	12.0000
EARTHQUAKE ACCELERATION	=	0.0000
EARTHQUAKE MAGNITUDE	=	5.0000
ETA	=	1.0000
TOTAL ANCHOR LENGTH	=	12.0000
FIXED ANCHOR LENGTH	=	4.0000
CAPACITY OF EACH ANCHOR	=	50.0000
COHESION-PLANE 1	=	2.4410
ANGLE OF FRICTION-PLANE 1	=	20.0000
COHESION-PLANE 2	=	4.8820
ANGLE OF FRICTION-PLANE 2	=	30.0000
DIP OF PLANE 1	=	45.0000
DIP DIRECTION OF PLANE 1	=	105.0000
DIP OF PLANE 2	=	70.0000
DIP DIRECTION OF PLANE 2	=	235.0000
DIP OF TOP PLANE	=	30.0000
DIP DIRECTION OF TOP PLANE	=	250.0000
DIP OF SLOPE PLANE	=	65.0000
DIP DIRECTION OF SLOPE PLANE	=	185.0000
DIP OF CRACK	=	80.0000
DIP DIRECTION OF CRACK	=	175.0000
PLUNGE OR DIP OF ANCHOR	=	30.0000
TREND OR DIP DIRECTION OF ANCHOR	=	5.0000
PLUNGE OR DIP OF EARTHQUAKE	=	0.0000
TREND OR DIP DIRECTION OF EARTHQUAKE	=	157.7324
DIP OF INTERSECTION 1 2	=	31.1965
DIP DIRECTION OF INTERSECTION 1 2	=	157.7324
APPARENT DIP OF INTERSECTION	=	34.2643
DEPTH OF TENSION CRACK	=	23.02
SEEPAGE PRESSURE ON JOINT PLANE 1	=	7.69
SEEPAGE PRESSURE ON JOINT PLANE 2	=	7.69
RESERVOIR WATER LEVEL ABOVE TOE	=	0.00
SATURATED WEIGHT OF WEDGE	=	17302.

***************************** STATIC CASE *********************************

FACTOR OF SAFETY	=	0.8433
SLIDING ALONG INTERSECTION 1 2		
TOTAL ANCHOR CAPACITY	=	3252.
FOR FACTOR OF SAFETY	=	1.2000
EARTHQUAKE FORCE	=	0.
WATER THRUST IN CRACK	=	2247.
SPACING OF ANCHOR	=	1.4685

************************ DYNAMIC CASE *********************************

FACTOR OF SAFETY	=	0.8433
SLIDING ALONG INTERSECTION 1 2		
TOTAL ANCHOR CAPACITY	=	3252.
FOR FACTOR OF SAFETY	=	1.2000
EARTHQUAKE FORCE	=	0.
WATER THRUST IN CRACK	=	2247.
SPACING OF ANCHOR	=	1.4685
PLUNGE OR DIP – EARTHQUAKE	=	0.0000
TREND OR DIP DIRECTION – EARTHQUAKE	=	157.7324
FOR UNREINFORCED WEDGE		
DYNAMIC SETTLEMENT IN METER	=	0.0000
CRITICAL EARTHQUAKE ACCELERATION	=	0.0000

REFERENCES

Deoja, B., Dhital, M., Thapa, B. and Wagner, A. 1991. *Mountain Risk Engineering Handbook*. Kathmandu, International Centre of Integrated Mountain Development, Part I, Chaps 10 and 13 and Part II, Section 24.9.

Jansen, R.B. 1990. Estimation of embankment dam settlement caused by earthquake. *Water Power and Dam Construction*, December, pp. 35–40.

Hoek, E. and Bray, J.W. 1981. *Rock Slope Engineering*. London, Institution of Mining and Metallurgy, Revised Third Edition, Chaps 5 and 8 and Appendix II – Comprehensive Solution

CHAPTER 11

Toppling failure – TOPPLE

"The most incomprehensible fact about nature is that it is comprehensible"
Albert Einstein

11.1 GENERAL

Toppling is a failure mode of slopes involving overturning of interacting columns.
In rock, such columns are comprised of regular bedding planes, cleavage or joints
which strike parallel to the slope crest and dip into the rock mass (Section 3.5 and
Fig.11.1). This is in contrast to a sliding mode of failure in which the controlling
discontinuities (surface supporting the sliding mass) dip towards slope face. The
toppling occurs when the centre of gravity of a unit block overhangs a pivot point
(edge) within the unit. Figure 11.1 shows a simple series of blocks as may be
found in nature.

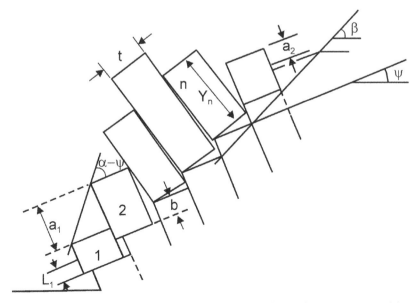

Figure 11.1. Model for limiting equilibrium analysis of toppling on a stepped base.

11.2   BASIC PRINCIPLES

The limit equilibrium method has been chosen as the method for analysing toppling failure. Three distinct zones can be identified in the slope.
1. The zones near the toe of the slope. This zone has blocks which are failing predominantly in sliding.
2. The intermediate zones in which the blocks undergo rotational movements, i.e. tend to topple.
3. The zones near the top which comprises of stable blocks.
It is important to note that the presence of above three zones is dependent on the geometry of the slope. It might be possible for a particular geometry that nearly all the blocks (including the toe blocks) tend to topple.

11.3   ASSUMPTIONS

The following assumptions are made while carrying out the limit equilibrium analysis:
• The thickness of the columns throughout the slope face is uniform.
• The columns contain a set of cross fractures which are close enough to form a regular stepped base.
• The inclination of the stepped base has to be assumed. The inclination of the stepped base has been generally found in the range of 20° to 30°.
• The inclination of the slope face is uniform.
• Because cross fractures are close and well developed, water table does not develop and consequently the effect of water pressure is not considered.
• The columns/blocks are rigid.
• A failure block either fails in toppling or sliding. No block is assumed to experience both sliding as well as toppling.
• The last block does not have any thrust acting on it from upslope direction.
In a slope susceptible to toppling failure, columns interact with one another and there are more degrees of freedom. Goodman et al. (1976) have obtained the kinematic condition for flexural slip prior to large mass movement. The state of stress in rock slope in uniaxial is with σ_1 parallel to the slope face. When the rock layers tend to slip past one another, the σ_1 stress will be inclined at an angle ϕ_j with the normal to layers, where ϕ_j is the inter column friction angle. If Ψ_f be the angle of the slope face and Ψ_c be the dip of the cross joints (both with respect to horizontal), the condition for inter layer slip for static case is

$$\Psi_f \geq 90 - \Psi_c + \phi_j \qquad (11.1)$$

In order to avoid the toppling failure the cut slope should be reduced to $90 - \Psi_c + \phi_j$. The reduction of angle of cut face is usually uneconomical in high slopes (H > 20 m) as large mass is excavated. Thus, in order to stabilise a potentially over-toppling slope, use of rock bolts/anchors become inevitable.

Goodman et al. (1976) have suggested the use of rock anchors to provide restoring force to the sliding toe block. Sakurai et al. (1985) have suggested the use of rock bolts throughout the slope to prevent excessive displacements.

The computer program TOPPLE has been developed on the basis of theory developed by Zabank (1983) which is a modified form of theory of Goodman et al. (1976). The theory has been further extended to account for seismic forces, as chances of toppling are more during earthquake. Thus, the computer program TOPPLE is better than Zabank's (1983) program TOPANL which ignores the seismic forces and the rock reinforcement.

11.4   USER'S MANUAL – TOPPLE

THIS PROGRAM IS FOR THE TOPPLING ANALYSIS OF ROCK SLOPE AND DESIGN OF REINFORCEMENT (FIG. 11.1)

NAME OF PROGRAM –   TOPPLE.FOR
INPUT FILE NAME –   ITOPPLE.DAT
OUTPUT FILE NAME –   OTOPPLE.OUT
UNITS USED –   TONNE – METER – DEGREE

GIVE INPUT DATA IN THE FOLLOWING SEQUENCE

TITLE OF PROBLEM IN ONE LINE (<80 CHARACTERS)
MM
KK (IGNORE THIS LINE IF MM = 2)
Y(I),I = 1,KK (IGNORE THIS LINE IF MM = 2)
PSI,TETA,ALF1,ALF2,THICK,HEIGHT,SPGRAV,PHICC,PHICB
ALFAH,ALFAV
A1BASE
RFORCE
BOLTLE
NUMB

DO YOU WANT HELP REGARDING DEFINITIONS OF VARIABLES USED
ENTER   0 FOR TERMINATION
              1 FOR FURTHER HELP
              2 FOR EXECUTION

MM	=	1 FOR SLOPE WITH KNOWN COLUMN DIMENSIONS
	=	2 FOR GENERAL SLOPE (The program will generate column dimensions)
KK	=	NUMBER OF COLUMNS
Y(I)	=	HEIGHT OF ITH COLUMN (I = 1 AT TOE OF SLOPE)
ALFAH	=	COEFF. OF HORIZONTAL ACCELERATION OF EARTHQUAKE
ALFAV	=	COEFF. OF VERTICAL ACCELERATION OF EARTHQUAKE
SPGRAV	=	UNIT WEIGHT OF ROCK
GAMAW	=	UNIT WEIGHT OF WATER
PSI	=	DIP ANGLE OF THE BASE PLANE
RFORCE	=	RESTRAINING FORCE AVAILABLE AT TOE
BOLTLE	=	MAXIMUM BOLT LENGTH AVAILABLE (in meters)

NUMB	=	0 IF SLOPE REINFORCEMENT IS NOT REQUIRED BY USER
	=	ANY NUMBER IF SLOPE REINFORCEMENT IS TO BE DESIGNED
A1BASE	=	HEIGHT OF THE FIRST COLUMN (measured at right angle to its base)
THETA	=	STEP ANGLE (= ARCTAN(b/t))
ALF1	=	SLOPE ANGLE
ALF2	=	DIP ANGLE OF THE TOP TERRACE OF THE ROCK SLOPE
THICK	=	THICKNESS OF COLUMNS
HEIGHT	=	SLOPE HEIGHT
PHICC	=	FRICTION ANGLE AT THE COLUMN TO COLUMN CONTACT
PHICB	=	FRICTION ANGLE AT THE COLUMN TO BASE CONTACT
b	=	STEP HEIGHT
t	=	COLUMN WIDTH
H	=	HEIGHT OF COLUMN
YN	=	HEIGHT OF NTH COLUMN
DX	=	THICKNESS OF COLUMN
MN	=	MOMENT ARM OF NTH INTERACTIVE FORCE (UPSLOPE SIDE)
LN	=	MOMENT ARM OF NTH INTERACTIVE FORCE (DOWNSLOPE SIDE)
PNT	=	NTH INTERACTIVE FORCE FOR TOPPLING FAILURE
PST	=	NTH INTERACTIVE FORCE FOR SLIDING FAILURE
PN	=	NORMAL FORCE ON NTH COLUMN FACE
RN	=	NORMAL FORCE AT THE BASE OF COLUMN
SN	=	TANGENTIAL FORCE AT THE BASE OF THE COLUMN

(a) Input file name: ITOPPLE.DAT

OVER-TOPPLING OF ROCK BLOCKS
2
30 15 56.6 0 2 25 2.5 35. 38
0 0
1
0
8
4

(b) Output file name: OTOPPLE.DAT

UNITS USED –   TONNE – METER – DEGREE

COMPUTER PROGRAM TITLE: TOPPLE.FOR
TITLE OF THE PROBLEM: OVER-TOPPLING OF ROCK BLOCKS

Slope Height	=	25.0 (meters)
Slope Angle	=	56.6 (degrees)
Dip Angle of Discontinuities	=	60.0 (degrees)
Dip Angle of Slope Top Surface	=	0.0 (degrees)
Dip Angle of Flat Base plane	=	30.0 (degrees)
Column Thickness	=	2.0 (meters)
Dip Angle of stepped Base plane	=	45.0 (degrees)
Step Height	=	0.536 (meters)

UNIT WEIGHT OF ROCK:
2.5 (Tons per cubic meter)

24.5 (Kilo Newton per meter cube)

FRICTION ANGLES:

At base-column contact	=	38.0 (degrees)
At column-column contact	=	35.0 (degrees)

COEFFICIENTS OF EARTHQUAKE ACCELERATION:

In horizontal direction (ALFAH)	=	0.000 (meter per second sq.)
In vertical direction (ALFAV)	=	0.000 (meter per second sq.)
NUMBER OF COLUMNS ON FLAT BASE	=	28
NUMBER OF COLUMNS ON STEPPED BASE	=	16

COL. NO.	YN	YN/ DX	MN	LN	PNT	PST	PN	RN	SN	SN/RN	COMMENTS
16	1.52	0.76	0.00	1.52	0.00	0.00	0.00	8.27	6.22	0.75	COLUMN IS STABLE
15	3.21	1.60	2.05	3.21	0.00	0.00	0.00	14.11	8.34	0.59	COLUMN IS STABLE
14	4.90	2.45	3.74	4.90	1.79	0.00	1.79	19.95	10.45	0.52	COLUMN TENDS TO TOPPLE
13	6.59	3.29	5.43	5.59	5.90	0.00	5.90	25.65	12.36	0.48	COLUMN TENDS TO TOPPLE
12	6.12	3.06	6.12	5.12	9.41	0.00	9.41	24.05	11.79	0.49	COLUMN TENDS TO TOPPLE
11	5.66	2.83	5.66	4.65	11.93	0.00	11.93	22.73	11.62	0.51	COLUMN TENDS TO TOPPLE
10	5.19	2.60	5.19	4.19	13.47	1.81	13.47	21.40	11.44	0.53	COLUMN TENDS TO TOPPLE
9	4.73	2.36	4.73	3.72	14.03	4.26	14.03	20.07	11.26	0.56	COLUMN TENDS TO TOPPLE
8	4.26	2.13	4.26	3.26	13.61	5.72	13.61	18.74	11.07	0.59	COLUMN TENDS TO TOPPLE
7	3.79	1.90	3.79	2.79	12.23	6.21	12.23	17.40	10.87	0.62	COLUMN TENDS TO TOPPLE
6	3.33	1.66	3.33	2.33	9.89	5.74	9.89	16.05	10.66	0.66	COLUMN TENDS TO TOPPLE
5	2.86	1.43	2.86	1.86	6.61	4.31	6.61	14.69	10.43	0.71	COLUMN TENDS TO TOPPLE
4	2.40	1.20	2.40	1.40	2.43	1.94	2.43	13.31	10.17	0.76	COLUMN TENDS TO TOPPLE
3	1.93	0.97	1.93	0.93	0.00	0.00	0.00	11.88	9.85	0.83	COLUMN IS STABLE

| 2 | 1.47 | 0.73 | 1.47 | 0.46 | 0.00 | 0.00 | 0.00 | 11.87 | 11.55 | 0.97 | COLUMN IS STABLE |
| 1 | 1.00 | 0.50 | 1.00 | 1.00 | 0.00 | 0.00 | 0.00 | 6.49 | 5.58 | 0.86 | COLUMN IS STABLE |

NATURE OF SLOPE:	Stable
FACTOR OF SAFETY IN SLIDING:	1.116

REFERENCES

Deoja, B., Dhital, M., Thapa, B. and Wagner, A. 1991. *Mountain Risk Engineering Handbook*. Kathmandu, International Centre of Integrated Mountain Development, Part I, Chaps 10 and 13 and Part II, Section 24.9.

Goodman, R.E. and Bray, J.W. 1976. Toppling of Rock Slopes. *Rock Engineering for Foundations and Slopes*, ASCE, pp. 201–234.

Hoek, E. and Bray, J.W. 1981. *Rock Slope Engineering*. London, Institution of Mining and Metallurgy, Revised Third Edition, Chap. 10.

Zabank C. 1983. Design charts for rock slopes susceptible to toppling. *ASCE J. Geotech. Eng.* 109(8): 1039–1062.

Sakurai, S. and Deeswasmongkoll, N. 1985. Study on rock slope protection of toppling failure by physical modelling. *26th U.S. Symposium on Rock Mechanics*, pp. 11–18.

CHAPTER 12

Circular slides

"Seek and you will find"
C.V. Raman, Nobel Laureate (1955)

12.1 CIRCULAR SLIDES

Rotational slides occur on slopes of homogeneous clay or shale, soil and crushed rocks. The slide movement is more or less rotational about an axis that is parallel to the contour of the slope. Highly altered and weathered rocks also tend to fail in this mode. The scarp due to the tension crack at the head may be almost vertical, while the toe bulges upwards and outwards and some time the slope mass flows out. Taylor (1948) suggested that actual slip surface may be replaced by a circular slip surface to get the factor of safety approximately. Hence rotational slides are assumed to be circular slides for the analytical purpose.

The problem of stability of natural slopes has long been a major field of interest to geotechnical engineers. In spite of considerable advancement made in understanding the soil behaviour during the last seventy years, the problem of evaluation of stability of slopes evades simple solution. The critical slip surface and the strength parameters of the soil are the two important pre-requisites for any fair evaluation of the stability of slopes.

12.2 THE PROGRAM SARC

Computer-orientated solutions can help an engineer in both these aspects. In this chapter, the computer program SARC (stability analysis of reservoir slopes with circular slide) is presented. Singh and Ramaswamy (1979) reported the method of seismic stability analysis. It is an extension of Bishop's (1954) method of slices, which has given excellent experience in case of circular slides. The program creates vertical slices below every point of the input values of coordinates along the profile of slope.

The unique feature of the SARC is that it computes the critical acceleration due to earthquake; and then predicts approximately the dynamic settlement of the slope at the top of the slope using Jansen's (1990) correlation. It is interesting to know that SARC in most of the cases has predicted only small dynamic settlements of slopes even due to earthquakes of high intensity (M > 8 on Richter's scale).

It is also realized that submerged rock/soil slopes are most vulnerable to fail during an earthquake. In fact, earthquake force acts on the saturated mass of the slope whereas the shearing resistance along the slip surface is governed by the effective stress or submerged weight of the water-charged slope. Consequently, the dynamic factor of safety of the submerged slope is reduced drastically. This assumption appears to be realistic.

It may be noted that the origin of axes should be the lowest point or toe of the slope as shown in Figure 12.4. The program SARC computes factors of safety of circular slip surface with exit point XEXITI, considering its various radii. Then program recalculates factors of safety of slip circle with exit point XEXIT + GAP with different radii. These calculations are repeated till the last exit point XEXITL. Finally the program writes the minimum factor of safety in the output file if NOPT = 0. Furthermore, many entry points along the slope may be considered so that the most critical slip surface is evaluated for the minimum factor of safety. Sometimes an error message is written in the output file that the

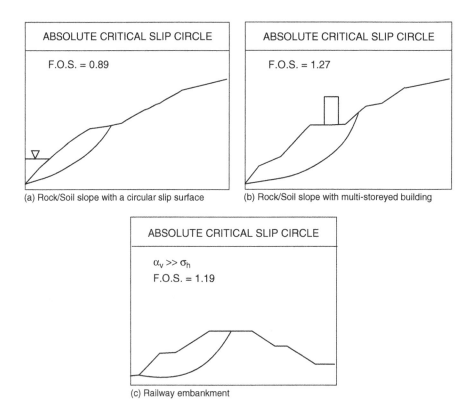

Figure 12.1. Applications of SARC software.

slip surface is above the profile of slope, especially in concave slopes. This program ignores such trial slip circles. So this error message may be ignored if the weight of wedge is not equal to zero. A major advantage of the program is that the centre of the critical slip circle is evaluated automatically and there is no need to give the probable grid of centres.

There are many applications of this software as shown in Figures 12.1 and 12.2. The brief descriptions are offered here.

1. Rock/soil slope with circular slip surface (Fig. 12.1a). It shows the failure of a slope submerged by a dam reservoir. The program assumes that the phraetic line is nearly horizontal up to the critical slip surface to enable easy calculation of the factor of safety.

2. Rock/soil slope with a multi-storied building (Fig. 12.1b). A multi-storied building may be substituted by an additional height of slice with equivalent weight. The mode of failure of the foundation of the building is seen in this figure as a circular slide. The design experience is that the actual factor of safety of slope and building is much less than the required factor of safety ($= 3$). One has to contend with a factor of safety of 1.3 to 1.5 at the best. It is important to note that a safe edge distance from the slope is the key to ensure stability of the heavy structures.

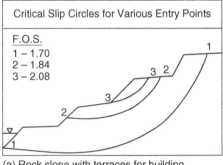

(a) Rock slope with terraces for building complexes

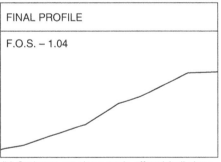

(b) Optimum cut slope angle off rock/soil slope

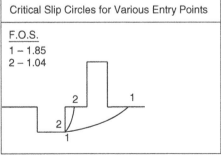

(c) Trench for excavation of foundation near a building founded on soil

Figure 12.2. Applications of SARC software.

3. Railway embankment (Fig. 12.1c). In the special case of a railway embankment, the vertical component of acceleration due to a moving railway engine is higher than the corresponding horizontal component. Actual acceleration should be measured and used in the design. The permissible dynamic factor of safety may be taken as 1.0 or as specified by the site engineers.

4. Rock slope with terraces for building complexes in hilly areas (Fig. 12.2a). Generally, terraces are made along hill slopes to support a building complex. It is necessary that stability of each terrace (say 2–2 and 3–3) and overall stability of slope (1–1) is to be checked up to be on safe side. The computer program SARC is developed specially to meet this requirement with easy input in a single computer run.

5. Optimum cut slope angle of rock/soil slope (Fig. 12.2b). The soil or rock slopes may be stabilized by excavating the excess mass above safe cut slope angle. Side slopes of dam abutments are excavated to become adequately stable. The program ASC gives the average optimum angle of the cut slope automatically. During construction, the slope should be excavated in the form of benches of stable heights along the average slope.

6. Trench for excavation of foundation near a building founded on soil (Fig. 12.2c). In urban areas, many owners wish to make buildings stand close together. This is seldom feasible as the side slope of a deep trench of foundation may collapse. The SARC program is also used to determine the safe gap between edges of two foundations.

7. Damage of foundation pit due to over-blasting (Fig. 12.3). The abutment of Gambhir Rail Bridge, JURL, India has been analysed considering high acceleration due to blasting. It is heartening to find that program SARC simulated accurately the position of cracks inside the pit of the foundation of abutment. This case history partially supports the pseudo-seismic stability analysis. The cracks have been sealed by cement grout and stitched by rock anchors which were installed across these cracks.

8. In development of sites in hilly areas, a safe edge distance is needed for locating heavy structures and high-rise building complexes. This edge distance is found by trial and error runs after increasing XEXITI with actual surcharge WI and XS = XEXITI (exit point of critical slip circle); until the desired static factor of safety, say 1.20, is achieved. Similar runs are made for seismic case to obtain XEXITI to get a dynamic factor of safety of 1.0. Finally contours are plotted on a building plan for safe edge distance (XEXITI) to locate heavy structures.

Another validation of SARC is that the position of tension crack was predicted realistically at the site of Power Grid at NJPC, HP, India. The average strength parameters obtained by field tests on the debris were used. The computed factor of safety was just less than 1.0, which also matched with the distressed condition of the steepest slope in that area. Later, a PCC retaining wall with PCC drain was designed to stabilize this rotational landslide using SARC.

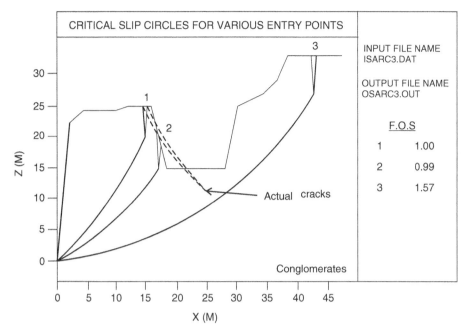

Figure 12.3. Dynamic stability of pit of foundation of abutment A2, Gambhir Bridge (Predicted crack match with actual due to overblasting).

12.3 USER'S MANUAL – SARC

THIS PROGRAM IS FOR THE STABILITY ANALYSIS OF RESERVOIR OR SUBMERGED SLOPES WITH CIRCULAR WEDGE MODE OF FAILURE [FIGS 12.4a, b]

NAME OF PROGRAM –   SARC.FOR
UNITS USED –   TONNE – METER – DEGREE

GIVE DATA IN THE FOLLOWING SEQUENCE

TITLE OF PROBLEM IN ONE LINE (<80 CHARACTERS)
N
X(I),Z(I),I = 1 to N
ROCK,RWL,XS,WI,ZC,ZWR
C,PHI,GAMA,GAMAW,BBAR,AH,AVR,EQM
NENP,(ENTX(I),ENTY(I),I = 1 to NENP)
Block 1.....................................
NEP,NOPT
XEXITI,XEXITL,GAP – This line is needed only when NEP = 0
XEXIT(I),I = 1 to NEP – This line is needed only when NEP > 0
End of block 1
Block 1 is repeated NENP number of times

ENTER   0 FOR TERMINATION

1 FOR FURTHER HELP REGARDING EXPLANATION OF TERMS IN INPUT DATA
2 FOR EXECUTION OF PROGRAM

N	=	Number of profile coordinates ($<$50)
(X,Z)	=	Coordinates of profile points ($X(I) < X(I+1)$)
ROCK	=	Reduced level of hard strata w.r.t. origin
RWL	=	Reduced level of gwt/reservoir water w.r.t. origin
XS	=	X-Coordinate of point from where surcharge starts
WI	=	Uniform surcharge intensity
ZC	=	Depth of tension crack
ZWR	=	Depth of water in tension crack/ZC
C	=	Cohesion of soil/rock
PHI	=	Angle of internal friction (Degree) of soil/rock
GAMA	=	Unit weight of soil/rock
GAMAW	=	Unit weight of pore water
BBAR	=	Pore water pressure/(GAMA* Average height of slices)
AH	=	Horizontal component of EQ. acceleration near crest of slope
AVR	=	Vertical component of EQ. acceleration/AH
EQM	=	Corresponding EQ. magnitude on Richter Scale
NENP	=	Number of entry points of slip circles ($<$10)
ENTX	=	X-Co-ordinate of entry point of circle
ENTY	=	Y-Co-ordinate of entry point of circle
NOPT	=	0, When only minimum factor of safety is required
	=	1, When all F.S. corresponding to all exit points are also required
NEP	=	Number of exit points ($<$50)
	=	0, When no individual point is given)
XEXITI	=	X-Co-ordinate of first exit point of circle
XEXITL	=	X-Co-ordinate of last exit point of circle
GAP	=	Horizontal distance between consecutive exit points
XEXIT	=	X-Co-ordinate of exit point of circle

(a) Input file name: ISARC.DAT

STABILITY ANALYSIS OF ROCK SLOPE WITH CIRCULAR SLIP SURFACE
15
0.0,0.0,100.,150.,150.,200.,200.,250.,250.,300.,300.
350.,400.,450.,500.,450.,600.,500.,700.,550.,800.,600.,
900.,625.,1000.,650.,1100.,700.,1200.,750.
−20.,0.0,0.0,0.0,2.0,1.0
30.0,35.0,2.6,1.0,0.0,0.1,0.5,8.
1,0.0,0.0
0,0
400.,1000.,50.

(b) Output file name: OSARC.DAT

STABILITY ANALYSIS OF ROCK SLOPE WITH CIRCULAR SLIP SURFACE

UNITS USED –	TONNE – METER – DEGREE
INPUT FILE NAME –	ISARC.DAT

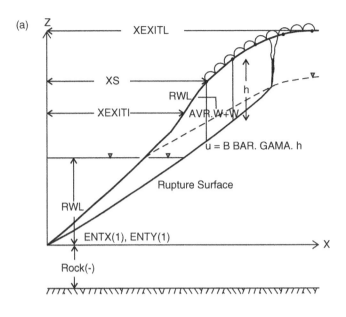

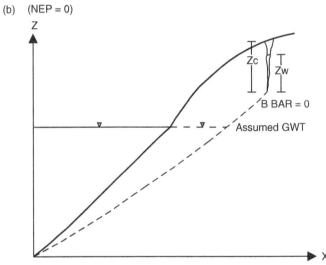

Figure 12.4. (a) Input variables for SARC. (b) Input variables for SARC (in case of no ground water table).

OUTPUT FILE NAME –   OSARC.DAT
COORDINATES OF POINTS ALONG SLOPE

X(1)	=	0.00000	Z(1)	=	0.00000
X(2)	=	100.00000	Z(2)	=	150.00000
X(3)	=	150.00000	Z(3)	=	200.00000
X(4)	=	200.00000	Z(4)	=	250.00000
X(5)	=	250.00000	Z(5)	=	300.00000

X(6)	=	300.00000	Z(6)	=	350.00000
X(7)	=	400.00000	Z(7)	=	450.00000
X(8)	=	500.00000	Z(8)	=	450.00000
X(9)	=	600.00000	Z(9)	=	500.00000
X(10)	=	700.00000	Z(10)	=	550.00000
X(11)	=	800.00000	Z(11)	=	600.00000
X(12)	=	900.00000	Z(12)	=	625.00000
X(13)	=	1000.00000	Z(13)	=	650.00000
X(14)	=	1100.00000	Z(14)	=	700.00000
X(15)	=	1200.00000	Z(15)	=	750.00000

ROCK	=	−20.000	RWL	=	0.000	XS	=	0.000	WI	=	0.000
ZC	=	2.000	ZWR	=	1.000						
C	=	30.000	PHI	=	35.000	GAMA	=	2.600	GAMAW	=	1.000
BBAR	=	0.000	AH	=	0.100	AVR	=	0.500	EQM	=	8.000

ENTX	=	0.000	ENTY	=	0.000			
NEP	=	0	NOPT	=	0			
XEXITI	=	400.000	XEXITL	=	1000.000	GAP	=	50.000

CHECK F.S. FOR -AVR ALSO

F.S.	DYN. DIS. (M)	WEIGHT OF WEDGE (T)	AH CRITICAL	COORDINATES OF CENTRE	COORDINATES OF EXIT POINT (XC, YC)	RADIUS R(M)
0.8606	0.712	0.10E + 06	0.014	(− 501.86, 954.11)	(450.00, 450.00)	1078.05

12.4   BACK ANALYSIS OF NATURAL SLOPES – BASC

12.4.1   General

The crisis in the analysis of landslides is what values of shear strength parameters should be used along potential deep-seated slip surfaces. Rarely, a slope is homogenous. Excavation of 4–5 m-deep pits show that the percentage of rock fragments increase with depth in thick debris or the extent of weathering decreases with depth in cases of rock slopes. It is, therefore, not surprising to observe that the in situ large scale shear tests on the surface of slopes have yielded in most cases lower strength parameters than those found from back-analysis of slopes in distress. Hence the practice of in situ shear tests have been discarded by The Indian Institute of Technology, Roorkee, India (previously named The University of Roorkee, India) in favour of back analysis.

Singh and Ramasamy (1979) have extended the concept of back analysis used to analyse failed slopes for the cases of stable slopes. The success of the back analysis depends on the following assumptions:

1. Potential slip surface is cylindrical and will closely follow the real rupture surface when failure takes place.
2. The assumed factor of safety is close to the real factor of safety.

3. The earthquake forces can be accounted for by the conventional method, i.e., by assuming additional static forces in vertical and horizontal directions.
4. Thin slices (at least 10 in numbers) are used in the analysis.

The real slopes are inhomogeneous and failure is three-dimensional which is initiated by the weakest zone. Hence, 2D back-analysis gives higher strength parameters of soil or rock mass. However, this error is compensated by 2D stability analysis of similar slopes by program SARC.

The method proposed herein to estimate ϕ and c involves the following steps:
1. The existing hill slopes may consist of slopes of varying height and slope angle. Choose the most critical slope (slope having least factor of safety).
2. Assign a factor of safety near unity.
3. Back analyse the slope to obtain the appropriate values of c for a given angle of internal friction ϕ.
4. Repeat 3 for various values of ϕ and slip surface. Select the appropriate values of ϕ and c.

The computer program BASC is developed based on the approach of Singh and Ramasamy (1979). This program gives back analysed values of cohesion for different anticipated values of the angle of internal friction. Sometimes, results in the output file are not obtained when centres of slip circles CX and CY are given wrongly. In that case, the program SARC may be used to find the center of the slip circle and cohesion by trial and error runs to get the desired factor of safety (1.0). Thus, coordinates of this center is fed back in the input file of BASC for detailed back analysis.

12.4.2   Selection of the critical slope

In cohesionless soils, the slope angle governs stability. In cohesive soils, both the height and the slope angle govern the stability. This should be kept in mind when choosing the critical slope. For example, in a soil having marginal cohesion, the steepest of the slopes may prove to be critical, whereas in a soil having considerable cohesion, the slope having the largest height may prove to be critical. In some situations, the critical slope may be obvious when the above criteria are applied and in others, more than one slope may have to be considered and only a back analysis can reveal which one of them is really critical.

12.4.3   Selection of the value of the factor of safety

The factor of safety should be assigned near unity depending upon to what extent the selected slope appears to be critical. The tilted trees and curved growth of trees and the age of trees are helpful in the assessment of the factor of safety. Appearance of surface cracks deeper than what can be expected due to the drying of the soil indicates that the slope is just stable. In these circumstances, a value of 1.0 to factor of safety may be more appropriate. In the absence of any of these, the slope may be stable with a factor of safety more than one and a higher value may be chosen. This needs engineering judgement and is a source of error also. The chosen critical slope is back analysed using Bishop's simplified method. Bishop's equation for factor of safety is modified to take into account the earthquake forces (Singh and Ramasamy, 1979).

12.4.4   Sensitivity studies

It is already pointed out that the factor of safety of the slope chosen for back analysis has to be assumed judiciously. In order to examine the possible effect of an error in the selection of the factor of safety on values of cohesion, the calculations for values of cohesion for various values of factor of safety are made for a slope.

A difference in factor of safety by 0.1 causes up to 30 percent change in the value of cohesion. However, even in laboratory testing of undisturbed samples, the order of error could be as high as 30 percent. Hence, an error of not more than 0.1 in the value of factor of safety may be tolerable which will keep the error in the estimation of c within acceptable limits.

In case two or more steep slopes of similar soil are back analysed, the highest value of cohesion for the anticipated internal angle of friction should be recommended. The reason is that the recommended strength parameter should give a factor of safety higher than 1.0 (or a realistic factor of safety).

12.4.5   Validation of method

Singh and Ramasamy (1979) back analysed a homogenous clayey slope at Srinagar, Garhwal, India. The back analysed value of cohesion for the actual angle of internal friction of soil is nearly the same. In case of a landslide on a non-homogeneous slope at Chilla, the back analysed strength parameters are almost equal to average of strength parameters which were obtained from laboratory direct shear tests on samples from different places. Eurocode 7 recommends a cautious estimate by adopting a value which is slightly less than the average of strength parameters found from a statistically significant number of laboratory or field tests in a homogenous soil mass (Ovesen, 2000). In general, back analysis gives higher strength parameters than those from field or laboratory tests on the surface as soil becomes denser with depth in many natural slopes.

12.5   USER'S MANUAL – BASC

THIS PROGRAM IS FOR THE BACK ANALYSIS OF SLOPE WITH CIRCULAR WEDGE MODE OF FAILURE TO DETERMINE STRENGTH PARAMETERS (C, PHI) [FIGS 12.5a, b]

NAME OF PROGRAM –   BASC.FOR

UNITS USED –   TONNE – METER – DEGREE

GIVE INPUT DATA IN THE FOLLOWING SEQUENCE

TITLE OF PROBLEM IN ONE LINE (<80 CHARACTERS)
NPHI
PHI(1),PHI(2),......,PHI(NPHI)
N
X1,Z1,X2,Z2,X3,Z3,.........XN,ZN.(profile data)
ROCK,XCUT,XHILL
NX,NY,NSET1
(Block 1) → Repeated NSET1 times.
CX,CY,ENTX,ENTY,DELX,DELY
End of block 1
NSET2
(Block 2) → Repeated NSET2 times...
GAMA,GAMAW,BBAR,ZC,ZWR,RWL,AH,AV
NFS,(FS(I),I = 1,NFS)
End of block 2.

ENTER   0 for termination
        1 for further help regarding definition of variables
        2 for execution of program

AH	=	Horizontal component of E.Q. acceleration
AV	=	Vertical component of E.Q. acceleration
BBAR	=	Pore water pressure/(Gama * Average height of slice)
C	=	Cohesion of soil/rock mass
CX,CY	=	Coordinates of center of slip circle
DELX	=	Increment in center in X direction
DELY	=	Increment in center in Y direction
ENTX	=	X-Coordinate of entry point of slip circle
ENTY	=	Y-Coordinate of entry point of slip circle
FS	=	Assumed factor of safety
GAMA	=	Unit weight of soil/rock mass
N	=	Number of profile coordinates (<50)
NPHI	=	Number of PHI values (<10)
NSET2	=	No of sets of → GAMA,GAMAW, BBAR,ZC,ZWR, RWL,AH,AV
NSET1	=	No. of sets of CX,CY,ENTX,ENTY,DELX,DELY
NFS	=	No. of sets of factor of safety(FS)
NX,NY	=	No. of center points in X and Y direction (in array)
PHI	=	Angle of internal friction (degree)
ROCK	=	Elevation of hard strata (Rock) w.r.t. origin
(X,Z)	=	Coordinates of slope profile
XCUT	=	X-Coordinate of a point before which about 10 slices should be there for proper accuracy
XHILL	=	X-Coordinate of steep/high hill (At the end of slope) exit point is not required after XHILL (Exit point of circle will be between XCUT & XHILL)
ZC	=	Depth of tension crack
ZWR	=	Depth of water in tension crack/ZC

(a) Input file name: IBASC.DAT

BACK ANALYSIS OF ROCK SLOPE WITH UNKNOWN CIRCULAR SLIP SURFACE
5
20.,25.,30.,35.,40.
17
0.0,0.0,100.,100.,150.,150.,200.,250.,250.,300.,300.,400.,
400.,425.,500.,500.,600.,500.,750.,500.,800.,525.,900.,525.,
1000.,550.,1100.,575.,1300.,625.,1500.,650.,1800.,750.
−20.,250.,1800.
3,3,2
−80.,710.,0.0,0.0,50.0,50.0
−150.,520.,0.0,0.0,50.0,50.0
1
2.6,1.,0.0,0.0,0.0,0.0,0.0,0.08,0.04
3,1.0,1.1,1.2

(b) Output file name: OBASC.DAT

BACK ANALYSIS OF ROCK SLOPE WITH UNKNOWN CIRCULAR SLIP SURFACE
UNITS USED –   TONNE – METER – DEGREE
INPUT FILE NAME –   IBASC.DAT
OUTPUT FILE NAME –   OBASC.DAT

NUMBER OF PHI VALUES = 5
VARIOUS VALUES OF PHI

20.00	25.00	30.00	35.00	40.00

PROFILE COORDINATES ALONG SLOPE

X(1)	=	0.00000	Z(1)	=	0.00000
X(2)	=	100.00000	Z(2)	=	100.00000
X(3)	=	150.00000	Z(3)	=	150.00000
X(4)	=	200.00000	Z(4)	=	250.00000
X(5)	=	250.00000	Z(5)	=	300.00000
X(6)	=	300.00000	Z(6)	=	400.00000
X(7)	=	400.00000	Z(7)	=	425.00000
X(8)	=	500.00000	Z(8)	=	500.00000
X(9)	=	600.00000	Z(9)	=	500.00000
X(10)	=	750.00000	Z(10)	=	500.00000
X(11)	=	800.00000	Z(11)	=	525.00000
X(12)	=	900.00000	Z(12)	=	525.00000
X(13)	=	1000.00000	Z(13)	=	550.00000
X(14)	=	1100.00000	Z(14)	=	575.00000
X(15)	=	1300.00000	Z(15)	=	625.00000
X(16)	=	1500.00000	Z(16)	=	650.00000
X(17)	=	1800.00000	Z(17)	=	750.00000

ROCK	=	− 20.000	XCUT	=	250.000	XHILL	=	1800.000
NX	=	3	NY	=	3	NSET1	=	2
CX(1)	=	−80.000	CY(1)	=	710.000	ENTX1	=	0.000
ENTY1	=	0.000	DELX1	=	50.000	DELY1	=	50.000
CX(2)	=	− 150.000	CY(2)	=	520.000	ENTX2	=	0.000

ENTY2	=	0.000	DELX2	=	50.000	DELY2	=	50.000

NSET 2 = 1

GAMA	= 2.600	GAMAW	= 1.000	BBAR	= 0.000	ZC	= 0.000
ZWR	= 0.000	RWL	= 0.000	AH	= 0.080	AV	= 0.040

FOS   = 1.000

**PHI*	COHESION	COORDINATE OF CRITICAL CENTER		*XEXITC*	*YEXITC*
	(MAX)	X	Y		
20.000	120.809	− 80.00	760.00	638.29	500.00
25.000	90.881	− 80.00	710.00	602.91	500.00
30.000	65.192	− 100.00	620.00	516.40	500.00
35.000	45.894	− 150.00	570.00	424.26	443.20
40.000	26.707	− 150.00	520.00	381.67	420.42

FOS = 1.100

**PHI*	COHESION	COORDINATE OF CRITICAL CENTER		*XEXITC*	*YEXITC*
	(MAX)	X	Y		
20.000	144.132	− 30.00	710.00	648.60	500.00
25.000	113.397	− 80.00	760.00	638.29	500.00
30.000	83.697	− 100.00	620.00	516.40	500.00
35.000	60.323	− 150.00	570.00	424.26	443.20
40.000	40.802	− 150.00	570.00	424.26	443.20

FOS = 1.200

**PHI*	COHESION	COORDINATE OF CRITICAL CENTER		*XEXITC*	*YEXITC*
	(MAX)	X	Y		
20.000	167.935	− 30.00	710.00	648.60	500.00
25.000	136.255	− 80.00	760.00	638.29	500.00
30.000	104.199	− 80.00	710.00	602.91	500.00
35.000	76.694	− 100.00	620.00	516.40	500.00
40.000	55.235	− 150.00	570.00	424.26	443.20

12.6   A SAFE ANGLE OF CUT SLOPE – ASC

Rotational slides may take place in rock slopes where all the joint sets are favorably oriented and rock mass is fractured extensively. It is essential that these rock and soil slopes are excavated at a safe angle of slope which ensures a desired factor of safety. Subsurface drainage of slope is equally important although it is very costly.

The computer program SARC has been modified as ASC to find out the safe angle of cut slope which is prone to circular slide. The output also gives the coordinates of a modified slope surface. Program ASC may be used for recommending a safe angle of cut slope in a soil slope also.

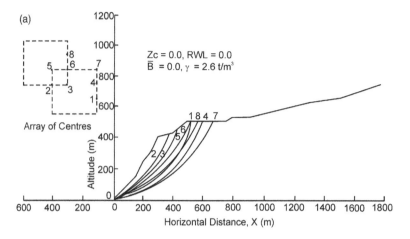

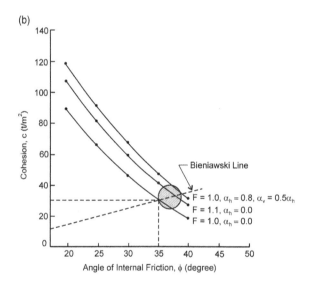

Figure 12.5. (a) Input variables for BASC. (b) Output results of BASC.

Abutment of all types of dams must be excavated to very safe angle of slope, even if all joint sets are favorably oriented and no rock wedge is sliding. The program ASC is specifically suitable for such analysis for dams.

Hoek and Bray (1981) have mentioned that the plan curvature of a rock slope has an important influence on the safe angle of cut slope. The angle of slope is significantly more in the case of a concave slope than that in a convex slope. Their recommendations have been incorporated in this computer program. Their observations have been found to be valid in slopes in soft rocks in lower Himalaya.

Extensive experience in the Himalayan region shows that a rock or soil slope is seldom homogeneous as generally assumed in the potentially circular slide.

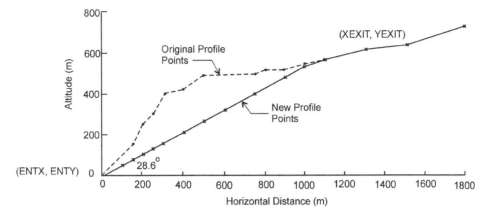

Figure 12.6. Optimum cut slope angle (c = 30 t/m², ϕ = 35°, γ = 2.6 t/m³, BBAR = 0.0, α_h = 0.25, α_v = 0.125, Zc = 0.0, RWL = 0.0, required factor of safety = 1.0).

The quality of rock mass improves with depth, due to a lesser degree of weathering and fracturing. As such, the strength parameters of rock mass are likely to be significantly more in the case of deeper slip circles. Based on the consultancy experience, it is recommended that design strength parameters should be derived from the back analysis of steepest and distressed rock slopes in similar rock masses, otherwise the computed angle of cut slope may be over-conservative.

12.7   USER'S MANUAL – ASC

CALCULATION OF OPTIMUM ANGLE OF CUT SLOPE WITH CIRCULAR WEDGE MODE OF FAILURE (FIG. 12.6)

NAME OF PROGRAM –   ASC.FOR
UNITS USED –   TONNE – METER – DEGREE

GIVE INPUT DATA IN THE FOLLOWING SEQUENCE

N,NCASE
X1,Z1,X2,Z2,X3,Z3,.........XN,ZN
XEXITL,AD
ENTX,ENTY,XS,WI
C,PHI,ROCK,GAMA,R1
Block 1. [Repeated NCASE times]
BBAR,AH,AVR,ZWR,GAMAW,ZC,RWL,FOS
End of block 1

DO YOU WANT HELP REGARDING DEFINITIONS OF VARIABLES USED
ENTER   0 For Termination
        1 For Help
        2 For Execution

AH	=	Horizontal component of EQ acceleration
AV	=	Vertical component of EQ acceleration
AVR	=	AV/AH
AD	=	Interval of exit points
BAR	=	Pore water pressure/(GAMA * Av. ht. of slice)
C	=	Cohesion of soil/rock
ENTX	=	X-Co-ordinate of entry point of slip circle
ENTY	=	Y-Co-ordinate of entry point of slip circle
FOS	=	Allowable Factor of safety of cut slope
GAMA	=	Unit weight of soil/rock
GAMAW	=	Unit weight of water
N	=	Number of profile co-ordinates (<50)
NCASE	=	Number of cases of line → BBAR,AH,AVR,ZWR,...
PHI	=	Angle of internal friction
R1	=	Radius of curvature of slope at base in plan (positive for concave)
ROCK	=	Reduced level of hard strata (Rock) w.r.t. origin
RWL	=	Reduced level of GWT/reservoir water w.r.t. origin
SAI	=	Angle of required cut slope (degree)
WI	=	Surcharge intensity
(X,Z)	=	Coordinates of slope profile
XEXITI	=	X-Co-ordinate of first exit point
XEXITL	=	X-Co-ordinate of last exit point
XS	=	X-co-ordinate of starting point of surcharge
ZC	=	Depth of tension crack
ZW	=	Depth of water in tension crack
ZWR	=	ZW/ZC

NOTE: Please give the profile data at close intervals especially in the case of soil/rock
of high strength where the cut slope angle is likely to be steep.

(a) Input file name: IASC.DAT

OPTIMUM CUT SLOPE ANGLE OF ROCK SLOPE WITH CIRCULAR SLIP SURFACE
20,1
0.0,0.0,100.,100.,150.,150.,200.,250.,250.,300.,300.,400.,400.,425.,500.,500.,600.,500.,750.,500.,
800.,525.,900.,525.,1000.,550.,1100.,575.,
1300.,625.,1500.,650.,1800.,750.,1900.,750.,2000.,750.,2200.,750.,
1800.,50.
0.0,0.0,0.0,0.0
30.,35., −20.,2.6,1000.
.0,.25,.5,0.0,1.0,0.0,0.0,1.

(b) Output file name: OASC.DAT

OPTIMUM CUT SLOPE ANGLE OF ROCK SLOPE WITH CIRCULAR SLIP SURFACE

UNITS USED –		TONNE – METER – DEGREE		
INPUT FILE NAME –		IASC.DAT		
OUTPUT FILE NAME –		OASC.DAT		
N	=	20	NCASE =	1
XEXITL	=	1800.000	AD =	50.000

ENTX	=	0.000	ENTY	=	0.000	XS	=	0.000		
WI	=	0.000								
C	=	30.000	PHI	=	35.000	ROCK	=	-20.000	GAMA	= 2.600
R1	=	1000.000								

Case No. 1

BBAR	=	0.000	AH	=	0.250	AVR	=	0.500	ZWR	=	0.000
GAMAW	=	1.000	ZC	=	0.000	RWL	=	0.000	FOS	=	1.000

CHECK FOS FOR -AVR ALSO
RECOMMENDED CUT SLOPE ANGLE = 28.600(Deg.) FOR FOS = 1.038

CO-ORDINATES OF POINTS ALONG SLOPE/CUT SLOPE

X(1)	=	0.000	Z(1)	=	0.000	ZL(1)	=	0.000
X(2)	=	100.000	Z(2)	=	100.000	ZL(2)	=	54.522
X(3)	=	150.000	Z(3)	=	150.000	ZL(3)	=	81.783
X(4)	=	200.000	Z(4)	=	250.000	ZL(4)	=	109.044
X(5)	=	250.000	Z(5)	=	300.000	ZL(5)	=	136.305
X(6)	=	300.000	Z(6)	=	400.000	ZL(6)	=	163.566
X(7)	=	400.000	Z(7)	=	425.000	ZL(7)	=	218.088
X(8)	=	500.000	Z(8)	=	500.000	ZL(8)	=	272.610
X(9)	=	600.000	Z(9)	=	500.000	ZL(9)	=	327.132
X(10)	=	750.000	Z(10)	=	500.000	ZL(10)	=	408.914
X(11)	=	800.000	Z(11)	=	525.000	ZL(11)	=	436.175
X(12)	=	900.000	Z(12)	=	525.000	ZL(12)	=	490.697
X(13)	=	1000.000	Z(13)	=	550.000	ZL(13)	=	545.219
X(14)	=	1100.000	Z(14)	=	575.000	ZL(14)	=	575.000
X(15)	=	1300.000	Z(15)	=	625.000	ZL(15)	=	625.000
X(16)	=	1500.000	Z(16)	=	650.000	ZL(16)	=	650.000
X(17)	=	1800.000	Z(17)	=	750.000	ZL(17)	=	750.000
X(18)	=	1900.000	Z(18)	=	750.000	ZL(18)	=	750.000
X(19)	=	2000.000	Z(19)	=	750.000	ZL(19)	=	750.000
X(20)	=	2200.000	Z(20)	=	750.000	ZL(20)	=	750.000

REFERENCES

Bishop, A.W. 1954. The use of slip circle in the stability analysis of slopes, *Geotechnique* 5(1): 7–17.

Deoja, B., Dhital, M., Thapa, B. and Wagner, A. 1991. *Mountain Risk Engineering Handbook*. Kathmandu, International Centre of Integrated Mountain Development, Part I, Chaps 10 and 13 and Part II, Section 24.9.

Jansen, R.B. 1990. Estimation of embankment dam settlement caused by earthquake, *Water Power and Dam Construction*, December, pp. 35–40.

Hoek, E. and Bray, J.W. 1981. *Rock Slope Engineering*. London, Institution of Mining and Metallurgy, Revised Third Edition, Chaps 5, 9 and 12.

Ovesen, N.K. 2000. *Workshop on Limit Analysis in Design Codes of Geotechnical Engineering*. University of Roorkee, February 24, 2000, p. 70.

Singh, Bhawani and Ramasamy, G. 1979. Back-analysis of natural slopes for evalution of strength parameters. *Int. Conf. on Computer Applications in Civil Engineering*, Roorkee, India, Vol. I, pp. VII-57–62.

Taylor, D.W. 1948. *Fundamentals of Soil Mechanics*. New Delhi, India, Asia Publishing House, Section 16.34, p. 700.

CHAPTER 13

Debris slide

*"O Mother Earth, enrich me with the wisdom so that I should not
damage or degrade you, wherever, I dig out, should quickly be
regenerated and covered with greenery"*
Atharv Veda

13.1 THE DEBRIS SLIDE

A vast majority of landslides are debris/talus slides on hill slopes. The debris
slides or talus slides occur generally along hill roads and rail lines after a long
rainfall. The mechanics of debris slides is fortunately simple. The loose debris/
talus/slope wash/colluvium gets deposited on hill slopes due to weathering of
rocks in the upper regions. A long rainfall for 5–10 days results in temporary
formation of a ground water table on the relatively impervious rock slope. In the
case of thin layers of debris, the ground water table may rise up to the slope
surface. Consequently, a thin layer of water-charged debris may slide down due
to high uplift seepage water pressure. It may be noted that the factor of safety of
cohesionless debris is reduced to half of that of the dry slope. A typical case is
shown in Figure 13.1. Contrary to popular belief, debris slides do not occur in
thick layers of debris because the water table may not rise close to the slope
surface in the same region.

One may not notice this temporary ground water table after a landslide has
taken place because ground water seeps out soon after rains. If 3−5 m-deep pits
are made on the slope surface, one may notice that there is slushy or wet soil
below relatively dry soil in each pit. The boundary between slushy and dry soil
represents the temporary (perched) ground water table (Fig.13.2).

It is inspiring to observe how Mother Nature manages the control of landslides.
Her choice is for the bio-engineering solution. Since the debris slides tend to take
place on thin layers due to water charging during long rains, the trees with
4−6 m-deep roots reinforce the debris into the rock slope. The grass is also
grown by Mother Nature on the surface of the slope. The humus layer generated
by grass acts as an impermeable membrane and slows down seepage of rain

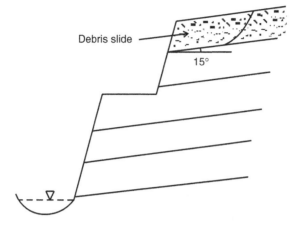

Figure 13.1. Bridge abutment in Pauri-Garhwal, India.

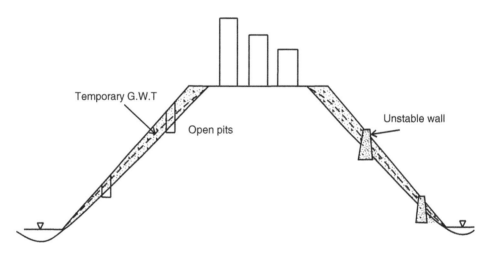

Figure 13.2. Paper mill site in Nagaland, India.

water inside the debris. Thus, the stability of thin debris slides is managed by nature. It is also interesting to note that chances of resonance of thin debris layers are low during earthquake according to Section 4.7. Unfortunately, her wonderful eco-regeneration is being destroyed by forest mafia. However, the solution still remains the same, which is the most economical, ecological, beautiful, effective, natural and sustainable.

13.2　THE PROGRAM SAST

The computer program SAST has been developed by Ranjan et al. (1984) to analyse debris slides and suggest automatically the most suitable remedial measure. The possible control measures are given by:

- Forestation and development of grass on the water-charged steeper slopes,
- Use of the sub-surface drains to lower the ground water table in flatter slopes,
- Excavation of excess thickness of layer of cohesive debris in steeper slopes.

The computer program assumes uniform thickness of layer of debris. In practice, average thickness of non-uniform debris layer is considered for the analysis. In the case of complex debris slides with non-uniform rock slope and non-uniform debris cover, computer software SANC discussed in Chapter 14 is recommended.

The details of a case of sub-surface pipe drainage system are shown in Figure 13.3. The network of sub-surface drains is according to the shape of debris slide. The cross-sections of the drains are also illustrated in this figure. Many such landslide stabilisations have been done in Nepal with success. This technique was invented by ITECO International, Switzerland. The authors recommend its use in other countries also.

Another case of the application of SAST is shown in Figure 13.2. To develop asite for a paper mill in Nagaland, India, a hill was cut at the top and the excavated material was dumped on the downhill slopes. During rains, the loose debris started sliding downwards, which aroused the fears of deep-seated instability of the entire hill. The investigation indicated that it was a case of debris slide. The stabilisation of the flatter slope was done by sub-surface trench drains. The stabilisation of the steeper slope was done by excavating the excess thickness of debris. Further grass cover and forestation was implemented to improve the stability of debris slides. It may be mentioned that retaining walls were tried earlier. These gave poor results, as earth pressure due to the debris slide was very high. The program SAST suggests the suitable depth of sub-surface drains for a given factor of safety.

13.3   SURFACIAL LANDSLIDES

Surfacial landslides have occurred in Southern California (Pradel and Raad, 1993), Nepal Himalaya (Deoja et al., 1991) and Indian Himalaya. Surfacial landslides are found to occur in soil slopes of silty soil up to a depth of 2 m during a prolonged period of heavy rains. The rain water seeps into the partially saturated soil slopes due to gravity and capillary action. The seepage water collects on a shallow impervious (silt/clay) layer/seam if any below the slope surface. The surfacial landslide takes place along the same due to increase in the uplift pressure and saturation of the silt/clay layer/seam. The depth (Z) of the impervious layer may be found from an electrical resistivity survey (Deoja et al., 1991). Then program SAST may be used to predict the factor of safety ($Z_w = 0$) of surfacial slides, considering strength parameters along the silt/clay layers.

Input variables of program SAST are given in Figure 13.4. The program SAST has been used extensively in consultancy in the Himalayas of India and Nepal with amazing results.

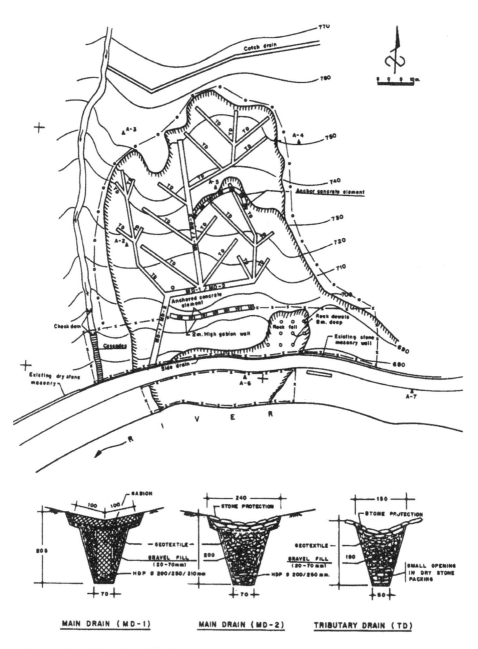

Source: Road Flood Rehabilitation Project, DOR, HMG Nepal, 1991

Figure 13.3. Landslide stabilization by sub-surface pipe drains (Deoja et al., 1991).

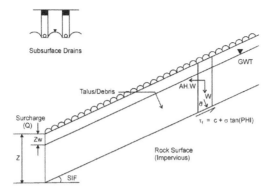

Figure 13.4. Input variables for SAST.

13.4 USER'S MANUAL – SAST

THIS PROGRAM IS FOR THE STABILITY ANALYSIS OF SLOPE WITH TALUS DEPOSIT
(FIG. 13.4)

NAME OF PROGRAM –   SAST.FOR
UNITS USED –   TONNE – METER – DEGREE
GIVE DATA IN THE FOLLOWING SEQUENCE
NO. OF SLOPES
TITLE OF PROBLEM IN ONE LINE (<80 CHARACTERS)
C,PHI,GAMA,GAMAW,Z,ZW,SIF,AH,Q,FS,EFFCY,EQM
(NUMBER OF ABOVE TWO LINES = NO. OF SLOPES)

ENTER   0 FOR TERMINATION
            1 FOR FURTHER HELP
            2 FOR EXECUTION

AV	=	Coefficient of vertical earthquake acceleration
AH	=	Coefficient of horizontal earthquake acceleration
EQM	=	Corresponding EQ. magnitude on Richter Scale
C	=	Cohesion of talus deposit
EFFCY	=	Efficiency of drains (generally observed to be 0.50. It is more for higher K(horz.)/K(vert.))
FS	=	Allowable factor of safety in static condition
GAMA	=	Unit weight of talus (saturated)
GAMAW	=	Unit weight of water
No	=	Number of cases to be analysed
PHI	=	Angle of internal friction of talus deposit
Q	=	Surcharge on slope(live) (for dead load increase Z and ZW by Q/GAMA)
SAI	=	Dip of slope face = dip of rock slope
Z	=	Average vertical depth of talus/debris deposit
ZW	=	Vertical depth of ground water during (worst) rainy season below slope surface
	=	Depth of wet soil below slope in off season

NOTE: THIS PROGRAM MAY ALSO BE USED FOR DEBRIS/DIP/REGOLITH SLOPES
         USING STRENGTH PARAMETERS OF PLANE OF SLIP. FOR SURFACIAL SLIDE
         IN SILTY SOIL DUE TO LONG SPELL OF RAINS, ASSUME ZW = 0.0 AND
         Z = OBSERVED DEPTH OF SATURATION BELOW SLOPE SURFACE.

(a) Input file name: ISAST.DAT

3
STABILITY ANALYSIS OF TALUS SLOPE OF DEPTH 4.87 M
.565,25.,2.0,1.0,4.87,0.00,16.5,0.12,0.0,1.30,0.50,6.
STABILITY ANALYSIS OF TALUS SLOPE OF DEPTH 4.87 M
.565,25.0,2.0,1.0,4.87,0.00,16.5,0.12,0.0,1.50,0.50,6.
STABILITY ANALYSIS OF TALUS SLOPE OF DE PTH 2.5 M
.565,25.0,2.0,1.0,2.50,1.50,31.0,0.12,0.0,1.50,0.50,6.

(b) Output file name: OSAST.DAT

STABILITY ANALYSIS OF TALUS SLOPE OF DEPTH 4.87 M

UNITS USED –	TONNE – METER – DEGREE
INPUT FILE NAME –	ISAST.DAT
OUTPUT FILE NAME –	OSAST.DAT

Case No. 1

C	=	0.565	PHI	=	25.000	GAMA	=	2.000	GAMAW	=	1.000
Z	=	4.870	ZW	=	0.000	SIF	=	16.500	AH	=	0.120
AV	=	– 0.060	EQM	=	6.000	Q	=	0.000	FS	=	1.300

FACTOR OF SAFETY WITH DIFFERENT CONDITIONS		CRITICAL ACCELERATION	DYNAMIC DISPLACEMENT(M)	
FS1(No Surcharge & E.Q., But Dry)	=	1.787		
FS2(With Surcharge & W.T., But No E.Q.)	=	1.000		
FS3(No Surcharge & E.Q., But W.T.)	=	1.000		
FS4(No Surcharge, With E.Q. & Dry)	=	1.217	0.221	0.00
FS5(No Surcharge, With E.Q. & W.T.[WORST])	=	0.632	0.000	1.94

Measure adopted to get required factor of safety →

SPACING OF DRAIN	=	18.55
DEPTH OF DRAIN	=	3.71
EFFICIENCY OF DRAIN	=	0.50
AVERAGE VERTICAL DEPTH OF W.T.	=	1.86

PROVIDE CARPET OF GREEN GRASS AND BUSHES OVER HILL TO REDUCE THE RATE OF INFILTRATION INSIDE THE SLOPE MATERIAL. BUSHES WILL ALSO REDUCE EROSION AND ARE EASY TO MAINTAIN. BUSHES OF ROOT DEPTH > Z SHOULD BE CHOSEN.

STABILITY ANALYSIS OF TALUS SLOPE OF DEPTH 4.87 M

UNITS USED –	TONNE – METER – DEGREE
INPUT FILE NAME –	ISAST.DAT
OUTPUT FILE NAME –	OSAST.DAT

Case No. 2

C	=	0.565	PHI	=	25.000	GAMA	=	2.000	GAMAW	=	1.000
Z	=	4.870	ZW	=	0.000	SIF	=	16.500	AH	=	0.120
AV	=	− 0.060	EQM	=	6.000	Q	=	0.000	FS	=	1.500

FACTOR OF SAFETY WITH DIFFERENT CONDITIONS		CRITICAL ACCELERATION	DYNAMIC DISPLACEMENT(M)
FS1(No Surcharge & E.Q.,But Dry)	= 1.787		
FS2(With Surcharge & W.T.,But No E.Q.)	= 1.000		
FS3(No Surcharge & E.Q., But W.T.)	= 1.000		
FS4(No Surcharge, With E.Q. & Dry)	= 1.217	0.221	0.00
FS5(No Surcharge, With E.Q. & W.T.[WORST])	= 0.632	0.000	1.94

Measures adopted to get required factor of safety →

DEPTH OF EXCAVATION REQUIRED = 3.415 FOR FACTOR OF SAFETY(3) = 1.50

STABILITY ANALYSIS OF TALUS SLOPE OF DEPTH 2.5 M

UNITS USED –	TONNE – METER – DEGREE
INPUT FILE NAME –	ISAST.DAT
OUTPUT FILE NAME –	OSAST.DAT

Case No. 3

C	=	0.565	PHI	=	25.000	GAMA	=	2.000	GAMAW	=	1.000
Z	=	2.5000	ZW	=	1.500	SIF	=	31.000	AH	=	0.120
AV	=	−0.060	EQM	=	6.000	Q	=	0.000	FS	=	1.500

FACTOR OF SAFETY WITH DIFFERENT CONDITIONS		CRITICAL ACCELERATION	DYNAMIC DISPLACEMENT(M)
FS1(No Surcharge E.Q., But Dry)	= 1.030		
FS2(With Surcharge & W.T.,But No E.Q.)	= 0.877		
FS3(No Surcharge & E.Q., But W.T.)	= 0.877		
FS4(No Surcharge, With E.Q. & Dry)	= 0.816	0.014	0.09
FS5(No Surcharge, With E.Q. &W.T.[WORST])	= 0.679	0.000	3.70

Measures adopted to get required factor of safety →

DEPTH OF EXCAVATION REQUIRED = 1.925 FOR FACTOR OF SAFETY(3) = 1.50

13.5   BACK ANALYSIS OF DEBRIS SLIDE – BAST

The problem of obtaining the values of strength parameters in analysis of a debris slope is equally disturbing. Evidently, back analysis of a nearby debris slide is a practical way of assessment of overall values of strength parameters. As such the computer program BAST has been developed to back analyse the relationship between cohesion and angle of internal friction. Sometimes the results of direct shear test on the saturated soil samples (fine material) from debris slides are superposed on c–ϕ curve for improving judgement of probable strength parameters of the debris (Fig.13.5). In case of cohesionless debris (colluvium), the selection of angle of internal friction as ϕ=0 is obvious.

Debris slides may occur along residual soils in hills. Knowledge of shear strength of residual soil is, therefore, helpful. Experience of back analysis suggests that residual soils are cohesive soils and their cohesion helps in stabilizing debris slides. Onalp (1988) has given analysis of 90 case histories and values of strength parameters.

The program BAST gives a set of values of c for various inputs of ϕ. The relationship between c and ϕ is shown in Figure 13.5. Although back analysis does not yield unique values of c and ϕ from a single debris slide, back analysis helps the young engineers from being over conservative.

Limited experience shows that the natural slopes are generally inhomogeneous and debris slopes are no exception. Therefore, direct shear tests on various samples from different places of debris provide widely different values of its strength parameters. The law of averages may be adopted here also and the average of c and ϕ represents the overall strength parameters of debris. In case of well-defined inhomogeneity of debris slopes, the computer program SANC is suggested.

It is really not necessary to mention that conditions leading to a debris slide should be simulated realistically in the program BAST. Generally,

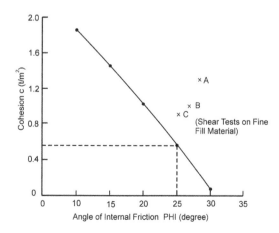

Figure 13.5. Output results of BAST (back analysis of talus slope).

debris slides take place during a long spell of continuous rain. Nevertheless, an earthquake of high intensity could also trigger off a debris slide even during mild rains.

13.6   USER'S MANUAL – BAST

THIS PROGRAM IS FOR THE BACK ANALYSIS OF SLOPE WITH TALUS DEPOSIT (FIG. 13.5)

NAME OF PROGRAM –   BAST.FOR
UNITS USED –   TONNE – METER – DEGREE

GIVE DATA IN THE FOLLOWING SEQUENCE

TITLE OF PROBLEM IN ONE LINE (< 80 CHARACTERS)
GAMA,GAMAW,Z,ZW,SIF,AH,F
NPHI
PHI(I),I = 1, NPHI

ENTER   0 FOR TERMINATION
              1 FOR FURTHER HELP
              2 FOR EXECUTION

AV	=	Coefficient of vertical earthquake acceleration
AH	=	Coefficient of horizontal earthquake acceleration
C	=	Cohesion of talus deposit
F	=	Assumed factor of safety of the talus slope
GAMA	=	Unit weight of talus (saturated)
GAMAW	=	Unit weight of water
NPHI	=	Number of PHI values
PHI	=	Angle of internal friction of talus deposit
SIF	=	Dip of slope face = dip of rock slope
Z	=	Average vertical depth of talus deposit
ZW	=	Vertical depth of perched water table during (worst) rainy season below ground level
	=	Depth of wet soil in off season

NOTE: THIS PROGRAM MAY ALSO BE USED FOR BACK ANALYSES OF DIP SLOPES AND REGOLITH

(a) Input file name: IBAST.DAT

BACK ANALYSIS OF A FAILED TALUS SLOPE AT NAGALAND
2.0,1.0,4.87,0.00,16.5,0.00,1.0
5
10.,15.,20.,25.,30.

(b) Output file name: OBAST.DAT

BACK ANALYSIS OF A FAILED TALUS SLOPE AT NAGALAND

UNITS USED –   TONNE – METER – DEGREE
INPUT FILE NAME –   IBAST.DAT

OUTPUT FILE NAME –   OBAST.DAT

UNIT WEIGHT OF TALUS	=	2.0000
UNIT WEIGHT OF WATER	=	1.0000
DEPTH OF TALUS DEPOSIT	=	4.8700
DEPTH OF WATER TABLE	=	0.0000
SLOPE ANGLE	=	16.500
COEF.OF HORZ.EARTHQUAKE ACCN	=	0.0000
FACTOR OF SAFETY-ASSUMED	=	1.0000

ANGLE OF FRICTION	COHESION
10.000	1.863
15.000	1.453
20.000	1.023
25.000	0.565
30.000	0.068

REFERENCES

Deoja, B., Dhital, M., Thapa, B. and Wagner, A. 1991. *Mountain Risk Engineering Handbook*, Part I, II and III. Kathmandu, International Centre of Integrated Mountain Development.

Onalp, A. 1988. Slope Stability problems on the south-eastern coast of Black Sea, *Proc. 5th Int. Sym. Landslides*, Vol. 1, pp. 275–278.

Pradel, D. and Raad, G. 1993. Effect of permeability on surfacial stability of homogeneous slopes. *J. Geotech. Eng., ASCE* 119(2): 315–332.

Ranjan, G., Singh, Bhawani, Saran, S., Viladkar, M.N. and Khazanchi, A.C. 1984. Stability analysis and protective measures for building complex on slopes. *Int. Symp. Landslide*, Toronto.

CHAPTER 14

A stability analysis of slopes with non-circular slip surface and non-vertical slices – SANC

"All things by immortal power near or far, hiddenly to each other are linked"
Francis Thompson (English Victorian Poet)

14.1   INTRODUCTION

Landslides are complex in natural and non-homogeneous slopes. For the analysis of stability of complex landslides, numerous methods are available, differing mainly in handling the degree of indeterminacy of the problem, shape of slip surface and slices. Sometimes, owing to the inherent structural weak planes in the sliding mass or other reasons, it becomes essential to consider slip surface as non-circular and slice as non-vertical and non-parallel. From observations it is known that sliding, depending on the main discontinuities of the rock, may take place along polygonal shaped surfaces (Fig. 14.1). For kinematic reasons, sliding on such 'external' polygonal surfaces involves failure within the sliding slope mass as well, i.e., a sufficient number of internal slip surfaces will develop (Kovari and Fritz, 1978). Taylor (1948) suggested to replace actual slip surface by circular slip surface but to use actual strength parameters for each slice. Most of the available methods cannot handle both the non-circular slip surface and non-vertical slices. Sarma's (1979) and Kovari and Fritz's (1978, 1984) methods can handle such problems, but these are very complex, especially for beginners and field engineers. Details of these methods are not given as these are used frequently and their computer programs are available (Hoek, 1987). Ramamurthy (1985) developed a variational calculus method of stability analysis and plotted charts for soil and rock slopes for the static case.

14.2   THE PROGRAM SANC

A new method is being presented here which makes use of simple equations of statics and which can handle all types of slip surfaces and shapes of slices. Above

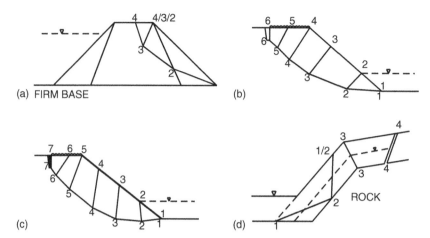

Figure 14.1. (a) Non-homogeneous earth dam with slices along potential planes of weakness or failure. (b) Rock slope with slices along pre-existing joints. (c) Circular failure with radial slices to give minimum factor of safety in homogeneous slopes. (d) Debris slide.

all, the proposed method is easily comprehensible. Owing to its generality, this method can be used for analysing the static and dynamic stability of non-homogenous earth dams, complex talus/debris slides, complex landslides and almost all types of slope failures due to sliding. The method is presented in the form of a computer program SANC.

Goodman and Kieffer (2000) have suggested a classification of rock slopes on the basis of complex mechanism of failure. The program SANC may be used to analyse many complex failures of rock slopes. The sophisticated computer program UDEC is recommended for non-linear dynamic analysis of other slopes. Experts should also be consulted.

The software SANC also computes the dynamic settlement of a slope approximately during an earthquake. In the proposed analysis, it is assumed that over-toppling of the slice does not occur. It fails only in sliding. The earthquake force is assumed to act on the saturated weight of the slice both above and below the ground water table. The program considers the pore water pressure as product of γ_w and the head of water, which is the difference in the elevation between GWT and the point under consideration.

Figure 14.1 shows some cases where the use of non-circular slip surface and non-vertical slice becomes inevitable.

1. Figure 14.1a. This is the case of a non-homogeneous earth dam with a firm base or foundation. For the stability of downstream slope, the slip surface 1-2-3-4 should be considered. Here 2-2 is the joint between the outer body and inner core, i.e. junction of non-homogeneity and hence it is the potential plane of sliding. Slip surface 1-2-3-4 is non-circular and the slope can be analysed by considering three slices and trying various inclinations of boundary 1-2-3-4

and finally choosing the one which gives a minimum factor of safety for the slope.

The recently built concrete-faced rockfill dams (CFRD) are performing very well all over the world. Here a reinforced concrete slab on the upstream face constitutes the impervious membrane, with the main body consisting exclusively of rockfill which remains dry. CFRD is more economical than earth and rockfill dams, even in seismic regions (Jatana, 1999). The program SANC may be used for preliminary design of CFRD considering slices parallel to the concrete slab.

2. Figure 14.1b. This is the case of rock slope in the rock mass with pre-existing joint sets 1-2, 2-3, 3-5 and 5-6 etc. Here 6-6 is a tension crack. In the proposed method, the inter-slice boundaries are not merely fictitious lines but they are actual sliding planes/joint planes. In this case, the inter-slice movement will occur along plane 2-2, 3-3, 4-4 and 5-5. The slip surface will be 1-2-3-4-5-6 (non-circular) and there will be five slices.

3. Figure 14.1c. Sarma (1979) has shown that the most critical slice has side inclinations which are approximately normal to the basal failure surface. Hence in the case of conventional circular slip surface, inter-slice boundaries should be considered approximately normal to the failure surface rather than by considering them vertical (Hoek, 1987).

4. Figure 14.1d. Generally, rock surface is undulating and talus or debris is also non-uniform. The slip surface is 1-2-3-4 and inter-slice boundaries are 2-2, 3-3 and non-vertical tension crack is 4-4.

Singh et al. (1996) derived all the equations for calculating the overall factor of safety of a complex landslide. The various equations are required to be solved iteratively. Hence, a computer program (SANC.FOR) has been developed by Shekhawat (1993) to solve different slope stability problems. A computer program (SARMA.BAS) written by Hoek (1987), based upon Sarma's (1979) method was also available. The results obtained by both the programs, i.e. SANC.FOR and SARMA.BAS were compared and a close comparison in the results was obtained as shown in Table 14.1.

The results are considered unacceptable when the normal stress at the base or side becomes tensile. Many slip surfaces and orientation of slices should be analysed by SANC to get the minimum factor of safety. Only this minimum factor of safety is true factor of safety of a slope.

14.3 APPLICATIONS

Three sites at Lal Bahadur Shastri Academy of Administration, Mussoorie, India were analysed using the SANC program to suggest remedial measures for the stability of buildings. These sites have steeper bedding planes in soft shale at the top terrace and flatter towards the toe of the hill. The computer program predicted negligible settlement due to earthquakes. Later, no settlement was

Table 14.1. Summary of results by SANC and SARMA methods.

S.No.	Type of slope	Comparison of results
1	Dry homogenous slope	Results from the two programs are in close agreement.
2	Submerged homogenous	Deviation in results: '0.05 to 0.08' in factors of safety and almost negligible in critical accelerations with results of SANC.FOR on lower side. The deviation is less than 5 percent on the safer side.
3	Submerged non-homogenous	Deviation in results: '0.1 to 0.2' in factor of safety and '0.1' in critical accelerations with results of SANC.FOR on lower side. The deviation is less than 10 percent on the safer side.

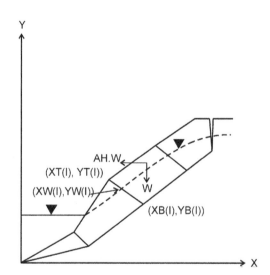

Figure 14.2. Input variables for SANC.

observed during the Uttarkashi earthquake in 1991. The program has been used at many other sites. Unlike the program SARMA.BAS, SANC fortunately did not give the problem of convergence and unrealistic results at any of the above three sites and for other slopes (Sharma, 1996; Deoja et al., 1991).

14.4   CONCLUDING REMARKS

On the basis of computations it may be concluded that the results obtained are reasonably accurate (maximum deviation is 10 percent for submerged non-homogeneous slopes) and the method is simple and easy to understand. Hence, the proposed program SANC is easy to use for analyzing the non-homogeneous earth dam, talus/debris slides, planar rock slides, complex landslides and other types of failures of slopes in seismic area.

14.5   USER'S MANUAL – SANC

PROGRAM FOR STABILITY ANALYSIS OF SLOPE WITH NON-CIRCULAR SLIP SURFACE AND NON-VERTICAL SLICES (FIG. 14.2)

NAME OF PROGRAM –	SANC.FOR
UNITS USED –	TONNE – METER – DEGREE

GIVE INPUT DATA IN FOLLOWING SEQUENCE

TITLE OF PROBLEM IN ONE LINE (<80 CHARACTERS)
N
AH AVR EQM GAMAW R(N1)
XT(I) YT(I) XB(I) YB(I) XW(I) YW(I) CS(I) PHIS(I)
[PLEASE REPEAT ABOVE LINE N + 1 TIMES FOR ALL N + 1 SIDES]
C(I) PHI(I) GAMA(I)
[PLEASE REPEAT ABOVE LINE N TIMES FOR BASES OF N SLICES]

N	=	NUMBER OF SLICES
N1	=	N + 1 = NUMBER OF SIDES
AH	=	COEFFICIENT OF HORIZONTAL ACCELERATION OF EARTHQUAKE NEAR CREST OF SLOPE
AVR	=	AH/AV
AV	=	COEFFICIENT OF VERTICAL EARTHQUAKE ACCELERATION
EQM	=	CORRESPONDING EARTHQUAKE MAGNITUDE (RICHTER SCALE)
GAMAW	=	UNIT WEIGHT OF WATER
R(N1)	=	THRUST OF WATER IN THE TENSION CRACK ON N + 1 SIDE
I	=	Ith SIDE OR SLICE STARTING FROM TOE OF SLOPE
XT(I)	=	X-COORDINATE OF TOP OF Ith SIDE
YT(I)	=	Y-COORDINATE OF TOP OF Ith SIDE
XB(I)	=	X-COORDINATE OF BOTTOM OF Ith SIDE
YB(I)	=	Y-COORDINATE OF BOTTOM OF Ith SIDE
XW(I)	=	X-COORDINATE OF WATER TABLE ON Ith SIDE
YW(I)	=	Y-COORDINATE OF WATER TABLE ON Ith SIDE
		COORDINATES MUST INCREASE FROM TOE TO TOP OF THE SLOPE
C(I)	=	COHESION AT BASE OF Ith SLICE
PHI(I)	=	ANGLE OF INTERNAL FRICTION AT BASE OF Ith SLICE
CS(I)	=	COHESION FOR Ith SIDE
PHIS(I)	=	ANGLE OF INTERNAL FRICTION FOR Ith SIDE
GAMA(I)	=	UNIT WEIGHT FOR Ith SLICE
R(I)	=	INTERSLICE FORCE FOR Ith SIDE
F(I)	=	FACTOR OF SAFETY FOR Ith SLICE

(a) Input File Name: ISANC.DAT

STABILITY ANALYSIS OF SLOPE WITH NON-CIRCULAR SLIP SURFACE/NON-VERTICAL SLICES
4
0.1 0.0 8. 1.0 0.0
6.00 7.40 4.00 1.40 6.00 12.00 0.00 26.00
29.80 10.00 29.80 5.80 29.80 12.00 0.00 26.00
35.00 15.00 35.00 8.00 35.00 12.00 0.00 26.00
39.40 20.00 39.40 10.00 39.40 12.00 0.00 26.00
50.00 20.00 50.00 14.80 50.00 12.00 0.00 26.00
0.00 26.00 2.00
0.00 26.00 3.10
0.00 26.00 2.86
0.00 26.00 2.40

(b) Output File Name: OSANC.DAT

STABILITY ANALYSIS OF SLOPE WITH NON-CIRCULAR SLIP SURFACE/NON-VERTICAL SLICES

UNITS USED –   TONNE – METER – DEGREE
INPUT FILE NAME –   ISANC.DAT
OUTPUT FILE NAME –   OSANC.DAT

SIDE NO.	XT	YT	XB	YB	XW	YW	C	PHI
1	6.00	7.40	4.00	1.40	6.00	12.00	0.00	26.00
2	29.80	10.00	29.80	5.80	29.80	12.00	0.00	26.00
3	35.00	15.00	35.00	8.00	35.00	12.00	0.00	26.00
4	39.40	20.00	39.40	10.00	39.40	12.00	0.00	26.00
5	50.00	20.00	50.00	14.80	50.00	12.00	0.00	26.00

SLICE NO.	COHESION (C)	PHI	GAMA	WEIGHT	WATER PRESSURE
1	0.000	26.00	2.00	245.960	8.400
2	0.000	26.00	3.100	90.272	5.100
3	0.000	26.000	2.860	106.964	3.000
4	0.000	26.000	2.400	193.344	2.500

RESULTS OF CALCULATIONS FOR STATIC CASE

AH = 0.000 AV = 0.000 EQM = 8.000 GAMAW = 1.000
WATER THRUST IN TENSION CRACK = 0.000

OVERALL FACTOR OF SAFETY = 1.25

FACTOR OF SAFETY OF WEDGE F(I)	INTERWEDGE REACTION R(I)	WEDGE NO. I
1.246	0.000	1
1.246	42.698	2
1.246	31.698	3
1.246	20.684	4

RESULTS OF CALCULATIONS FOR DYNAMIC CASE

CRITICAL ACCELERATION	=	0.057
DYNAMIC SETTLEMENT(M)	=	0.177
AH = 0.100 AV = 0.000 EQM = 8.000 GAMAW = 1.000		
WATER THRUST IN TENSION CRACK	=	0.000
OVERALL FACTOR OF SAFETY	=	0.87

FACTOR OF SAFETY OF WEDGE, F(I)	INTERWEDGE REACTION, R(I)	WEDGE NO. I
0.869	0.000	1
0.869	36.492	2
0.869	25.604	3
0.869	16.890	4

REFERENCES

Deoja, B., Dhital, M., Thapa, B. and Wagner, A. 1991. *Mountain Risk Engineering Handbook*. Kathmandu, International Centre of Integrated Mountain Development, Part I, Chap. 10 and 13 and Part II, Section 24.9.

Goodman, R.E. and Kieffer, D.C. 2000. Behaviour of rock in slopes. *J. Geotech. Geoenviron. Eng.*, ASCE 126(8): 675–684.

Hoek, E. 1987. General two-dimensional slope stability analysis, Ch.3, pp. 95–128, *Analytical and Computational Methods in Engineering Rock Mechanics* (ed. Brown, E.T.). London, Allen and Unwin.

Jatana, B.L. 1999. Fail-safe large dams in earthquake prone himalayan region, 19th ISET annual lecture. *ISET J. Earthquake Technology*. Department of Earthquake Engineering, University of Roorkee, Roorkee, India, 36(1): 1–13.

Kovari, Kalman and Fritz, P. 1978. Slope stability analysis with plane wedge and polygonal sliding surfaces. *Int. Symp. Rock Mechanics Related to Dam Foundations*, Rio-de-Janerio, Brazil.

Kovari, K. and Fritz, P. 1984. Recent developments in the analysis and monitoring of rock slopes. *4th Int. Symp. on Landslides*, Toronto, I.S.L.

Ramamurthy, T. 1985. *Stability of rock mass*, VIII Annual IGS Lecture, IGC, Roorkee, India, pp. 37–44.

Sarma, S.K. 1979. Stability analysis of embankments and slope. *J. Geotech. Engg. Div., ASCE*, GT12: 1511–1524.

Sharma, S. 1996. *Engineering geological investigations of Lakhwar Dam Project, Garhwal, Himalaya*, U.P. Ph.D. Thesis. Department of Earth Sciences, University of Roorkee, Roorkee.

Shekhawat, R.K. 1993. *Stability analysis of slopes with non-circular slip surface and non-vertical slice*. M.E. Dissertation, Civil Engineering Department, University of Roorkee, Roorkee.

Singh, Bhawani, Samadhiya, N.K. and Shekhawat, R.K. 1996. Stability analysis of slopes with non-circular slip surface and non-vertical slices. *Indian Geotech. J.* 26(4): 417–429.

Taylor, D.W. 1948. *Fundamentals of Soil Mechanics*. Asia Publishing House, New Delhi, India, Section 16.34, p. 700.

CHAPTER 15

Waves due to a landslide in a reservoir — WAVE

15.1 BACKGROUND

The safety of a dam also depends upon the height of the wave caused by a nearby deep-seated landslide in the dam reservoir. In one case, a rockslide generated wave over-toppled the Vjönt concrete dam in Italy in 1960 killing many villagers downstream. The cause of the rockslide was the submergence of a rock slope which was very near to the dam site. The rockslide was catastrophic due to the brittle failure of rocks. The concrete dam is still standing as a monument to history but it has not been filled. Thus, due caution must be shown in case of deep-seated landslides by the side of a dam.

Millions of landslides tend to take place along the rim of a dam reservoir, but only deep-seated landslides/rockslides have the ability to generate a very high wave in the reservoir. It is needless to mention that submerged rotational slides may be triggered by a major earthquake during a rainy season. The requirement of design provided the motivation to prepare software packages for analysis of landslides in reservoirs. Fortunately, we have now this complete package of programs of landslide to predict the weight of potential landslide in each case. There is nothing to worry about if the weight of the landslide mass is small such as in cases of 3D wedge failures, toppling failure and debris slides.

15.2 THE PROGRAM WAVE

The Vjönt case history outlined above has provided a motivation to study in detail the chances of the over-toppling of dams due to landslides in the entire dam reservoir. Slingerland and Voight (1979) obtained an approximate correlation for the maximum height of a wave due to a landslide in the reservoir on the basis of case histories. They have also given an equation of how the height

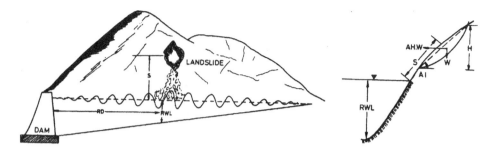

Figure 15.1. Input variables for WAVE.

of a wave decreases with travel towards the dam site. The software WAVE has been written on these findings to predict the height of landslide-generated waves at the dam site (Fig.15.1).

There is a need for more research in this field. More field and model tests should be conducted in dam reservoirs. There are many parameters affecting the height of wave such as shape of reservoir, depth of water, type of landslide, cause of landslide, besides mass of landslide and height of slide with respect to reservoir, etc.

The University of Roorkee (now known as The Indian Institute of Technology, Roorkee) has carried out studies on the stability of rims of the dam reservoirs of Tehri, Koteshwar, Lakhwar and Kishau (India). Fortunately, the site of potential deep-seated landslides due to submergence are far away from the respective dam sites and no danger of over-toppling is predicted by the software WAVE.

In the case of the dam reservoir of the Columbia River in U.S.A., the landslide by the side of the dam is being stabilized by an underground drainage system. Although it may be difficult to increase the factor of safety of this rotational landslide by more than 6 percent through a good drainage system, the rate of movement of landslide has been arrested to a permissible limit of 5 mm per year. Mountains move more than this limit in that region due to tectonic forces (Hoek, 1998). This indicates that regular monitoring of the rate of displacement of slopes should be carried out near the dam site. Moreover, the safety of the dams should be reviewed every 5 years as it is done in the U.S.A.

15.3   USER'S MANUAL – WAVE

THIS PROGRAM CALCULATES THE HEIGHT OF WAVE GENERATED IN A RESERVOIR DUE TO LANDSLIDE (FIG. 15.1)

GIVE DATA IN THE FOLLOWING SEQUENCE:
NCASE (= NUMBER OF PROBLEMS)
TITLE OF PROBLEM IN ONE LINE (<80 CHARACTERS)
FS,PHI,W,H,AI,S,RWL,G,GAMAW,RD
THE ABOVE TWO LINES ARE REPEATED NCASE TIMES

```
ENTER   0 FOR TERMINATION
        1 FOR FURTHER HELP
        2 FOR EXECUTION
```

KE	=	KINETIC ENERGY OF LANDSLIDE (DIMENSIONS)
W	=	TOTAL WEIGHT OF LANDSLIDE MASS
AI	=	DIP OF PLANE OR INTERSECTION OF JOINT PLANES
	=	AVERAGE DIP OF CIRCULAR WEDGE/DEBRIS SLIDE
FS	=	STATIC RESIDUAL FACTOR OF SAFETY OF THE SLOPE DURING LANDSLIDE. (C = 0.0 & PHI = PHIR/2)
G	=	ACCELERATION DUE TO GRAVITY
GAMAW	=	UNIT WEIGHT OF WATER
H	=	HEIGHT OF SLOPE
PHI	=	ANGLE OF SLIDING FRICTION (= PHIR/2)
RWL	=	MEAN DEPTH OF RESERVOIR
RD	=	DISTANCE OF LANDSLIDE FROM DAM
S	=	DISTANCE OF MOVEMENT OF LANDSLIDE FROM RWL

(a) Input File Name: IWAVE.DAT

```
2
HEIGHT OF WAVE DUE TO LAND SLIDE
0. 14. .375E+08 295. 25. 696. 100. 9.81 1. 1000.
HEIGHT OF WAVE DUE TO LAND SLIDE
0. 14. .375E+08 295. 25. 0.00 100. 9.81 1. 1000.
```

(b) Output File Name: OWAVE.DAT

HEIGHT OF WAVE DUE TO LAND SLIDE

UNITS USED –	TONNE – METER – DEGREE	
INPUT FILE NAME –	WAVE.DAT	
OUTPUT FILE NAME –	OWAVE.DAT	

Case No. 1

RESIDUAL FACTOR OF SAFETY	=	0.5347
RESIDUAL ANGLE OF FRICTION	=	14.0000
WEIGHT OF WEDGE (TOTAL)	=	0.375E + 08
HEIGHT OF SLOPE	=	295.0000
DIP OF LAND SLIDE	=	25.0000
DISTANCE OF SLOPE MOVEMENT	=	696.0000
HEIGHT OF WATER ABOVE TOE	=	100.0000
ACCELERATION DUE TO GRAVITY	=	9.8100
UNIT WEIGHT OF WATER	=	1.0000
MAXIMUM WAVE HEIGHT DUE TO LAND SLIDE	=	92.1166
HEIGHT OF WAVE	=	36.8466
AT DISTANCE FROM LAND SLIDE	=	1000.00
MAXIMUM VELOCITY OF LAND SLIDE	=	51.8205
KINETIC ENERGY OF LAND SLIDE	=	51.3257

HEIGHT OF WAVE DUE TO LAND SLIDE

UNITS USED –	TONNE – METER – DEGREE	
INPUT FILE NAME –	IWAVE.DAT	
OUTPUT FILE NAME –	OWAVE.DAT	

Case No. 2		
RESIDUAL FACTOR OF SAFETY	=	0.5347
RESIDUAL ANGLE OF FRICTION	=	14.0000
WEIGHT OF WEDGE (TOTAL)	=	0.375E + 08
HEIGHT OF SLOPE	=	295.0000
DIP OF LAND SLIDE	=	25.0000
DISTANCE OF SLOPE MOVEMENT	=	467.6797
HEIGHT OF WATER ABOVE TOE	=	100.0000
ACCELERATION DUE TO GRAVITY	=	9.8100
UNIT WEIGHT OF WATER	=	1.0000
MAXIMUM WAVE HEIGHT DUE TO LAND SLIDE	=	69.4622
HEIGHT OF WAVE	=	27.7849
AT DISTANCE FROM LAND SLIDE	=	1000.0000
MAXIMUM VELOCITY OF LAND SLIDE	=	42.4787
KINETIC ENERGY OF LAND SLIDE	=	34.4885

REFERENCES

Hoek, E. 1998. Lecture at Tehri Hydro Development Corporation Ltd., Rishikesh, India, on 4th April 1998. Abstract Reported in *J. Rock Mech. Tunnel. Technol.* 4(2): 156–157.

Slingerland, R.L. and Voight, B. 1979. Occurrence, properties and predictive models and landslide generated waves, rockslides avalanches. *Develop. Geotech. Eng. Series* 14 B.

CHAPTER 16

Retaining walls and breast walls – RETAIN

"Two roads diverged in a yellow wood,
And sorry I could not travel both . . . . . .
I took the one less travelled by,
And that has made all the difference."
Robert Frost

16.1   GENERAL

The function of retaining walls is to retain the backfill below (i) the hill road (ii) rail lines or (iii) terraces for building complexes. Table 16.1 illustrates types and choice of appropriate retaining wall. Table 16.2 shows various types of breast walls and the selection of an appropriate breast wall. The purpose of a breast wall is to stabilize a cut slope in soils and very poor rock masses and to prevent erosion of the cut slopes during the rainy season. Experience suggests that a breast wall embedded in rock is not just ornamental but a structural support, even if its thickness is less.

16.2   DESIGN CONSIDERATIONS

Experience shows that retaining walls and breast walls with their bases tilted towards a hill, as shown in Tables 16.1 and 16.2, are the most suitable and economical in the seismic hilly areas. This is because earth pressures are very high due to seismic forces and a tilted base gives extra resistance to sliding and overturning. Similarly, if the back of a retaining/breast wall is tilted towards a hill (Table 16.2), the wall is highly economical (Arya and Gupta, 1983). In fact the earth between the vertical line and the back of the wall tends to act as part of the wall. That is why only a nominal thickness of a breast wall is needed in case it is founded on the rock mass. The walls are designed as simple rigid walls. The design criteria are specified below.
1. Static factor of safety against overturning   >2.0 (static forces)
   about toe (F_o)
   Dynamic factor of safety against overturning   >1.5 (earthquake forces)
   about toe ($F_{o\ dyn}$)

Table 16.1. Selection of retaining walls (Deoja et al., 1991).

Type		Timber crib	Dry stone	Banded stone/masonry dry	Cement masonry	Gabion Low	Gabion High	Reinforced earth
Diagrammatic cross-section								
C O N S T R U C T I O N	Top width	2 m	0.6 – 1.0 m	0.6 – 1.0 m	0.5 – 1.0 m	1 m	1 – 2 m	4 m or 0.7 – 0.8 H
	Base width	–	0.5 – 0.7 H	0.6 – 0.65 H	0.5 – 0.65 H	0.6 – 0.75 H	0.55 – 0.65 H	4 m or 0.7 – 0.8 H
	Front batter	4 : 1	vertical	varies	10 : 1	6 : 1	6 : 1	3 : 1
	Back batter	4 : 1	varies	vertical	varies	varies	varies	3 : 1
	Inward dip of foundation	1 : 4	1 : 3	1 : 3	Horizontal or 1 : 6	1 : 6	1 : 6	Horizontal
	Foundation depth below drain	0.5 – 1 m	0.5 m	0.5 – 1 m	0.5 – 1 m	0.5 m	1 m	0.5 m
	Range of height	3 – 9 m	1 – 6 m	6 – 8 m	1 – 10 m	1 – 6 m	6 – 10 m	3 – 25 m
	Hill slope angle	<30°	<35°	20°	35° – 60°	35° – 60°	35° – 60°	<35°
	Toe protection in case of soft rock/soil	Boulder pitching	Boulder pitching	Boulder pitching	Boulder pitching			No
N O T E S	General	Timbers 15cm φ with stone rubble well packed behind timbers. 10% of all headers to extend into fill. Eco-logically unaccep-table.	Set stones along foundation bed. Use long bond stones. Hand packed stones in back fill.	Cement masonry bands of 50cm thickness at 3m c/c. Other specifications as for dry stone wall.	Weep holes 15x15cm size at 1–2m c/c. 50cm rubble backing for drainage.	Stones to be hand packed. Stone shape important, blocky preferable to tabular. Specify maximum/minimum stone size. No weathered stone to be used. Compact granular back fill in layers (<15cm). Use B type gabion wall front face nearly vertical.		Granular back fill preferred. Use geo-grid for H<4m and tensar grid for H>4m. Provide drainage layer in case of seepage problems. Specify spacing of reinforcement grids.

1. Foundations to be stepped up if rock encountered
2. All walls require durable rock filling of small to medium size
3. Provide 15cm thick gravel layer for drainage in case of clayey foundation

Least Durable — Most durable — Can take differential settlement and slope movement — Huge potential, used more as stable reinforced fill.

Non ductile structures most susceptible to earthquake damage — Very flexible structures

Application: Platform for road rather than preventive method of slope support.

1. Design as conventional retaining walls. Assume surcharge on road of 2T/sq.m
2. Use d both as cut slope and fill slopes support. Breast wall is more economical for cut slopes
3. Choice of wall depends on local resources, local skill, hill slope angle, foundation conditions and also shape of back fill wedges

Table 16.2. Selection of breast walls (Deoja et al., 1991).

Type	Breast Walls/Retaining Walls					Notes
	Dry stone	Banded dry stone/masonry	Cement masonry	Gabion	Horizontal drum walls	
Diagrammatic cross-section						1. Wall construction requires special skills and practical labour. Curing of masonry walls generally not feasible in hill due to paucity of water. 2. The typical dimensions shown rely both on well-drained back fill and good foundation condition. 3. Detailed design is necessary in case of soil slopes and walls higher than 6m and poor foundation conditions. 4. Gabion walls should be used in case of poor foundation / seepage conditions. They can take considerable differential settlement and some slope movement. 5. Other measures should also be taken, e.g., check drains, turfing, benching of cut slopes in soft rocks sealing of cracks, etc. All preventive measures should be implemented in one season. Total system of measures is far more effective than individual measures.
CONSTRUCTION — Top width	0.5 m	0.5 m	0.5 m	2.0 m	1.0 m	
Base width	0.29 H / 0.3 H / 0.33 H		0.23 H	2.0	1.0	
Back batter	3:1 / 4:1 / 5:1	3:1	3:1	3 to 5:1	3:1	
Inward dip of foundation	1:3 / 1:4 / 1:5	1:3	1:3	1:5	1:3	
Foundation depth below drain	0.5 m / 0.5 m / 0.5 m	0.5 m	0.5 m	0.5–1 m	0.25 m	
Range of height	6m / 4m / 3m	3-8 m	1-10 m	1-8 m	<2.2 m	
Hill slope angle	35°–60°	35°–60°	35°–70°	35°—60°	35°	
Toe protection in case of soft rock/soil	No pitching	No	No	No	No	
NOTES — General	Pack stones along foundation bed. Use long stones. Specify minimum stone size.	Cement masonry (1:6) bands of 50cm thickness at 3m c/c.	Weep holes 15x15cm size at 1.5–2m c/c and grade 1:10. Cement sand mortar (1:6)	Step in front face 20–50cm wide. Otherwise as for retaining walls.	Use vertical single drum for 0.7 m height. Anchor drum walls on sides. Fill debris material.	
	Revetment walls have uniform section of 0.5m/0.75m thickness for batter of 2:1 or more. Section shaped to suit variation and overbreak in rock cut slope.					
Application	Least durable / economical	Little used	Most durable / costly	Quite durable / costlier	Promising / most economical	
	Nonductile structures most susceptible to earth-quake damage			Very flexible	Flexible	
	Revetments are used to prevent only major erosion, rock fall, slope degradation particularly where vulnerable structures are at risk					

2. Static factor of safety against sliding >1.5 (static forces)
   at base (F_s)
   Dynamic factor of safety against >1.0 (earthquake forces)
   sliding at base ($F_{s\ dyn}$)
3. Highest base pressure, F_{max} (static) $<q_a$ (allowable bearing
                                                           capacity)

   Lowest base pressure, F_{min} (static or dynamic) >0
   Highest dynamic base pressure, $<1.33\,q_a$
   F_{max} (dynamic)
4. If the hill slope itself is not stable, the wall cannot be stable, as the slip surface may pass below the toe and heel of the retaining wall. So the stability of the slope with a retaining wall should also be checked up. Indirectly, the criteria in (3) above account for an unstable slope as the allowable bearing pressure will be very low in that case near the toe (Chapter 17).

Furthermore, the peak strength parameters of soil should be used in the designs of retaining/breast walls, as walls do not permit large deformations of soils.

Retaining walls of high damping coefficient should be used in highly seismic areas. The reinforced earth retaining walls may, therefore, survive during earth-quake. There should be an inverted filter, about 30 cm thick, behind retaining and breast walls. The inverted filter should consist of three layers of coarse sand, pebbles and boulders so that there is no flow of fines from soil backfill. The best strategy is to have free draining backfill of boulders only. Furthermore, there should be weep holes of 15 cm diameter at spacing of about 1.5 m centre to centre to drain out the seepage water. Finally, there should be a drain at the toe of the breast wall. One should not forget to place a 30 cm-thick layer of impervious silt all over the free draining backfill and inverted filter. This will prevent ingress of rainwater into the backfill and the filter. Deoja et al. (1991) have given details of drainage and construction of walls.

16.3   THE PROGRAM RETAIN

The software RETAIN has been developed by Professor M.N. Viladkar on the basis of wedge theory (Viladkar et al., 1985; Deoja et al., 1991) and design criteria (1), (2) and (3) above. The last criterion may be checked by other software packages like SARC or SANC. The program RETAIN automatically gives the optimum base width of a retaining wall or breast wall for given input after satisfying design criteria (1), (2) and (3). This software also points out probable errors in the input data and prints them out. The design experiences mentioned above are off-shoots of this program.

The program RETAIN has been used extensively in the design of walls in development of sites for building complexes in Himalaya. Figure 16.1 shows three design of breast walls along cut slopes in debris, sand rock and clay shale along Jammu-Udhampur Rail Line (JURL), India. These designs were done with the

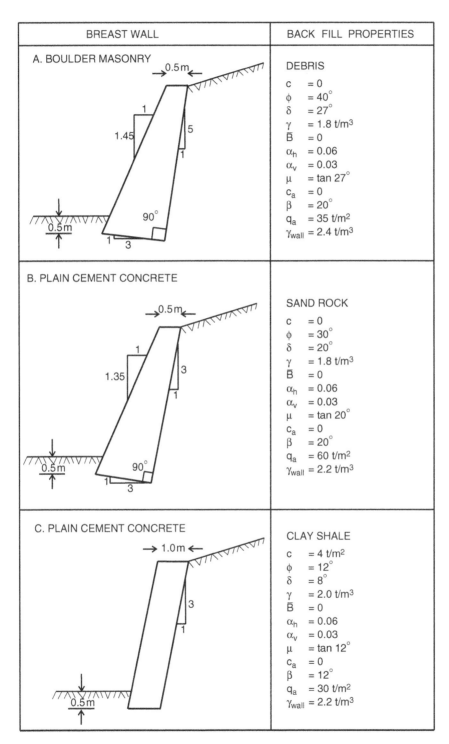

Figure 16.1. Design of breast walls by computer program RETAIN for JURL, India.

help of RETAIN. In the case of shale, the program indicated stability of the cut slope. So a nominal thickness of 1 m of breast wall was recommended. These walls have been built and have been functioning satisfactorily since 1995.

Most of the failures of walls have been due to construction of inadequate sections of the walls. Unfortunately, a series of small retaining walls have been made to retain unstable high slopes. The result is obviously disastrous. A large number of smaller walls are not equal to wall of the height of the unstable slope. Moreover, debris slides cannot be stabilized by the breast walls without sub-surface drainage.

There has been a failure of retaining walls without weep holes on the sides of the storm water drains in cities during heavy rains. Obviously, the side walls should be designed for higher value of pore water pressure coefficient $\bar{B}$. The backfill should also be compacted densely to reduce settlement during rains.

The modern trend is for the use of reinforced earth retaining walls in seismic hilly areas also. Japanese experience in the Kobe earthquake shows that reinforced earth retaining walls have survived where RCC walls have failed. It appears that the damping coefficient of a reinforced earth retaining wall is quite high to dampen

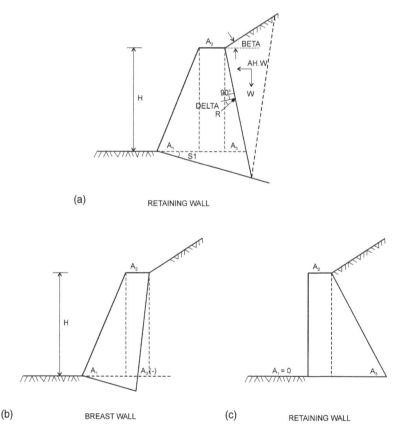

Figure 16.2. Input variables for RETAIN.

out the effect of resonance. This topic is out of scope of this chapter and therefore not covered here.

16.4   USER'S MANUAL – RETAIN

THIS PROGRAM IS FOR DESIGN OF RETAINING AND BREAST WALLS USING COULOMB'S WEDGE THEORY UNDER STATIC AND SEISMIC CASES (FIG. 16.2)

NAME OF PROGRAM –   RETAIN.FOR
UNITS USED   –   TONNE – METER – DEGREE

GIVE DATA IN THE FOLLOWING SEQUENCE

NPROB
TITLE OF PROBLEM IN ONE LINE (<80 CHARACTERS)
C,PHI,GS,BETA
GW,DELTA,AMUE,CA,BBAR,QA,SI
AH,AV,QS
H,A1,A2,A3
(THE ABOVE 5 LINES ARE REPEATED NPROB TIMES)

ENTER   0 FOR TERMINATION
        1 FOR FURTHER HELP
        2 FOR EXECUTION

AH	=	Coefficient of horizontal earthquake acceleration
AMUE	=	Base friction angle
AV	=	Coefficient of vertical earthquake acceleration
A1,A2,A3	=	Projection of toe slope, top width and projection of heel slope on the horizontal plane respectively
BBAR	=	Pore water pressure/(GS * h)
BETA	=	Angle of stable slope of the backfill (<PHI) (+ for upward and − for downward backfill/hillslope)
C	=	Cohesion of backfill/soil
CA	=	Adhesion at the base of the wall
DELTA	=	Angle of wall friction
GS	=	Unit weight of backfill soil
GW	=	Unit weight of wall material
H	=	Height of the retaining wall
NPROB	=	Number of problems of retaining walls
PHI	=	Angle of internal friction of the backfill/soil
QA	=	Allowable bearing pressure (it assumes 1.33 * QA for seismic case)
SI	=	Downward dip of base with horizontal (degree)
QS	=	Surcharge intensity on the backfill
F1	=	Base pressure at toe of wall <QA
F2	=	Base pressure at heel of wall <QA >0.1
FSS	=	Factor of safety of wall against sliding
	=	1.5 for static case and = 1.0 for seismic case
FSO	=	Factor of safety of wall against overturning
	=	2.0 for static case and = 1.5 for seismic case

(a) Input File Name: IRETAIN.DAT

```
1
OPTIMUM SEISMIC DESIGN OF RETAINING WALL WITH INCLINED BASE
0.0,30.0,2.0,0.0
2.2,20.0,20.0,0.0,0.0,20.0,6.0
0.08,0.04,0.0
6.0,2.0,0.5,0.0
```

(b) Output File Name: ORETAIN.DAT

OPTIMUM SEISMIC DESIGN OF RETAINING WALL WITH INCLINED BASE

UNITS USED –	TONNE – METER – DEGREE
INPUT FILE NAME –	IRETAIN.DAT
OUTPUT FILE NAME –	ORETAIN.DAT

COHESION	=	0.000
ANGLE OF INTERNAL FRICTION	=	30.000
BBAR	=	0.000
UNIT WEIGHT OF WALL	=	2.200
UNIT WEIGHT OF BACKFILL	=	2.000
SLOPE ANGLE OF BACKFILL	=	0.000
ANGLE OF WALL FRICTION	=	20.000
ANGLE OF BASE FRICTION	=	20.000
ADHESION AT THE BASE OF WALL	=	0.000
ALLOWABLE BEARING PRESSURE	=	20.000
SURCHARGE INTENSITY ON BACKFILL	=	0.000
HORZ. COMPONENT OF EQ. ACCELERATION	=	0.080
VERT. COMPONENT OF EQ. ACCELERATION	=	0.040
INCLINATION OF BASE WITH HORIZONTAL	=	6.000

WALL HEIGHT = 6.000   A1 = 2.000   A2 = 0.500   A3 = 0.000

OUTPUT DATA

STATIC ANALYSIS
CRITICAL ANGLE IN DEGREES = 56.00000
MAXIMUM EARTH PRESSURE = 11.65754

A1	TOP WIDTH	A3	BASE WIDTH	FSO	FSS	F1	F2
2.00	0.50	0.00	2.51	2.18	1.11	17.12	3.17
2.05	0.55	0.00	2.61	2.36	1.17	15.72	4.60
2.10	0.60	0.00	2.71	2.55	1.24	14.48	5.85
2.15	0.65	0.00	2.82	2.75	1.30	13.39	6.97
2.20	0.70	0.00	2.92	2.96	1.37	12.42	7.95
2.25	0.75	0.00	3.02	3.18	1.44	11.55	8.84
2.30	0.80	0.00	3.12	3.40	1.51	10.78	9.63

EARTHQUAKE ANALYSIS WITH DOWNWARD VERTICAL COMPONENT

CRITICAL ANGLE IN DEGREES	=	52.00000
MAXIMUM EARTH PRESSURE	=	14.50387

A1	TOP WIDTH	A3	BASE WIDTH	FSO	FSS	F1	F2
2.30	0.80	0.00	3.12	2.24	1.02	18.35	3.68

EARTHQUAKE ANALYSIS WITH UPWARD VERTICAL COMPONENT

CRITICAL ANGLE IN DEGREES	=	52.00000
MAXIMUM EARTH PRESSURE	=	13.56671

A1	TOP WIDTH	A3	BASE WIDTH	FSO	FSS	F1	F2
2.30	0.80	0.00	3.12	2.25	0.99	16.92	3.48
2.35	0.85	0.00	3.22	2.38	1.03	15.95	4.45

REFERENCES

Arya, A.S. and Gupta, V.P. 1983. Retaining wall for hill roads, *J. Indian Roads Congr.* 44–1: 356.

Deoja, B., Dhital, M.,Thapa, B. and Wagner, A. 1991. *Mountain Risk Engineering Handbook*, Part I, Chaps 9 and 17 and Part II, Section 24.6. Kathmandu, International Centre of Integrated Mountain Development.

Viladkar, M.N., Ranjan, G. and Singh, Bhawani 1985. Site development in himalayan region and protective measures – A case history. *Proc. Indian Geotech. Conf.*, IGC-85, Roorkee, Vol. 1, pp. 365–370.

CHAPTER 17

Footing on slopes

17.1 GENERAL

The bearing capacity of a footing on soil or rock slope depends upon (i) slope angle, (ii) depth of foundation, (iii) width and length of footing (iv) eccentricity of load and (v) inclination of load. Figure 17.1 shows examples of footing/raft where the resultant load is eccentric and inclined. Brinch and Hansen (in Bowles, 1977) have developed a theory for the bearing capacity of footings, taking into account the above parameters. The theory clearly proves that the bearing capacity of footing is drastically affected by the angle of slope. Consequently, the design of a foundation is a challenging job in hilly areas. Another serious problem with hills is that hillside footings rest on rock, whereas valley side footings may rest on filled up soil below the terrace. As such, masonry buildings crack due to significant differential settlement in hilly areas. RCC raft foundation has performed well in partly soil and partly rock mass foundation in hilly areas.

17.2 THE PROGRAM QULT

The software QULT has been written to predict the ultimate bearing capacity of shallow footings on the basis of Brinch and Hansen's theory for spot calculations and decisions. A shallow footing is defined if the depth (D) of foundation is less than the short width (B) of foundation. It is assumed that slope is homogeneous and obeys Coulomb's theory of shear strength (Fig. 17.2). Furthermore, no seepage erosion takes place such as in unprotected slopes of sand rocks, conglomerates, soils, etc. The engineering practice is that the net bearing capacity is recommended and it is equal to the ultimate bearing capacity minus overburden weight above footing (γD). The allowable net bearing capacity is the net ultimate bearing capacity divided by factor of safety ($=3$). It is suggested that the factor of safety may be reduced to the permissible factor of safety of slopes in steep slopes (1.5 in soils).

The net allowable bearing pressure also depends upon allowable settlement of footings. The same is determined from plate load tests in open pits both for soils and rocks. The criterion of settlement for sand may be applied to jointed

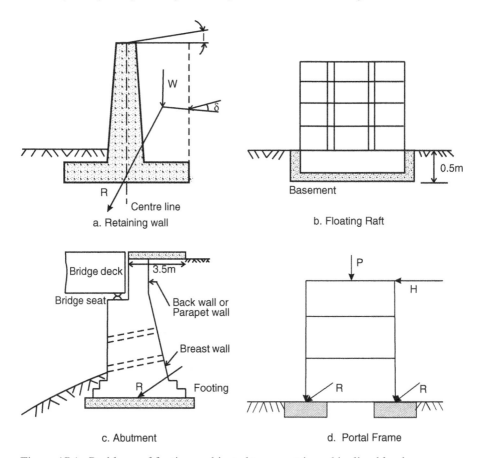

a. Retaining wall

b. Floating Raft

c. Abutment

d. Portal Frame

Figure 17.1. Problems of footings subjected to eccentric and inclined loads.

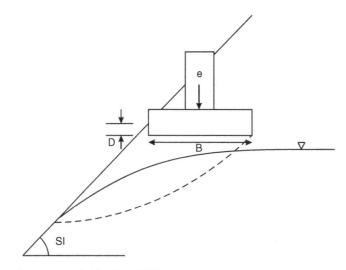

Figure 17.2. Input variables for QULT.

rock masses also where allowable settlement of footings is only 12 mm. This engineering practice has served us well in India since 1987 (IS: 12070-1987 and IS: 13063-1991). Section 23.9 presents more knowledge on the same.

The key to success is in the choice of type of structure for a building complex. Field experience suggests that a RCC frame with all round heavy plinth beam is a near ideal choice for a building on a terrace. The heavy/stiff RCC plinth beam reduces the extent of differential settlements of a building. There is little cracking of masonry walls as they are built on the plinth beam. The extra cost of stiff plinth beam is worth its advantages. Furthermore, ductile RCC buildings have performed better than brittle masonry buildings during earthquakes from the point of view of safety of life of its inhabitants.

17.3   USER'S MANUAL – QULT

ULTIMATE BEARING CAPACITY OF A RECTANGULAR FOOTING ON HILL SOIL SLOPE (FIG. 17.2)

(a) Input File Name: QULT.DAT

ULTIMATE BEARING CAPACITY OF A RECTANGULAR FOOTING ON HILL SOIL SLOPE
0 40. 5. 5. 2. 39. 1.6 .79 .79 27.5 275. 0. 1. 10.

PLEASE TYPE THE NAME OF SITE (<80 CHARACTERS)
PLEASE TYPE THE FOLLOWING PARAMETERS
COHESION (C) AND ANGLE OF INTERNAL FRICTION (PHI)
LENGTH (L), WIDTH (B < L) AND DEPTH OF FOOTING (D)
SLOPE ANGLE (SI), UNIT WEIGHT OF SOIL (GAMA)
ECCENTRICITY ALONG LENGTH AND WIDTH OF FOUNDATION (EL, EB)
LOAD COMPONENTS PARALLEL AND PERPENDICULAR TO FOOTING (H, V)
DIP OF FOOTING (ETA) TOWARDS HILL SIDE (−VE SLOPE SIDE)
UNIT WEIGHT AND DEPTH OF G.W.T. BELOW FOOTING (GAMAW, DW)

(b) Output File Name: QULT.OUT

ULTIMATE BEARING CAPACITY OF A RECTANGULAR FOOTING ON HILL SOIL SLOPE

THE INPUT PARAMETERS ARE AS FOLLOWS

COHESION (C) AND ANGLE OF INTERNAL FRICTION (PHI)
LENGTH (L), WIDTH (B < L) AND DEPTH OF FOOTING (D)
SLOPE ANGLE (SI), UNIT WEIGHT OF SOIL (GAMA)
ECCENTRICITY ALONG LENGTH AND WIDTH OF FOUNDATION (EL, EB) LOAD COM-
PONENTS PARALLEL, PERPENDICULAR TO FOOTING H, V
DIP OF FOOTING (ETA) TOWARDS HILL SIDE (−SLOPE SIDE)
UNIT WEIGHT AND DEPTH OF G.W.T. BELOW FOOTING GAMAW, DW

C	=	0.00000	PHI	=	40.00000	L	=	5.00000
B	=	5.00000	D	=	2.00000	SI	=	39.00000
GAMA	=	1.60000	EL	=	0.79000	EB	=	0.79000
H	=	27.50000	V	=	275.00000	ETA	=	0.00000
GAMAW	=	1.00000	DW	=	10.00000			

BEARING CAPACITY FACTORS (AFTER BRINCH AND HASEN, 1970)

NQ	=	64.51383	NC	=	75.64967	NGAMA	=	9.98727
SC	=	1.85280	SQ	=	1.83958	SGAMA	=	0.60000
DC	=	1.16000	DQ	=	1.08560	DGAMA	=	1.00000
IQ	=	0.77378	IC	=	0.77022	IGAMA	=	0.69569
GC	=	0.73469	Q	=	0.07450	GGAMA	=	0.07450
BC	=	1.00000	BQ	=	1.00000	BGAMA	=	1.00000

ULTIMATE BEARING CAPACITY IS   = 33.7
NET ULTIMATE BEARING CAPACITY IS   = 30.5

PLEASE CHECK STABILITY OF SLOPE

REFERENCES

Bowles, J.E. 1977. *Foundation analysis and design.* McGraw-Hill Kogakusha Ltd., Sections 4.3, 4.4, 4.5 and 4.6.

IS: 12070-1987. *Code of practice for design and construction of shallow foundations on rock.* New Delhi, India, Bureau of Indian Standards.

IS: 13063-1991. *Code of practice for structural safety of buildings on shallow foundations on rock.* New Delhi, India, Bureau of Indian Standards.

CHAPTER 18

Slope mass rating (SMR)

18.1 THE SLOPE MASS RATING (SMR)

For evaluating the stability of rock slopes, Romana (1985) proposed a classification system called slope mass rating (SMR) system. Slope mass rating (SMR) is obtained from Bieniawski's rock mass rating (RMR) by subtracting adjustment factors of the joint-slope relationship and adding a factor depending on method of excavation,

$$SMR = RMR_{basic} + (F_1 \cdot F_2 \cdot F_3) + F_4 \qquad (18.1)$$

where RMR_{basic} is evaluated according to Bieniawski (1979, 1989) by adding the ratings of five parameters (Singh and Goel, 1999). The F_1, F_2, and F_3 are adjustment factors related to joint orientation with respect to slope orientation and F_4 is the correction factor for method of excavation. These are defined below:

F_1 depends upon parallelism between joints and slope face strikes. It ranges from 0.15 to 1.0. It is 0.15 in case of the angle between the critical joint plane and the slope face is more than 30° and the failure probability is very low, whereas it is 1.0 when both are near parallel.

The value of F_1 was initially established empirically. Subsequently, it was found to match approximately the following relationship:

$$F_1 = (1 - \sin A)^2 \qquad (18.2)$$

where A denotes the angle between the strikes of the slope face (α_s) and that of the joints (α_j) that is ($\alpha_s - \alpha_j$).

F_2 refers to joint dip angle (β_j) in the planar failure mode (Fig. 18.1). Its values also vary from 0.15 to 1.0. It is 0.15 when the dip of the critical joint is less than 20° and 1.0 for joints with dip greater than 45°. For the toppling mode of failure, F_2 remains equal to 1.0.

$$F_2 = \tan \beta_j \qquad (18.3)$$

F_3 refers to the relationship between the slope face and joint dips.

In planar failure, F_3 refers to a probability of joints 'day lighting' in the slope face. Conditions are called fair when the slope face and the joints are parallel. If the slope dips $10°$ more than the joints, the condition is termed very unfavourable. For the toppling failure, unfavourable conditions depend upon the sum of dips of joints and the slope $\beta_j + \beta_s$.

Values of adjustment factors F_1, F_2, and F_3 for different joint orientations are given in Table 18.1.

F_4 pertains to the adjustment for the method of excavation. It includes the natural slope, or the cut slope excavated by pre-splitting, smooth blasting, normal blasting, poor blasting and mechanical excavation (see Table 18.2 for adjustment rating F_4 for different excavation methods).

Natural slopes are more stable, because of long time erosion and built in protection mechanism (vegetation, crust desiccation), **$F_4 = +15$.**

Normal blasting applied with sound methods does not change slope stability conditions and therefore **$F_4 = 0$.**

Deficient blasting or poor blasting damages the slope stability, therefore **$F_4 = -8.0$.**

Table 18.1. Values of adjustment factors for different joint orientations (Romana, 1985).

Case of slope failure		Very favourable	Favourable	Fair	Unfavourable	Very unfavourable
P T W	$\|\alpha_j-\alpha_s\|$ $\|\alpha_j-\alpha_s-180°\|$ $\|\alpha_i-\alpha_s\|$	$>30°$	$30-20°$	$20-10°$	$10-5°$	$<5°$
P/W/T	F_1	0.15	0.40	0.70	0.85	1.00
P W	$\|\beta_j\|$ $\|\beta_i\|$	$<20°$	$20-30°$	$30-35°$	$35-45°$	$>45°$
P/W	F_2	0.15	0.40	0.70	0.85	1.00
T	F_2	1.0	1.0	1.0	1.0	1.0
P W	$\|\beta_j-\beta_s\|$ $\|\beta_i-\beta_s\|$	$>10°$	$10-0°$	$0°$	$0-(-10°)$	$<-10°$
T	$\|\beta_j+\beta_s\|$	$<110°$	$110-120°$	$>120°$	–	–
P/W/T	F_3	0	-6	-25	-50	-60

Notations: P – planar failure; T – toppling failure; W – wedge failure; α_s – slope strike; α_j – joint strike; α_i – plunge direction of line of intersection; β_s – slope dip and β_j – joint dip (see Fig. 18.1); β_i – plunge of line of intersection.

Table 18.2. Values of adjustment factor F_4 for method of excavation (Romana, 1985).

Method of excavation	Value of F_4
Natural slope	$+15$
Pre-splitting	$+10$
Smooth blasting	$+8$
Normal blasting or mechanical excavation	0
Poor blasting	-8

Mechanical excavation of slopes, usually by ripping, can be done only in soft and/or very fractured rock and is often combined with some preliminary blasting. The plane of slope is difficult to finish. The method neither increases nor decreases slope stability and therefore $F_4 = 0$.

The minimum and maximum values of SMR from Equation 18.1 are 0 and 100 respectively. It is needless to mention here that the slope stability problem is not found in areas where the discontinuities are steeper than the slope. Therefore, this condition is not considered in the empirical approach.

Romana (1985) used planar and toppling failures for his analysis. The wedge failures have been considered as a special case of plane failures and analyzed in forms of individual planes and the minimum value of SMR is taken for assessing the rock slopes. Experience shows that dip β_i and dip direction α_i of the intersection of these planes should be taken as β_j and α_j respectively, i.e., $\beta_j = \beta_i$ and $\alpha_j = \alpha_i$ where wedge failure is likely to occur (Fig. 18.2).

The effect of future weathering on the slope stability cannot be assessed with rock mass classification as it is a process which depends mostly on the mineralogical conditions of rock, and the climate. In certain rock masses, e.g., some marls, clays and shales, slopes are stable when excavated but fail sometime afterwards (usually one to two years later). In such conditions, it is suggested that the classification should be applied twice: initially and afterwards for weathered conditions. It is always prudent to check SMR against adjoining stable rock slopes before applying it to rock slopes in distress.

In recent times Hack (1998) has developed slope stability probability classification (SSPC) system for weathered and unweathered soil and rock slopes under European climatic conditions. He has given chart for assessment of the probability of failure of a slope. He has also found correlations for sliding angle of friction ϕ along joints. SSPC is enjoying popularity in hilly regions of Europe but needs to be tested in Himalaya and in other climatic conditions.

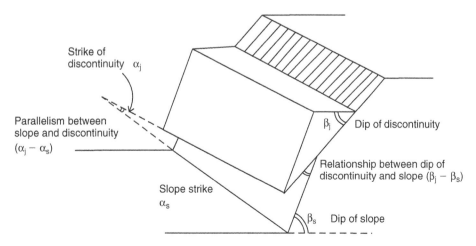

Figure 18.1. Planar failure.

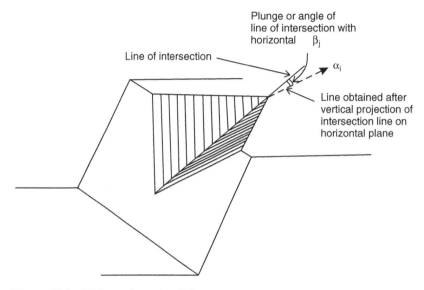

Figure 18.2. Wide angle wedge failure.

Water conditions govern the stability of many slopes which are stable in summer and fail in winter because of heavy raining or freezing. The worst possible water conditions must be assumed for analysis. This technique is not applicable to mountains which are covered by snow most of the times. Moreover, freezing and thawing of water in rock joints cause rock slides in these regions.

18.2   SLOPE STABILITY CLASSES

According to the SMR values, Romana (1985) defined five stability classes. These are described in Table 18.3. It is inferred from Table 18.3 that the slopes with SMR value below 20 may fail very quickly. No slope has been registered with SMR value below 10 because such slopes would not be physically existing.

The stability of slope also depends upon length of joints along slope. Table 18.3 is found to over-estimate SMR where the length of joint along slope is less than 5 percent of the affected height of the landslide. SMR is also not found to be applicable to opencast mines because heavy blasting creates new fractures in the rock slope and depth of cut slope is also large.

Slope mass rating is being used successfully for landslide zonation in rocky and hilly areas. Detailed studies should be carried out where SMR is less than 40 and life and property is in danger and slopes should be stabilized accordingly. Otherwise, a safe cut slope angle should be determined to increase SMR to 60.

Table 18.3. Various stability classes as per SMR values (Romana, 1985).

Class No.	V	IV	III	II	I
SMR value	0–20	21–40	41–60	61–80	81–100
Rock mass description	Very bad	Bad	Normal	Good	Very good
Stability	Completely unstable	Unstable	Partially stable	Stable	Completely stable
Failures	Big planar or soil like or circular	Planar or big wedges	Planar along some joint and many wedges	Some block failure	No failure
Probability of failure	0.9	0.6	0.4	0.2	0

Table 18.4. Suggested supports for various SMR classes.

SMR classes	SMR values	Suggested supports
Ia	91–100	None
Ib	81–90	None, scaling is required
IIa	71–80	(None, toe ditch or fence), spot bolting
IIb	61–70	(Toe ditch or fence nets), spot or systematic bolting
IIIa	51–60	(Toe ditch and/or nets), spot or systematic bolting, spot shotcrete
IIIb	41–50	(Toe ditch and/or nets), systematic bolting/anchors, systematic shotcrete, toe wall and/or dental concrete
IVa	31–40	Anchors, systematic shotcrete, toe wall and/or concrete (or re-excavation), drainage
IVb	21–30	Systematic reinforced shotcrete, toe wall and/or concrete, re-excavation, deep drainage
Va	11–20	Gravity or anchored wall, re-excavation

Less popular support measures are given in brackets.

18.3   SUPPORT MEASURES

Many remedial measures can be taken to support a slope. Both detailed study and good engineering sense are necessary to stabilize a slope. Classification systems can only try to point out the normal techniques for each different class of supports as given in Table 18.4.

In a broader sense, the SMR range for each group of support measures is as follows:

SMR 65–100   None, scaling
SMR 30–75   Bolting, anchoring
SMR 20–60   Shotcrete, concrete
SMR 10–30   Wall erection, re-excavation

As pointed out by Romana (1985), wedge failure has not been discussed in his SMR classification separately. To overcome this problem, Anbalagan et al. (1992)

has modified SMR to make it applicable for the wedge mode of failure also. This modification is presented in the following paragraphs.

18.4   A MODIFIED SMR APPROACH

Though the SMR accounts for planar and toppling failures in rock slopes, it takes into consideration different planes forming the wedges and analysing the different planes individually in the case of wedge failure. The unstable wedge is a result of combined effect of the intersection of various joints (Fig. 18.2). Anbalagan et al. (1992) considered plane and wedge failures as different cases and presented a modified SMR approach for slope stability analysis.

In the modified SMR approach, the same method is applicable for planar failures and the strike and the dip of the plane are used for the analysis. But in the case of wedge failures, the plunge and the direction of line of intersection of the unstable wedge are used. Thin wedges with low angle are likely to be stable and should not be considered. In Table 18.1, adjustment ratings for F_1, F_2, and F_3 are also given in the case of wedge failure as suggested by Anbalagan et al. (1992).

For example: Consider two joint sets having dips of 45° and 35° and dip directions of 66° and 325° respectively. The inclination of slope is N10°/50°. The plunge and trend of line of intersection of these two joints forming wedge are 28° and 4° respectively (Fig. 18.3).

According to the SMR approach, the SMR value for the above two joint sets are worked out separately and the critical value of SMR is adopted for classification

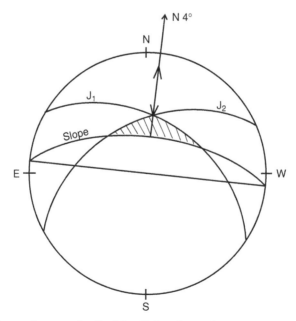

Figure 18.3. Usage of stereo plot for identifying the wedge.

purpose. According to this approach, adjustment factor ($F_1 \cdot F_2 \cdot F_3$) for the first joint set and the slope works out as -6.4 (Table 18.5). Similarly, considering the second joint set and slope, the adjustment factor works out as -6.3 (Table 18.5). Now, if we consider the plunge and the trend of the wedge formed by the two joint sets and the slope, the adjustment factor works out as -20.4. This clearly shows that the SMR calculated for the third case is more critical than the first and the second cases. Therefore, it is more logical and realistic to use the plunge and the trend of line of intersection for potential wedge failure.

18.5 A CASE STUDY OF STABILITY ANALYSIS

Anbalagan, Sharma and Raghuvanshi (1992) have analysed 20 different slopes using a modified SMR approach along the Lakshmanjhula-Shivpuri road in the lesser Himalayas of Distt. Garhwal, U.P., India.

18.5.1 Geology

The Lakshmanjhula-Shivpuri road section area forms the northern part of Garhwal syncline. The road section has encountered Infra-Krol formation. Krol 'A', Krol 'B', Krol 'C + D' formations, lower Tal formation, upper Tal formation and Blaini formation. The rocks are folded in the form of a syncline called the Narendra Nagar syncline. The axis of the syncline is aligned in a NE–SW

Table 18.5. Calculations for adjustment factors F_1, F_2, and F_3.

A. Details of geological discontinuities

	Dip direction	Dip
Joint J_1	N60°	45°
Joint J_2	N325°	35°
Slope	N10°	50°

B. Details of line of intersection of J_1 and J_2

Trend = 4°	See Figure 18.3
Plunge = 28°	

C. Adjustment factor F_1, F_2, and F_3 for different conditions

No.	Condition	F_1	F_2	F_3	Adjustment factor ($F_1 \cdot F_2 \cdot F_3$)
1	Considering joint J_1 and slope	0.15	0.85	-50	-6.4
2	Considering joint J_2 and slope	0.15	0.70	-60	-6.3
3	Considering the plunge and trend of line of intersection of J_1 and J_2 and the slope (modified SMR approach)	0.85	0.40	-60	-20.4

direction so that the sequence of Blaini and Tal formations from Lakshmanjhula are repeated again to the north of the syncline axis.

The Infra Krol formation mainly consists of dark grey shales while Krol A consists of shaly limestones and Krol B includes red shales. The Krol C + D comprises gypsiferous limestones. The lower Tal formation consists of shales, whereas the upper Tal comprises of quartzites. The rocks of Blaini formation are exposed near Shivpuri including laminated shales.

18.5.2 Rock slope analysis

Twenty rock slopes along the road were chosen such that they cover different rock types (Fig. 18.4). The RMR_{basic} for different rock types were estimated (Table 18.6). The graphical analysis is performed for the joints to deduce the mode

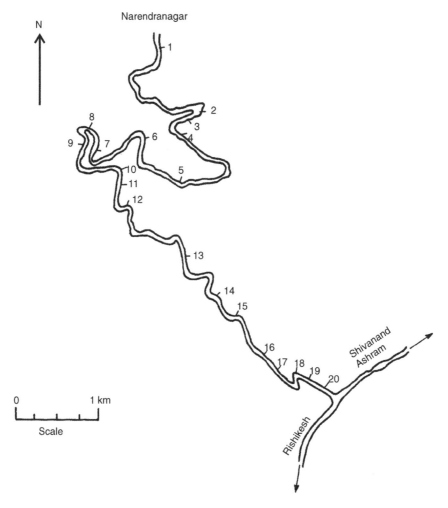

Figure 18.4. Location map of slope stability study.

Table 18.6. Rock mass rating (RMR) for various rock types of Lakshmanjhula-Shivpuri area (Anbalagan et al., 1992).

Rock type	Uniaxial compressive strength	RQD from J_v	Joint spacing	Joint condition	Ground water condition	RMR_{basic}
Infra Krol shales	7	13	8	22	15	65
Krol 'A' shaly limestones	12	13	8	22	15	70
Krol 'B' shales	12	13	8	22	15	70
Krol 'C + D' limestones	12	13	8	22	15	70
Lower Tal shales	7	13	8	22	15	65
Upper Tal quartzites	12	17	10	22	15	76
Blaini shales	7	13	8	22	15	65

Table 18.7. Slope stability analysis along Lakshmanjhula-Shivpuri area (Anbalagan et al., 1992).

Location No. (Fig. 18.4)	SMR value	Class No.	Slope description	Stability	Observed failure
1	44.2	III	Normal	Partially stable	Wedge failure
2	47.8	III	Normal	Partially stable	Wedge failure
3	36.3	IV	Bad	Unstable	Planar failure
4	32.4	IV	Bad	Unstable	Planar failure
5	18.0	V	Very bad	Completely unstable	Big wedge failure
6	24.0	IV	Bad	Unstable	Planar or big wedge failure
7	26.0	IV	Bad	Unstable	Wedge failure
8	40.6	III	Normal	Partially stable	Planar failure
9	56.8	III	Normal	Partially stable	Planar failure
10	30.0	IV	Bad	Unstable	Planar failure
11	69.6	II	Good	Stable	Some block failure
12	55.2	III	Normal	Partially stable	Planar failure
13	51.6	III	Normal	Partially stable	Planar failure
14	36.6	IV	Bad	Unstable	Wedge failure
15	60.9	II	Good	Stable	Some block failure
16	24.0	IV	Bad	Unstable	Planar failure
17	61.8	II	Good	Stable	Some block failure
18	57.0	III	Normal	Partially stable	Wedge failure
19	22.65	IV	Bad	Unstable	Planar failure
20	18.5	V	Very bad	Completely unstable	Big planar failure

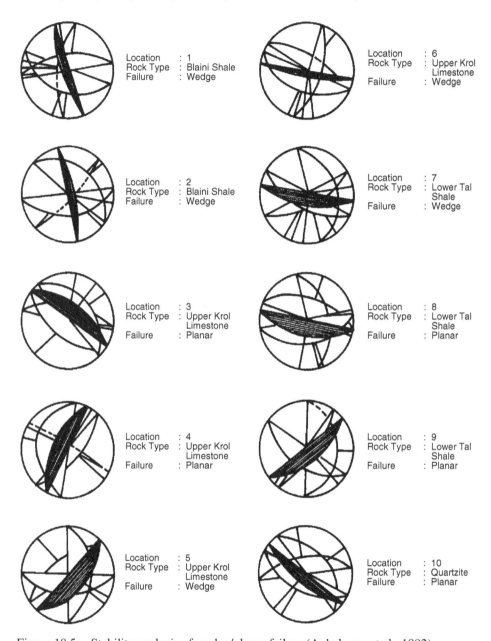

Figure 18.5a. Stability analysis of wedge/planar failure (Anbalagan et al., 1992).

of failure. In this method, the poles of discontinuities were plotted on an equal area stereonet and contours were drawn to get the maxima of pole concentrations. The probable failure patterns were determined by studying the orientation of various joints and the intersection and comparing the same with the slope. The graphical analysis of individual slope has been shown in Figures 18.5a and b. The result of the SMR approach has been given in Table 18.7.

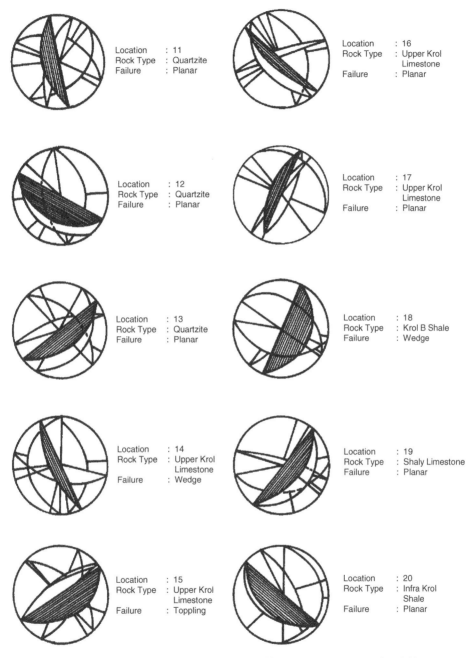

Figure 18.5b. Stability analysis of wedge/planar failure (Anbalagan et al., 1992).

It may be noted that the modified approach for wide angle wedge failure appears to be valid as SMR predictions are matched with the observed failure modes. However, for identifying potentially unstable wedges, one should use ones discretion.

REFERENCES

Anbalagan, R., Sharma, Sanjeev and Raghuvanshi, T.K. 1992. Rock mass stability evaluation using modified SMR approach. *Proc. 6th Nat. Sym. Rock Mech.*, Bangalore, India. pp. 258–268.

Bieniawski, Z.T. 1979. The geomechanics classification in rock engineering applications. Reprinted from: *Proc. 4th Cong. Int. Society for Rock Mech.*/Comptes-rendus/ Berichte-Montreux, Suisse, 2–8 September 1979. 2208pp., 3 Vols. Rotterdam, Netherlands, A.A. Balkema.

Bieniawski, Z.T. 1989. *Engineering Rock Mass Classifications*. John Wiley, p. 251.

Hack, R. 1998. Slope Stability Probability Classification – SSPC, International Institute for Aerospace Survey and earth Sciences (ITC), Kanaalweg 3, 2628 EB Delft, The Netherlands, 2nd Edition, ITC Publication No. 43, p. 258.

Romana, M. 1985. New Adjustment Ratings for Application of Bieniawski Classification to Slopes. *Int. Sym. Role Rock Mech.*, Zacatecas, pp. 49–53.

Singh, Bhawani and Goel, R.K. 1999. *Rock Mass Classification: A Practical Approach in Civil Engineering*, Chapter 17. Reproduced and Revised, Permission from Elsevier Science Ltd., U.K., p. 267.

CHAPTER 19

Landslide hazard zonation

"Landslide is a mountain cancer. It is cheaper to cure than to endure it"
Anonymous

19.1   INTRODUCTION

A landslide hazard zonation (LHZ) map is an important tool for designers, field engineers and geologists. It is used classify the land surface into zones of varying degrees of hazards based on the estimated significance of causative factors which influence the stability (Anbalagan, 1992). The LHZ map is a rapid technique of hazard assessment of the land surface (Anbalagan and Singh, 2001; Gupta and Anbalagan, 1995 and Gupta et al., 2000). It is useful for the following purposes:

1. The LHZ maps help the planners and field engineers to identify the hazard prone areas and therefore enable one to choose favourable locations for site development schemes. In case the site cannot be changed and it is hazardous, the zonation before construction helps to adopt proper precautionary measures to tackle the hazard problems.
2. These maps identify and delineate the hazardous area of instability for adopting proper remedial measures to check further environmental degradation of the area.
3. Geotechnical monitoring of structures on the hills should be done specially in the hazardous areas by preparing a contour map of displacement rates. Landslide control measures and construction controls may be identified accordingly for safety of buildings on the hilly areas.
4. Tunnels should be realigned to avoid regions of deep-seated major land-slides to eliminate risks of high displacement rate. The tunnel portals should be relocated in the stable rock slope. The outlet of the tailrace tunnel of a hydro-electric project should be much above flood level in the deep gorges which are prone to landslide.

Based on the scale of LHZ maps, these are classified into three categories.

1. Mega – Regional – Scale of 1:50,000 or more
2. Macro – Zonation and Risk Zonation – 1:25,000 to 1:50,000
3. Micro – Zonation – Scale of 1:2,000 to 1:10,000

Methodology of preparing the LHZ map is described in the following paragraphs with an example to show the method of applying a LHZ mapping technique in the field for demarcating the landslide prone areas.

19.2   LANDSLIDE HAZARD ZONATION MAPS: THE METHODOLOGY

19.2.1   Factors

The technique of landslide hazard zonation has been developed by Anbalgan (1992). Many researchers have developed various methods of landslide zonation but they are not based on causative factors. The main merit of Anbalagan's method is that it considers causative factors in a simple way. His method has become very popular in India, Italy, Nepal and other countries. The technique in a broader sense, classifies the area into five zones on the basis of the following six major causative factors:

1. Lithology – Characteristics of rock and land type
2. Structure – Relationship of structural discontinuities with slopes
3. Slope Morphometry
4. Relative Relief – Height of slope
5. Land Use and Land Cover
6. Ground Water Condition

These factors have been called as Landslide Hazard Evaluation Factors (LHEF). Ratings of all the Landslide Hazard Evaluation Factors (LHEF) are given in Table 19.1, whereas the maximum assigned rating to each LHEF is given in Table 19.2. The basis of assigning ratings in Table 19.1 is discussed parameter-wise below.

Lithology
The erodibility or the response of rocks to the processes of weathering and erosion should be the main criterion in awarding the ratings for lithology. The rock types such as unweathered quartzites, limestones and granites are generally hard and massive and more resistant to weathering, and therefore form steep slopes. But ferruginous sedimentary rocks are more vulnerable to weathering and erosion. The phyllites and schists are generally more weathered close to the surface. Accordingly, higher rating, i.e., LHEF ratings, should be awarded (Table 19.1).

In the case of soil-like materials, the genesis and age are the main considerations in awarding the ratings. The older alluvium is generally well compacted and has high strength whereas slide debris is generally loose and has low shearing resistance. Nearly vertical slopes of inter-locked sand are stable for several decades in lesser Himalaya. Gupta et al. (2000) observed in Garhwal Himalaya that maximum landslides were found in rocks with large amount of talc minerals as well as in riverbed material.

Table 19.1. Landslide hazard evaluation factor (LHEF) rating scheme (Gupta and Anbalagan, 1995).

S.No.	Contributory factor	Category	Rating	Remarks
1.	**Lithology**			
	(a) Rock type	*Type – I*		*Correction factor for weathering:*
		– Quartzite and Limestone	0.2	(a) Highly weathered – rock
		– Granite and Gabbro	0.3	discolored joints open with
		– Gneiss	0.4	weathering products, rock
				fabric altered to a large extent;
		Type – II		correction factor C_1
		– Well cemented ferruginous	1.0	
		sedimentary rocks, dominantly		
		sandstone with minor beds of		(b) Moderately weathered – rock
		claystone		discolored with fresh rock
		– Poorly cemented ferruginous	1.3	patches, weathering more
		sedimentary rocks, dominantly		around joint planes but rock
		sandstone with minor clay shale		intact in nature; correction
		beds		factor C_2
		Type – III		(c) Slightly weathered – rock
		– Slate and phyllite	1.2	slightly discolored along joint
		– Schist	1.3	planes, which may be
		– Shale with interbedded clayey and	1.8, 2.0	moderately tight to open, intact
		non-clayey rocks		rock; correction factor C_3
		– Highly weathered shale, phyllite and	0.8	
		schist, any rock with talc mineral		
	(b) Soil type	– Older well compacted fluvial fill	1.0	**The correction factor for**
		material/RBM (alluvial)		**weathering should be a multiple**
		– Clayey soil with naturally formed	1.4	**with the fresh rock rating to get**
		surface (alluvial)		**the corrected rating**
		– Sandy soil with naturally formed	1.2	
		surface (alluvial)		

(*continued*)

Table 19.1. (*continued*)

S.No.	Contributory factor	Category	Rating	Remarks
		– Debris comprising mostly rock pieces mixed with clayey/sandy soil (colluvial)	2.0	*For rock type I* $C_1 = 4, C_2 = 3, C_3 = 2$ *For rock type II* $C_1 = 1.5, C_2 = 1.25, C_3 = 1.0$
		I. older well compacted		
		II. younger loose material		
2.	**Structure**			
	(a) *Parallelism between the slope and discontinuity.** PLANAR ($\alpha_j - \alpha_s$) WEDGE ($\alpha_i - \alpha_s$)	I. >30°	0.2	α_j = dip direction of joint α_i = direction of line of intersection of two discontinuities
		II. 21–30°	0.25	
		III. 11–20°	0.3	
		IV. 6–10°	0.4	
		V. <5°	0.5	α_s = direction of slope inclination
	(b) *Relationship of dip of discontinuity and inclination* PLANAR ($\beta_j - \beta_s$) WEDGE ($\beta_i - \beta_s$)	I. >10°	0.3	β_j = dip of joint β_i = plunge of line of intersection β_s = inclination of slope
		II. 0–10°	0.5	
		III. 0°	0.7	
		IV. 0–(−10°)	0.8	
		V. <−10°	1.0	
	(c) *Dip of discontinuity* PLANAR (β_j) WEDGE (β_i)	I. <15°	0.2	*Category* I = very favourable II = favourable III = fair IV = unfavourable V = very unfavourable
		II. 16–25°	0.25	
		III. 26–35°	0.3	
		IV. 36–45°	0.4	
		V. >45°	0.5	
3.	**Slope morphometry**			
	– *Escarpment/cliff*	>45°	2.0	
	– *Steep slope*	36–45°	1.7	
	– *Moderately steep slope*	26–35°	1.2	
	– *Gentle slope*	16–25°	0.8	
	– *Very gentle slope*	<15°	0.5	

Table 19.1. (*continued*)

S.No.	Contributory factor	Category	Rating	Remarks
4.	**Relative relief**			
	Low	<100 m	0.3	
	Medium	101–300 m	0.6	
	High	>300 m	1.0	
5.	**Land use and land cover**			
	– *Agriculture land/populated flat land*		0.65	
	– *Thickly vegetated area*		0.90	
	– *Moderately vegetated*		1.2	
	– *Sparsely vegetated with lesser ground cover*		1.2	
	– *Barren land*		2.0	
	– *Depth of soil cover*	<5 m	0.65	
		6–10 m	0.85	
		11–15 m	1.3	
		16–20 m	2.0	
		>20 m	1.2	
6.	**Ground water condition**	Flowing	1.0	
		Dripping	0.8	
		Wet	0.5	
		Damp	0.2	
		Dry	0.0	

* Discontinuity refers to the planar discontinuity or the line of intersection of two planar discontinuities, whichever is important concerning instabilities.
Note: In regions of low seismicity (1, 2, and 3 zones), the maximum rating for relative relief may be reduced to 0.5 times and that of hydrogeological conditions be increased to 1.5 times (Table 19.1). For high seismicity (4 and 5 zones), no corrections are required.

Table 19.2. Proposed maximum LHEF rating for different contributory factors for LHZ mapping (Gupta and Anbalagan, 1995).

Contributory factor	Maximum LHEF rating
Lithology	2
Structure – relationship of structural discontinuities with slopes	2
Slope morphometry	2
Relative relief	1
Land use and land cover	2
Ground water condition	1
Total	10

Structure

This includes primary and secondary rock discontinuities, such as bedding planes, foliations, faults and thrusts. The discontinuities in relation to slope direction has greater influence on the slope stability. The following three types of relations are important:

1. The extent of parallelism between the directions of discontinuity or the line of intersection of two discontinuities and the slope.
2. Steepness of the dip of discontinuity or plunge of the line of intersection of two discontinuities.
3. The difference in the dip of discontinuity or plunge of the line of intersection of two discontinuities of the slope.

The above three relations are same as that of F_1, F_2 and F_3 of Romana (1985) and discussed in Chapter 18. Various sub-classes of the above conditions are also more or less similar to Romana (1985). It may be noted that the inferred depth, in case of soil, should be considered for awarding the ratings.

Slope morphometry

Slope morphometry defines the slope categories on the basis of frequency of occurrence of particular slope angle. Five categories representing the slopes of escarpment/cliff, steep slope, moderately steep slope, gentle slope and very gentle slope are used in preparing slope morphometry maps. On regional basis, the angle can be obtained from topo sheets for initial study.

Relative relief

A relative relief map represents the local relief of maximum height between the ridge top and the valley floor within an individual facet. Three categories of slopes of relative relief, namely low, medium and high, should be used for hazard evaluation purposes. A facet is a part of hill slope which has more or less similar characters of slope showing consistent slope direction and inclination.

Land use and land cover

The nature of land cover is an indirect indication of hill slope stability. Forest cover, for instance, protects slopes from the effects of weathering and erosion.

Table 19.3. Classification of landslide hazard zonation LHZ (Gupta and Anbalagan, 1995).

Zone	Value of TEHR	Description of LHZ	Practical significance
I	<3.5	Very Low Hazard (VLH)	Safe for development schemes
II	3.5–5.0	Low Hazard (LH)	
III	5.1–6.0	Moderate Hazard (MH)	Local vulnerable zones of instabilities
IV	6.1–7.5	High Hazard (HH)	Unsafe for development schemes
V	>7.5	Very High Hazard (VHH)	

A well-developed and spread root system increases the shearing resistance of the slope material. The barren and sparsely vegetated areas show faster erosion and greater instability. Based on the vegetation cover and its intensity, therefore, ratings for this parameter have been awarded. (A review of literature shows that extra cohesion due to root reinforcement is seldom more than 5 T/m^2.) Thus, continuous vegetation and grass cover on an entire hill slope is not fully responsible in landslide control because of root reinforcement, but a drastic decrease in the infiltration rate of rain water through a thin humus layer on account of grass cover is more beneficial. It may be noted that, in the case of thickly populated areas, smaller facets of rock slopes may be taken into consideration.

Ground water conditions
Since the ground water in hilly terrain is generally channeled along structural discontinuities of rocks, it does not have uniform flow pattern. The observational evaluation of the ground water on hill slopes is not possible over large areas. Therefore, for quick appraisal, surface indications of water such as damp, wet, dripping and flowing are used for rating purposes. It is suggested that studies should be carried out soon after the monsoon season.

Other factors
A 100 m- to 200 m-wide strip on either side of major faults and thrusts and intra-thrust zones may be awarded an extra rating of 1.0 to consider higher landslide susceptibility depending upon intensity of fracturing.

19.2.2   Landslide hazard zonation

Ratings of all the parameters are added to obtain a total estimated hazard rating (TEHR). Various zones of landslide hazard have subsequently been classified on the basis of TEHR as given in Table 19.3.

19.2.3   Presentation of LHZ maps

The results should be presented in the form of maps. The terrain evaluation maps are prepared in the first stage showing the nature of facet-wise distribution of parameters. The terrain evaluation maps are superimposed and TEHR is estimated

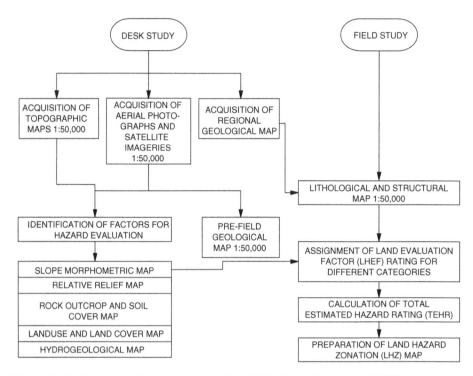

Figure 19.1. Procedure for macro-regional landslide hazard zonation (LHZ) mapping.

for individual facets. Subsequently, LHZ maps are prepared based on facet-wise distribution of TEHR values. For this exercise, two types of studies are performed – (i) Desk or laboratory study and (ii) Field study. The general procedures of LHZ mapping techniques have been outlined in the form of a flow chart (Fig. 19.1). This method has been adopted by Bureau of Indian Standards (IS 14496 Part 2: 1998). This technique is not applicable to mountains which are covered by snow most of the times. Moreover, freezing and thawing of water in rock joints cause rock slides in these regions. A case history has been presented to clarify the LHZ methodology and to develop confidence amongst users.

19.3   A CASE HISTORY (GUPTA AND ANBALAGAN, 1995)

The present investigation covers the Tehri-Pratapnagar area falling between Latitude (30°22′15″–30°30′5″) and Longitude (78°25′–78°30′) (Fig. 19.2).

19.3.1   The geology of the area

The study area lies in the Tehri District of Uttar Pradesh in India. The rock masses of the area belong to the Damtha, Tejam and Jaunsar Groups. The stratigraphic sequence of the area and its vicinity is as follows (Valdiya, 1980).

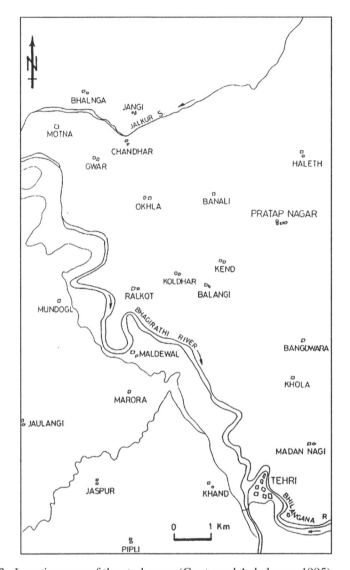

Figure 19.2. Location map of the study area (Gupta and Anbalagan, 1995).

Nagthat–Berinag Formation	
Chandpur formation	Jaunsar group
Deoban formation	Tejam group
Rautgara formation	Damtha group

The area has been mapped on 1:50,000 scale for studying the lithology and structure. The rocks exposed in the area include phyllites of Chandpur formation interbedded with sublitharenites of Rautgara formation, dolomitic limestone of Deoban formation and quartzites of Nagthat–Berinag formation. The phyllites

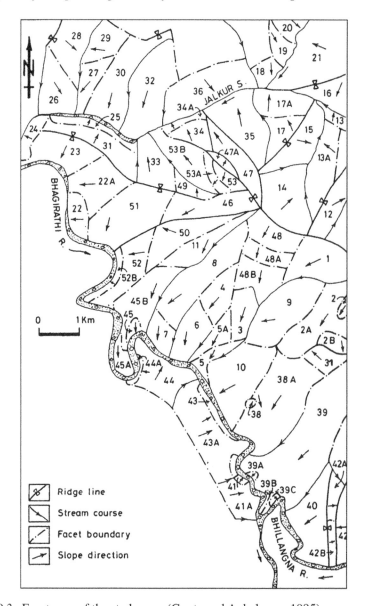

Figure 19.3. Facet map of the study area (Gupta and Anbalagan, 1995).

are grey and olive green interbedded with metasiltstones and quartzitic phyllites. The Rautgara formation comprises purple, pink and white coloured, medium grained quartzites interbedded with medium grained grey and dark green sublitharenites and slates as well as metavolcanics. The Deoban formation consists of dense, fine-grained dolomites of white and light pink colours with minor phyllitic intercalations. They occupy topographically higher ridges. The Nagthat–Berinag formation includes purple, white and green coloured quartzites interbedded with greenish and grey slates as well as grey phyllites.

The Chandpur Formation is delimited towards the north by a well defined thrust called North Almora thrust trending roughly northwest–southeast and dipping southwest. Moreover, the Deoban and the Nagthat–Berinag Formations have a thrusted contact, the thrust trending parallel to North Almora thrust and dipping northeast. The thrust is called Pratapnagar thrust. The rocks are badly crushed in the thrust zones.

19.3.2   Landslide hazard zonation mapping

The LHZ map of this area has been prepared on a 1:50,000 scale using a LHEF rating scheme for which a facet map of the area has been prepared (Fig. 19.3). A facet is a part of hill slope which has more or less similar characters of slope, showing consistent slope direction and inclination. The thematic maps of the area, namely a lithological map (Fig. 19.4), a structural map (Fig. 19.5), a slope morphometry map (Fig. 19.6), a land use and land cover map (Fig. 19.7), a relative relief map (Fig. 19.8), and a ground water condition map (Fig. 19.9) have been prepared using the detailed LHEF rating scheme (Table 19.2).

19.3.3   Lithology (Fig. 19.4)

Lithology is one of the major causative factors for slope instability. The major rock types observed in the area include phyllites, quartzites and dolomitic limestones. In addition, fluvial terrace materials are present in abundance to the right of river Bhagirathi all along its course.

Phyllites are exposed on either bank close to the Bhagirathi river. Though older terrace materials are present at lower levels, thick alluvial and colluvial soil cover are present at places in the upper levels on the right bank. On the left bank, the phyllites are generally weathered close to the surface and support thin soil cover. At places, the thickness of soil cover increases up to 5 m.

The North Almora thrust separates the Chandpur phyllites on the South from the quartzites of the Rautgara formation. The Rautgara quartzites interbedded with minor slates and metavolcanics are pink, purple and white coloured, well jointed and medium grained. The rocks and soil types in the area have the following distribution: phyllites – 44.17 percent, quartzites – 27.41 percent, marl/limestones – 12.48 percent, metabasics 0.25 percent, river terrace material 6.11 percent, phyllites with thin alluvial soil cover 6.16 percent and quartzites with thin soil cover 3.41 percent of the study area.

19.3.4   Structure (Fig. 19.5)

Major structural features seen in the area are the North Almora thrust and the Pratapnagar thrust which form part of the Berinag thrust. The structures used for landslide hazard zonation mapping include beddings, joints and foliations. The dispositions of the structures have been plotted in a stereonet for individual facets. The interrelation of the structural discontinuity with slope is studied carefully to award ratings.

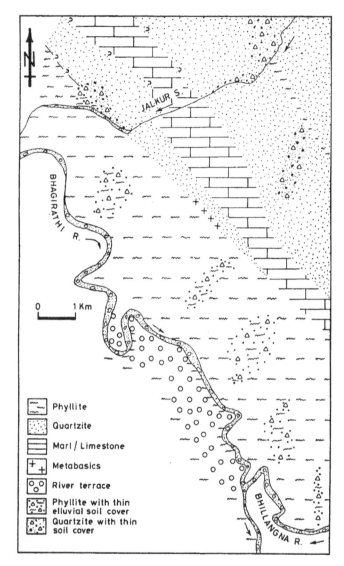

Figure 19.4. Lithological map (Gupta and Anbalagan, 1995).

19.3.5   Slope morphometry (Fig. 19.6)

A slope morphometry map represents the zones of different slopes, which have a specific range of inclination. The area of study has a good distribution of slope categories. The area to the west of the Bhagirathi river, mainly occupied by terrace deposits, falls in the category of a very gentle slope. Gentle slopes are mainly confined to the agricultural fields. There is a good distribution throughout the area of study. Moderately steep slopes mainly occur in the central and eastern part of the area. Steep slopes mainly occur in the central and the eastern parts of the area. Very steep slopes occur in the northern part of the study area adjoining the Jalkur stream.

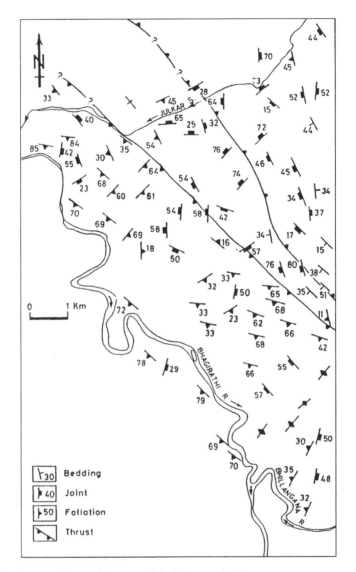

Figure 19.5. Structural map (Gupta and Anbalagan, 1995).

In fact, the Jalkur stream flows through a tight, narrow, V-shaped gorge in this reach. Very steep slopes/escarpments occur in small patches, mainly close to the watercourses possibly because of toe erosion. The area has the following distribution – 6.14 percent, 31.92 percent, 42.32 percent, 11.37 percent and 8.27 percent of very gentle slope, gentle slope, moderately steep slope, steep slope and very steep slope/escarpment respectively.

19.3.6   Land and land cover (Fig. 19.7)

Vegetation cover generally smoothens the action of climatic agents and protects the slope from weathering and erosion. The nature of land cover may indirectly

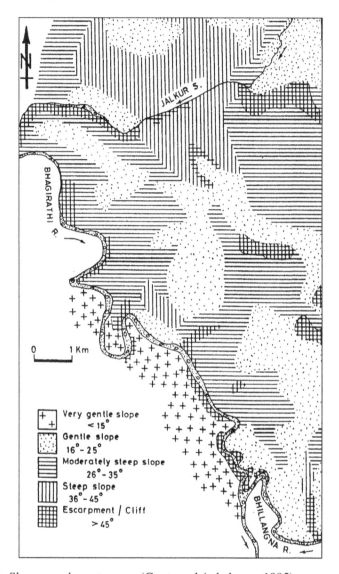

Figure 19.6. Slope morphometry map (Gupta and Anbalagan, 1995).

indicate the stability of hill slopes. Agriculture lands/populated flat lands are extensively present in the central, south-eastern, southern and parts of north-eastern areas. Thickly vegetated forest areas are seen in the Pratapnagar–Bangdwara area. Moderately vegetated areas are mainly present in small patches to the west of thickly vegetated areas. Sparsely vegetated and barren lands are mainly confined to quartzitic and dolomitic limestone terrain where steep to very steep slopes are present. These types of slopes are seen along the Bhagirathi valley adjoining the river course generally on steep slopes. The five categories of land use and land cover namely agricultural lands/populated flat lands, thickly vegetated forest area,

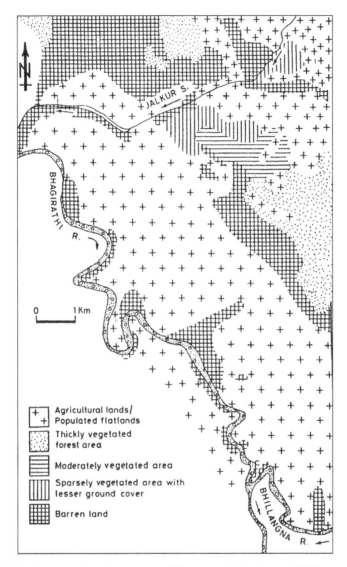

Figure 19.7. Land use and land cover map (Gupta and Anbalagan, 1995).

moderately vegetated area, sparsely vegetated area and barren land have the distribution of 65.44 percent, 5.94 percent, 1.73 percent, 3.78 percent and 23.10 percent
respectively in the study area.

19.3.7   Relative relief (Fig. 19.8)

Relative relief is the maximum height between the ridge top and the valley floor
within an individual facet. The three categories of relative relief, namely high
relief, medium relief and low relief, occupy 75.53 percent, 15.96 percent and 8.74
percent of the study area respectively.

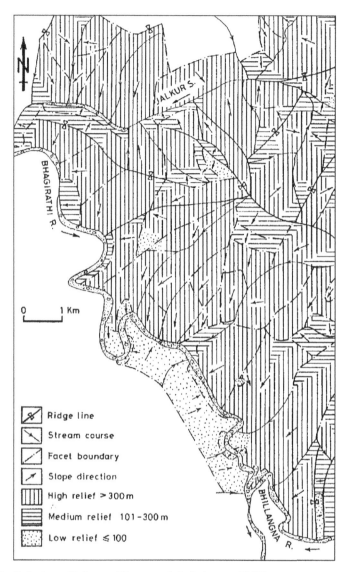

Figure 19.8. Relative relief map (Gupta and Anbalagan, 1995).

19.3.8   Ground water condition (Fig. 19.9)

The surface manifestation of ground water, such as wet, damp and dry have been observed in the study area. The area dominantly shows dry condition in about 54.86 percent of the area, damp condition in about 40.96 percent of the area and 4.8 percent of the study area is covered by wet ground water condition. Dry condition is mainly observed in the northern part and well distributed in rest of the study area. Damp and wet conditions are present in a number of facets in the southern, eastern and central part of the study area.

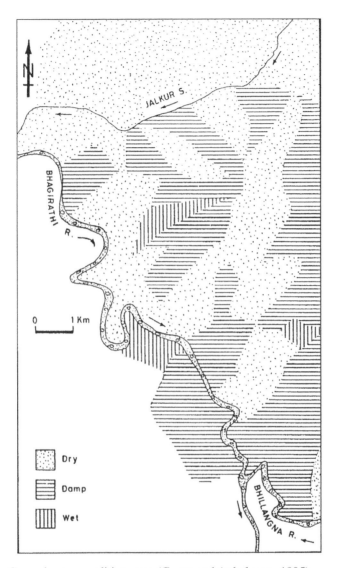

Figure 19.9. Groundwater condition map (Gupta and Anbalagan, 1995).

19.3.9 Landslide hazard zonation (Fig. 19.10)

The sum of all causative factors within an individual facet gives the total estimated hazard rating (TEHR) for a facet. The TEHR indicates the net probability of instability within an individual facet. Based on the TEHR value, facets are divided into different categories of hazard zones (Anbalagan, 1992).

The five categories of hazards, namely, very low hazard (VLH), low hazard (LH), moderate hazard (MH), high hazard (HH) and very high hazard (VHH) are found to be present in the study area. The areas showing VLH and LH constitute about 2.33 percent and 43.27 percent of the study area respectively. They are well

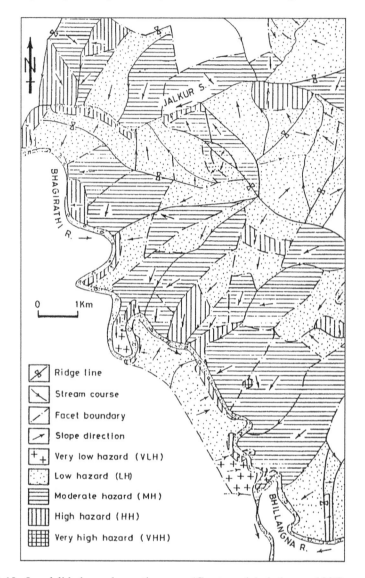

Figure 19.10. Landslide hazard zonation map (Gupta and Anbalagan, 1995).

distributed within the area. MH zones are mostly present in the immediate vicinities to the east of the Bhagirathi river. HH and VHH zones occur as small patches, mostly close to the watercourses. They represent areas of greater instability where detailed investigations should be carried out.

Some difficulty was experienced in zonation at the boundary lines. The visual inspection matched with Figure 19.10 for more than an 85 percent area. As such, Anbalagan's technique may be adopted in all mountainous terrains with minor adjustments in his ratings. For rocky hill areas, SMR is preferred.

19.4   A PROPOSITION FOR TEA GARDENS

Tea gardens are recommended in medium and high hazard zones because of suitable soil and climatic conditions in this area. The tea gardens will reduce infiltration of rainwater into the debris significantly and thereby stabilize landslide prone areas. Tea gardens will also provide job opportunities to local people and remove their poverty.

19.5   GEOGRAPHIC INFORMATION SYSTEMS

Geographic Information Systems (GIS) are software tools that are used to store, analyse, process, manipulate and update information in layers where geographic location is an important characteristic or critical to the analysis (Aronoff, 1989). Landslide hazard zonation mapping as described in Section 19.2 can be done efficiently by using Geographic Information System (GIS). The landslide hazard evaluation factors (LHEF) can be used as layers of information to GIS using various input devices. For example, Figures 19.2, 19.3, 19.4, 19.5 19.6, 19.7, 19.8 and 19.9 can be used as the layers of information to a GIS using input devices such as a digitiser, scanner, etc. to carry out landslide hazard zonation mapping of the considered area and thus provide an output like Figure 19.10. Amin et al. (2001) have developed a software package GLANN using GIS, neural network analysis and genetic algorithm for automatic landslide zonation. They used Anbalagan's LHZ system successfully.

While handling and analysing data that are referenced to a geographic location are the key capabilities of a GIS, the power of the system is most apparent when the quantity of data involved in mapping of landslide hazard zonation is too large to be handled manually. Over and above the main causative factors mentioned in Section 19.2.1, there may be many other features which may considered for LHZ based on site specific conditions and many factors associated with each feature or location. These data may exist as maps, table of data or even list of names (Fig. 19.2). Such a large volume of data can not be efficiently handled by manual methods. However, when those data have been inputted into a GIS, it can be easily processed and analyzed efficiently and economically.

Fortunately, GPS has been used successfully in monitoring landslides with an accuracy of nearly 2 mm (Brunner et al., 2000). It is a practical method of landslide hazard zonation by plotting contours of rates of displacement per year.

REFERENCES

Amin, S., Gupta, R., Saha, A.K., Arora, M.K. and Gupta, R.P. 2001. Genetic algorithm based neural network for landslide hazard zonation: some preliminary results, *Workshop on Application of Rock Engineering in Nation's Development* (In honour of Prof. Bhawani Singh), Roorkee, India, pp. 203–216.

Anbalagan, R. 1992. Terrain evaluation and landslide hazard zonation for environmental regeneration and land use planning in mountainous Terrain. *Int. Sym. Landslides*, Christ church, New Zealand, pp. 861–868.

Anbalagan, R. and Singh, Bhawani 2001. Landslide hazard and risk mapping in the Himalaya. An ICIMOD Publication on Landslide Hazard Mitigation in the Hindu Kush Himalayas, Edited by Li Tianchi et al., Kathmandu, pp. 163–188.

Aronoff, S. 1989. *Geographic information systems: a management perspective*. Ottawa, Canada, WDL Publications.

Brunner, F.K., Hartinger, H. and Richter, B. 2000. Continuous Monitoring of Landslides using GPS : A Progress Report; Publication on Geophysical Aspects of Mass Movements, Austrian Academy of Sciences (Eds. S.J. Bauer and F.K. Weber), Vienna, pp. 75–88.

Gupta, P. and Anbalagan, R. 1995. Landslide hazard zonation, mapping of Tehri-Pratapnagar area, Garhwal Himalayas. *J. Rock Mech. Tunnelling Technol*, India, 1(1): 41–58.

Gupta, P., Jain, Neelam, Anbalagan, R. and Sikdar, P.K. 2000. Landslide hazard evaluation and geostatistical studies in Garhwal Himalaya, India. *J. Rock Mech. Tunnel. Technol.* 6(1): 41–60.

IS 14496 (Part 2): 1998. Indian Standard Code on Preparation of Landslide Hazard Zonation Maps in Mountainous Terrains – Guidelines – Part 2 Macro-zonation.

Singh, Bhawani and Goel, R.K. 1999. *Rock Mass Classification: A Practical Approach in Civil Engineering*, Chapter 18. Reproduced and Revised, Permission from Elsevier Science Ltd., U.K., p. 267.

Valdiya, K.S. 1980. *Geology of Kumaon Lesser Himalaya*. Dehradun, India, Wadia Institute of Himalayan Geology, p. 291.

PART 2: TUNNELLING

CHAPTER 20

An introduction to software packages on underground openings in rock mass

"I believe that the engineer needs primarily the fundamentals of mathematical analysis and sound methods of approximation"
Von Karman

There is a concern in the mind of civil engineers for a conservative and durable solution of geological problems that are of an unclear nature. However, the mining engineers are more concerned about the safe and profitable production of ores at a faster rate. The past interaction among civil, mining engineers and geologists needs to be further developed to improve the state of the art. As such, six computer programs on approximate analysis of stresses and stability of underground opening in rock mass have been presented in this part of the book. The programs may also be used to design support system to stabilise the underground openings. The aim of this package is speedy and on-spot crisis management. Simple geological input data is enough. You are welcome to use the following simple and field-tested software packages.

20.1   A LIST OF SOFTWARE PROGRAMS AND THEIR FEATURES

20.1.1   UWEDGE

- Developed by Hoek and Brown (1980) and modified to design support systems.
- Analyses stability of wedges in roof and walls of an arched underground opening or cavern in rock mass, with and without support system (rock bolts/rock anchors/cable anchors, welded mesh and shotcrete/steel fibre reinforced shotcrete, SFRS).
- Wedge theory gives conservative design of support system as it does not account for interlocking of wedges and restrained dilatancy in tunnels and caverns.
- It suggests special specifications for tunnelling according to the site conditions.
- This program is also useful for the design of a support system for mine roadways with a flat roof.

- Analyses stability of wedges in both vertical and inclined shafts.
- Analyses stability of wedges in portals and half-tunnels also.
- Output is roof and wall support pressures and factors of safety of all possible both unsupported and supported wedges. The aim should be to stitch all the wedges.
- It is a great time saver to enable a fast rate of tunnelling in non-squeezing rock masses and on-spot crisis management.

20.1.2 SQUEEZE

- Predicts support pressure in tunnels in squeezing ground.
- Based on elasto-plastic theory initially proposed by Daemen (1975).
- Assumes multi-axial strength criterion of jointed rock mass in the elastic zone.
- Assumes residual and sympathetic failure criterion in the broken zone.
- Gives ground response curve.
- Field experience is encouraging in the instrumented tunnels.

20.1.3 TM

- Suggests special specifications for tunnelling according to the site conditions.
- Adopts semi-empirical design of rock bolt/rock anchor and shotcrete/steel fibre reinforced shotcrete and steel ribs and grouted rock arch (theory based on Q-System).
- It takes into account seepage pressure, spacing of fractures in rock mass and earthquake.
- Applicable to caverns and tunnels in the squeezing ground conditions also.
- Based on concept of the self-stable reinforced rock arch in tunnels and the reinforced rock frame in caverns.
- Output is the capacity of entire support system, thickness of shotcrete and the optimum angle of roof arch in the caverns.
- It is a great time saver to enable fast rate of tunnelling in rock and on-spot crisis management in various ground conditions.

20.1.4 LINING

- Analyses stresses in solid or cracked PCC lining of pressure tunnels in rock mass.
- It estimates automatically the number of cracks in PCC lining and gaps in the cracks.
- Analyses RCC lining also.
- Takes into account the possible radial cracks in concrete and rock mass due to low in situ stresses.
- Checks if rock cover is adequate on PCC lining.
- Analyses steel liner and calculates spacing of stiffners to prevent buckling of the liner.

- Estimates the seepage loss in rock mass.
- It leads to design of a PCC cracked lining without hoop reinforcement and thus one can save in construction time and on millions of dollars spent on steel.
- Experience shows that performance of cracked lining in hydroelectric projects in India is excellent even after continuous use since 1970.

20.1.5   BEM

- Developed by Hoek and Brown (1980).
- Analyses stresses and displacement around underground openings in homogeneous, isotropic and elastic rock mass (such as in dry rocks with good rock mass quality).
- Useful for study of flow of principal stress trajectories around openings.
- Facilitates in finding out shapes and spacing between series of openings.
- Output is the zones of shear strength failure and tensile failure in rock mass.

20.1.6   MSEAMS

- Developed by Prof. S. L. Crouch, University of Minnesota, U.S.A.
- Analyses subsidence and tilt above horizontal mines in anisotropic and elastic rock mass with characteristic *low shear modulus*.
- Computes stresses on mine pillar.
- Tilt, horizontal strains and subsidence of building should be within safe limits.
- Maximum normal stress on edges of a mine pillar should not exceed a permissible strength of rock mass in hard rock mines.

The emphasis is on safe design of the support system and not on rigorous analysis. Thus simple geological data is enough. Programs will automatically identify critical joint planes in the rock mass. The strength parameters obtained from the back analysis and experience ensure that the design of a support system is not over-conservative. The advantage of loading this software on PC Note Book is that one may find a number of technical solutions or alternative designs to select finally the optimum design on the basis of economy, field experience, its feasibility at the site and minimum need for its maintenance. The joy of designing remedial measures cannot be described in words.

Extensive experience suggests that rock masses behave much better than our expectations in the cases of arched underground openings. Earthquakes do not affect the stability of underground openings beyond a 10 m depth below ground, if the opening is away from thick plastic seams, faults and thrusts and intra-thrust zone. Earthquake effects have not thus been considered in these programs. In the cases of complex geological conditions, computer programs such as UDEC or 3DEC developed by the research group of Professor C. Fairhurst, Professor Peter Cundall and others should be used to design a support system. Sharma et al. (1999) have published a book on their practical applications.

The uncertainties in geology, rock parameters and environmental conditions may be managed by a high factor of safety and redundancy (additional support systems e.g. shotcreting, etc.). Thus, it is suggested that 2 or 3 lines of defences or support types be built to manage uncertainties in explorations, testing and our assumptions in complex geological conditions. Experience suggests that reinforced rock arch or reinforced rock frame is self-stable and a robust support system capable of withstanding unexpected non-uniform support pressures. Furthermore, spot bolting may be done locally where the rock is intensely fractured. Recent tunnelling experience shows that steel fibre-reinforced shotcrete (SFRS) is an ideal choice for tunnelling through difficult ground conditions. Beginners are advised to consult internationally reputed consultants in cases of complex geological environments.

Many civil engineers, mining engineers and geologists have used these programs since 1985. The programs are not intended to take authority from the decision-makers, but to be used as aids. After all, we are working for mutual benefits. The software packages are vital to give an innovative design of rock structures where empirical solutions are not valid.

20.2   THE USE OF COMPUTER PROGRAMS

All the computer programs discussed in Part 2 are available in the directory SOFT-WARE/TUNNEL in the enclosed compact disc. The source programs are also given in the directory SOFTWARE/TUNNEL/SOURCE. All these source files in Fortran may be translated by F2C software in source files of C++ without any difficulty. The programs are written in Fortran 77 and EXE files work in DOS environment. The attached CD also has a user's manual file for almost all the programs. A user's manual is also presented in each chapter discussing a particular program.

For more clarifications of users, typical input data files are given beginning with I. Similarly, corresponding output files are also given, beginning with O. File details, for example, are as follows (considering Program UWEDGE):

File Name	*Details*
Uwedge.txt	Users Manual
Iuwedge.dat	Typical input data file
Ouwedge.dat	Typical output data file

The typical computer commands to work in DOS environment are (for example considering program UWEDGE):

Command	*Operation*
EDIT IUWEDGE.DAT	(TO PREPARE INPUT FILE)
UWEDGE	(TO RUN EXE FILE OF PROGRAM UWEDGE)
IUWEDGE.DAT	(TYPE INPUT FILE NAME)
OUWEDGE.DAT	(TYPE OUTPUT FILE NAME)
2	(TYPE 2 FOR EXECUTION)

EDIT OUWEDGE.DAT   (FOR VIEWING AND PRINTING OUTPUT
                     FILE OUWEDGE.DAT)

The use of DOS Editor has been suggested to create an input data file. All the data files are in ASCII format. Users may use other file names for different sites. However, SQUEEZE and TM (Chapters 25 and 26 respectively) are user friendly programs with SQUEEZE.OUT and TM.OUT as the output files.

A new file, say Ixyz.DAT, may be created by command EDIT Ixyz.DAT. After this, the input data is typed as per the sequence shown in the users manual. Similarly, a new output file Oxyz.DAT should be created and the empty file closed. Finally, program XYZ is run with the commands as shown above. After the execution of the program, a temporary empty output file is created which should be deleted.

Experience shows that an input file is very good choice for frequent refinements in input data for optimizing design of support system based on the experience, intuition and common sense. Appendix 4 may be referred for running these softwares/programs on WINDOWS environment.

The next three chapters are devoted to rock mass quality, tunnelling hazards and deformation modulus and in situ stresses to provide the information on selection of input parameters for various software packages discussed subsequently.

REFERENCES

Daemen, J.J.K. 1975. *Tunnel support loading caused by rock failure*. Ph.D. Thesis, University of Minnesota, Minneapolis, U.S.A.

Hoek, E. and Brown, E.T. 1980. *Underground Excavations in Rock*. Institution of Mining and Metallurgy, England, Revised Edition, Appendix 6: Underground wedge analysis, p. 52.

Sharma, V.M., Saxena, K.R. and Woods, R.D. 1999. *Distinct Element Modelling in Geomechanics*, Chapters 6, 7 and 8. New Delhi, Oxford and IBH Publishing House, p. 222.

CHAPTER 21

Rock mass quality (Q)-SYSTEM

"Genius is 99 percent perspiration and 1 percent inspiration"
Bernard Shaw

21.1   THE Q-SYSTEM

Barton, Lien and Lunde (1974) of the Norwegian Geotechnical Institute (NGI) originally proposed the Q-system of rock mass classification based on about 200 case histories of tunnels and caverns (particularly civil engineering cases). They have defined the rock mass quality Q as follows:

$$Q = [RQD/J_n][J_r/J_a][J_w/SRF] \qquad (21.1)$$

where

$$
\begin{aligned}
RQD &= \text{Deere's Rock Quality Designation} \geqslant 10, \\
&= 115 - 3.3\,J_v \\
J_n &= \text{Joint set number,} \\
J_r &= \text{Joint roughness number for the critically oriented joint set,} \\
J_a &= \text{Joint alteration number for critically oriented joint set,} \\
J_w &= \text{Joint water reduction factor,} \\
SRF &= \text{Stress reduction factor, and} \\
J_v &= \text{Volumetric joint count.}
\end{aligned}
$$

For various rock conditions, the ratings (numerical value) to these six parameters are assigned. The six parameters given in Equation 21.1 are defined in tables of Barton et al. (1974) and Singh and Goel (1999). The value of J_w should correspond to the future ground water condition where seepage erosion or leaching of chemical can alter the permeability of rock mass significantly. In case the rock mass quality varies from Q_{min} to Q_{max}, the average rock mass quality of $(Q_{max} \times Q_{min})^{1/2}$ may be assumed in the design calculations. The wall factor (Q_w) is obtained after multiplying Q by a number which depends on the magnitude of Q as given below:

Range of Q	Wall Factor Q_w	Remarks
>10	5.0 Q	p_{wall} is observed to be
0.1–10	2.5 Q	negligible in tunnels
<0.1	1.0 Q	

21.2   THE WIDTH OF SELF-SUPPORTING TUNNELS

According to Barton et al. (1974), the width (B_s) of a self-supporting tunnel is as follows

$$B_s = 2ESR \, Q^{0.4} \tag{21.2}$$

where ESR is excavation support ratio which is equal to 1 for important structures in rock mass. It may be noted that support pressure is zero if the width of the tunnel is less than or equal to B_s.

21.3   SUPPORT PRESSURE IN AN OPENING WITH AN ARCHED ROOF

21.3.1   Correlation by Singh et al. (1992)

It may be mentioned that Q, referred to in the above correlations, is actually the post-excavation quality of a rock mass, because, in tunnels, the geology of the rock mass is usually studied after blasting and an on-the-spot decision is taken on support density.

a. Short-term support pressure
Vertical or roof support pressure.   The observed roof support pressure is related to the short-term rock mass quality (Q_i) in 30 instrumented tunnels by the following empirical correlation

$$p_v = \frac{0.2}{J_r} \cdot Q_i^{-1/3} \cdot f \cdot f' \cdot f'' \quad \text{MPa} \tag{21.3}$$

$$f = 1 + (H - 320)/800 \geqslant 1 \tag{21.4}$$

where

$$
\begin{aligned}
Q_i &= 5\,Q = \text{short-term rock mass quality,} \\
p_v &= \text{short-term roof support pressure in MPa,} \\
f &= \text{correction factor for overburden,} \\
f' &= \text{correction factor for tunnel closure in the squeezing ground} \\
&\quad \text{condition (H} > 350\,Q^{1/3}\,\text{m and } J_r/J_a < 1/2), \\
&= 1 \text{ in non-squeezing,} \\
f'' &= \text{correction factor for the time after excavation and support} \\
&\quad \text{erection (Eqn. 21.5); and} \\
H &= \text{overburden above crown or tunnel depth below ground level in} \\
&\quad \text{metres.}
\end{aligned}
$$

While developing Equation 21.3, the correction factors have been applied in steps. Firstly, the correction factor for tunnel depth has been applied, afterwards the correction for tunnel closure and finally the correction for time (Singh et al., 1992).

Values of correction factors for tunnel closure (f′) can be obtained from Figure 21.1 based on the design value of tunnel closure in case of the squeezing ground. It may be noted that Figure 21.1 represents normalised observed ground response (reaction) curves for a tunnel roof and walls respectively in the squeezing ground.

The correction factor f″ for time was found as

$$f'' = \log (9.5\, t^{0.25}) \tag{21.5}$$

where t is time in months after support installation.

It may be noted that dilatant joints or J_r values play a dominant role in the stability of underground openings. Consequently, support capacities may be independent of the opening size (size effect was believed by Terzaghi, 1946) and evident from observations in Figure 21.2. In Figure 21.2, p_r^{obsd} is the observed roof support pressure and $p_{ir} = (0.2/Jr)Q_i^{-1/3}$. The observed short-term wall support pressure is insignificant generally in non-squeezing rock conditions. It is, therefore, recommended that these may be neglected in the case of tunnels in rock masses of good quality.

It should be noted however that although the wall support pressure would be negligible in non-squeezing ground conditions, high wall support pressure is common in poor ground or squeezing ground conditions. Therefore, invert struts with steel ribs should be used where the estimated wall support pressure requires the use of wall support in exceptionally poor rock conditions and highly squeezing or swelling ground conditions. Otherwise, NATM or NMT are better choices.

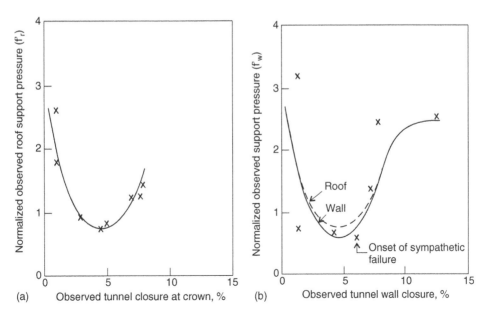

Figure 21.1. Correction factor for (a) roof closure and (b) wall closure under squeezing ground condition (Singh et al., 1992).

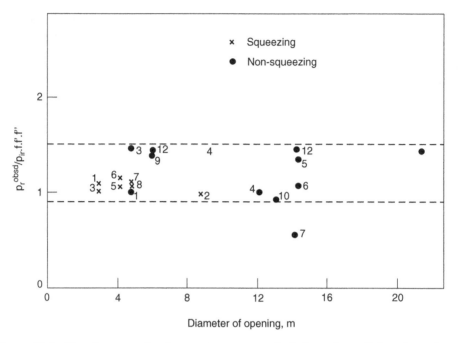

Figure 21.2. Showing normalised support pressure to be independent of size of openings in rock masses.

b. Ultimate support pressure
Replacing Q_i by Q in Equation 21.3, Singh et al. (1992) proposed the following correlation for ultimate tunnel support pressure p_{ult}:

$$p_{ult} = \frac{0.2}{J_r} \cdot Q^{-1/3} \cdot f \cdot f' \quad MPa \tag{21.6}$$

Long-term monitoring at the Chhibro cavern of the Yamuna hydroelectric project in India has enabled researchers to study the support pressure trend with time and with saturation. The study on the basis of 10 years monitoring has shown that the ultimate support pressure for water charged rock masses with erodible joint fillings may increase up to 6 times the short-term support pressure (Mitra, 1991; Mitra and Singh, 1997).

On extrapolating the values of support pressure for 100 years, a study of Singh et al. (1992) has shown that the ultimate support pressure would be about 1.75 times the short-term support pressure under non-squeezing ground conditions. But, in squeezing ground conditions, Jethwa (1981) has estimated that the ultimate support pressure would be 2 to 3 times the short-term support pressure.

In the case of water conductor systems, the rock mass is charged with water all around the concrete lining. In argillaceous rocks, the post-construction saturation reduces the modulus of deformation significantly. Consequently, high support pressure is likely to develop on the concrete lining. Verman (1993) has derived the following expression for extra support pressure.

$$p_{ult} = \left[1 - \frac{E_{sat}}{E_{dry}}\right]\gamma H$$

(21.7)

where

E_{sat} = modulus of deformation after saturation, and
E_{dry} = modulus of deformation in natural condition.

21.4   THE EFFECT OF EARTHQUAKES ON SUPPORT PRESSURE

Dowding and Rozen (1978) have compiled the seismic response of 71 tunnels. This study showed that the tunnels are less susceptible to damage than the surface structures. The peak acceleration of less than 0.2 g magnitude at the surface did no damage to the tunnels. The acceleration between 0.2 and 0.5 g did only minor damage. The damage was found to be significant only when the peak ground acceleration exceeded 0.5 g. In such cases most of the damage that occurred was located near portals. One may say that the portals are essentially surface structures. Barton (1984) found that dynamic increment in the support pressure was about 25 percent of the static ultimate support pressure.

An earthquake of 6.6 magnitude on the Richter scale occurred on 20th October, 1991, with an epicenter at Uttarkashi in India, about 70 km away from the Chhibro powerhouse cavern. This earthquake devastated the entire Uttarkashi area. The recorded damage in the region was closest to the shear and the fault zones. There was no damage, however, in the sections of the Chhibro powerhouse complex away from the shear and the fault zones. In Uttarkashi itself, there was no damage reported to the Maneri-Uttarkashi power tunnel.

A further example is the Chamoli earthquake in Indian Himalaya. This earthquake of 6.8 magnitude on the Richter scale occurred on 29th March, 1999. The epicenter of this earthquake was only 80 km from Tehri town and 90 km from Uttarkashi town. There was no reported damage in the Tehri Project underground openings and the Maneri-Uttarkashi power tunnel.

An analysis by Mitra and Singh (1997) shows that the dynamic support pressures may be 25 percent more than the long-term support pressures in the roof of the chamber near the shear and the fault zones due to accumulated strains in the nearby rock mass. It is understood that the size of natural caves, tunnels and caverns is smaller than the quarter-wave length of seismic waves. Hence, openings are not noticed by seismic waves and so there is no resonance of openings.

The dynamic increment of support pressure in rail tunnels may also be assumed to be same as for the seismic case. However, where overburden is less than 2B (where B is the width of the opening), the vertical support pressure is taken to be equal to the overburden pressure. The conservative practice is due to errors inherent in the survey of hilly terrain.

It is, therefore, recommended that underground earthquake shelters should be built in cities prone to earthquakes. These shelters will also be useful in other natural disasters and wars to some extent.

21.5   DESIGN OF SUPPORTS

Since the early 1980s, wet mix steel fibre reinforced shotcrete (SFRS) together with rock bolts have been the main components of a permanent rock support in underground openings in Norway. Based on this experience, Grimstad and Barton (1993) suggested a new support design chart using the steel fibre reinforced shotcrete (SFRS) as shown in Figure 21.3. This chart is recommended for tunnelling in poor rock conditions. The excavation support ratio (ESR) is 1.0 for important openings, which have a long life as in civil engineering practice.

Example 1
In a major hydro-electric project in the dry quartzitic phyllite, the rock mass quality is found to be in the range of 6 to 10. The joint roughness number J_r is 1.5 and the joint alteration number J_a is 1.0 for a critically oriented joint in the underground machine hall. The width of the cavern is 25 m and its height is 50 m and the roof is arched. The overburden is 450 m. Suggest the design of the support system.

Solution
The average rock mass quality is $(6 \times 10)^{1/2} = 8$ (approx.). The overburden above the crown is less than $350 \, (8)^{1/3} = 700$ m. Hence the rock mass is non-squeezing. The correction factor for overburden $f = 1 + (450 - 320)/800 = 1.16$. The correction for tunnel closure $f' = 1.0$. Short-term support pressure in roof from Equation 21.3 is ($f'' = 1.0$),

$$= (0.2/1.5)(5 \times 8)^{-1/3} \, 1.16 = 0.045 \, \text{MPa}$$

Short-term wall support pressure is

$$= (0.2/1.5)(5 \times 2.5 \times 8)^{-1/3} \, 1.16 = 0.033 \, \text{MPa (practically negligible)}$$

Ultimate support pressure in roof from Equation 21.6 is given by

$$p_{\text{roof}} = (0.2/1.5)(8)^{-1/3} \, 1.16 = 0.077 \, \text{MPa}$$

Ultimate wall support pressure is (see Section 21.1) given by

$$p_{\text{wall}} = (0.2/1.5)(2.5 \times 8)^{-1/3} \, 1.16 = 0.057 \, \text{MPa}$$

The modulus of deformation of the rock mass is (see Table 23.1 in Chapter 23) given by

$$E_d = (8)^{0.36} \, (450)^{0.2} = 7.0 \, \text{GPa}$$

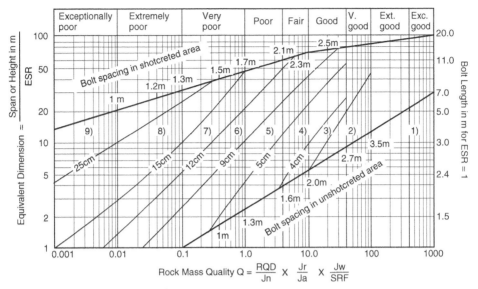

REINFORCEMENT CATEGORIES

1) Unsupported
2) Spot bolting, sb
3) Systematic bolting, B
4) Systematic bolting (and unreinforced shotcrete, 4-10cm, B(+S)
5) Fiber reinforced shotcrete and bolting, 5-9cm, Sfr+B

6) Fiber reinforced shotcrete and bolting, 9 - 12cm, Sfr+B
7) Fiber reinforced shotcrete and bolting, 12 - 15cm, Sfr+B
8) Fiber reinforced shotcrete > 15cm, reinforced ribs of shotcrete and bolting, Sfr, RRS+B
9) Cast concrete lining,CCA

Figure 21.3. Chart for the design of steel fibre reinforced shotcrete SFRS support (Grimstad and Barton, 1993).

The excavation support ratio is 1.0 for important structures. Figure 21.3 gives the following support system in the roof,

Bolt length $= 9$ m

Bolt spacing $= 2.5$ m

Thickness of steel fibre reinforced shotcrete $= 90$ mm

Figure 21.3 is also useful in recommending the following wall support system of the cavern ($Q = 2.5 \times 8 = 20$, ESR $= 1$, Height $= 50$ m)

Bolt length $= 11$ m

Bolt spacing $= 2.5$ m

Thickness of steel fibre reinforced shotcrete $= 70$ mm

REFERENCES

Barton, N., Lien, R. and Lunde, J. 1974. Engineering classification of rock masses for the design of tunnel support. *Rock Mechanics*, Vol. 6. Springer-Verlag, pp. 189–236.

Barton, N. 1984. Effects of rock mass deformation on tunnel performance in seismic regions. *Tunnel Technology and Surface Use* 4(3): 89–99.

Dowding, C.H. and Rozen, A. 1978. Damage to rock tunnels from earthquake shaking. *J. Geotech. Eng., ASCE* 104(2): 175–191.

Grimstad, E. and Barton, N. 1993. Updating of the Q-system for NMT. *Int. Symp. on Sprayed Concrete – Modern Use of Wet Mix Sprayed Concrete for Underground Support*, Fagernes (Editors Kompen, Opsahll and Berg. Norwegian Concrete Association, Oslo).

Jethwa, J.L. 1981. *Evaluation of rock pressures in tunnels through squeezing ground in lower Himalayas*. Ph.D. Thesis. Department of Civil Engineering, University of Roorkee, India, p. 272.

Mitra, S. 1991. *Studies on long-term behaviour of underground powerhouse cavities in soft rocks*. Ph.D. Thesis, University of Roorkee, India, p. 194.

Mitra, S. and Singh, Bhawani 1997. Influence of geological features on long-term behaviour of underground powerhouse cavities in lower Himalayan region – A case study. *J. Rock Mech. Tunnel. Technol.* 3(1): 23–76.

Singh, Bhawani, Jethwa, J.L., Dube, A.K. and Singh, B. 1992. Correlation between observed support pressure and rock mass quality. *Int. J. Tunnel Underground Space Technol.* 7(1): 59–74.

Singh, Bhawani and Goel, R.K. 1999. *Rock Mass Classification: A Practical Approach in Civil Engineering*. Part of Chapter 8, Permission from Elsevier Science Ltd., p. 268.

Terzaghi, K. 1946. Rock defects and load on tunnel supports. *Introduction to Rock Tunnelling with Steel Supports* by Proctor, R.V. and White, T.L. Youngstown, Ohio, Commercial Shearing and Stamping Company.

Verman, M.K. 1993. *Rock mass-tunnel support interaction analysis*. Ph.D. Thesis, Department of Civil Engineering. University of Roorkee, Roorkee, India, p. 258.

CHAPTER 22

Tunnelling hazards

"The most incomprehensible fact about nature is that it is comprehensible"
Albert Einstein

22.1   INTRODUCTION

The knowledge of potential tunnelling hazards plays an important role in the selection of excavation method and designing a support system for underground openings. The tunnelling media could be stable/competent (and or non-squeezing) or failing/squeezing depending upon the in situ stress and the rock mass strength. A weak over-stressed rock mass would experience squeezing ground condition, whereas a hard and massive over-stressed rock mass may experience rock burst condition. On the other hand, when the rock mass is not over-stressed, the ground condition is termed as stable or competent (non-squeezing).

Tunnelling in the competent ground conditions can again face two situations – (i) where no supports are required, i.e., a self-supporting condition and (ii) where supports are required for stability; let us call this a non-squeezing condition. The squeezing ground condition has been divided into three classes on the basis of tunnel closures by Singh et al. (1995) as mild, moderate and high squeezing ground conditions (Table 22.1).

The worldwide experience is that tunnelling through the squeezing ground condition is a very slow and hazardous process because the rock mass around the opening loses its inherent strength under the influence of in situ stresses. This may result in mobilization of high support pressure and tunnel closures. Tunnelling under the non-squeezing ground condition, on the other hand, is comparatively safe and easy because the inherent strength of the rock mass is maintained. Thus, the first important step is to assess whether a tunnel would experience a squeezing ground condition or a non-squeezing ground condition. This decision controls the selection of the excavation method and the support system. For example, a large tunnel could possibly be excavated full face with light supports. Under the non-squeezing ground condition, it may have to be excavated by heading and benching method with a flexible support system under the squeezing ground condition.

Table 22.1. Classification of ground conditions for tunnelling (Singh and Goel, 1999).

S.No.	Ground condition class	Sub-class	Rock behaviour
1	Competent self-supporting	–	Massive rock mass require no support for tunnel stability
2	Incompetent Non-squeezing	–	Jointed rock mass requires supports for tunnel stability. Tunnel walls are stable and do not close
2	Ravelling	–	Chunks or flakes of rock mass begin to drop out of the arch or walls after the rock mass is excavated
3	Squeezing	Mild squeezing ($u_a/a = 1–3$ percent) Moderate squeezing ($u_a/a = 3–5$ percent) High squeezing ($u_a/a < 5$ percent)	Rock mass squeezes plastically into the tunnel both from the roof and the walls and the phenomenon is time dependent; rate of squeezing depends upon the degree of over-stress; may occur at shallow depths in weak rock masses like shales, clay, etc.; hard rock masses under high cover may experience slabbing/poping/rock burst
4	Swelling	–	Rock mass absorbs water, increases in volume and expands slowly into the tunnel, (e.g. in montmorillonite clay).
5	Running	–	Granular material becomes unstable within steep shear zones
6	Flowing/ sudden flooding	–	A mixture of soil like material and water flows into the tunnel. The material can flow from invert as well as from the face crown and wall and can flow for large distances completely filling the tunnel and burying machines in some cases. The discharge may be 10–100 l/sec which can cause sudden flood. A chimney may be formed along thick shear zones and weak zones
7	Rock burst	–	A violent failure in hard (brittle) and massive rock masses of Class II* type when subjected to high stress

Notations: u_a = radial tunnel closure; a = tunnel radius; u_a/a = normalised tunnel closure in percentage; * = UCS test on Class II type rock shows reversal of strain after peak failure.

Non-squeezing ground conditions are common in most of the projects. The squeezing conditions are common in the Lower Himalaya in India, the Alps and in other parts of the world where the rock masses are weak, highly jointed, faulted, folded and tectonically disturbed and the overburden is high.

22.2   TUNNELLING HAZARDS

Various tunnelling conditions encountered during tunnelling have been summarized in Table 22.1. Table 22.2 suggests the method of excavation, the type of supports and precautions for various ground conditions. The Commission on Squeezing Rocks in Tunnels of International Society for Rock Mechanics (ISRM) has published *Definitions of Squeezing* which is quoted here (Barla, 1995).

> Squeezing of rock is the time dependent large deformation, which occurs around a tunnel and other underground openings, and is essentially associated with creep caused by exceeding shear strength. Deformation may terminate during construction or continue over a long time period.

This definition is complemented by the following additional statements:
- Squeezing can occur in both rock and soil as long as the particular combination of induced stresses and material properties pushes some zones around the tunnel beyond the limiting shear stress at which creep starts.
- The magnitude of the tunnel convergence associated with squeezing, the rate of deformation, and the extent of the yielding zone around the tunnel depend on the geological conditions, the in situ stresses relative to rock mass strength, the ground water flow and pore pressure and the rock mass properties.
- Squeezing of rock masses can occur as squeezing of intact rock, as squeezing of infilled rock discontinuities and/or along bedding and foliation surfaces, joints and faults.
- Squeezing is synonymous of over-stressing and does not comprise deformations caused by loosening as might occur at the roof or at the walls of tunnels in jointed rock masses. Rock bursting phenomena do not belong to squeezing.
- Time dependent displacements around tunnels of similar magnitudes as in squeezing ground conditions, may also occur in rocks susceptible to swelling. While swelling always implies volume increase due to penetration of the air moisture into the rock, squeezing does not, except for rocks which exhibit a dilatant behaviour. However, it is recognized that in some cases squeezing may be associated with swelling.
- Squeezing is closely related to the excavation, support techniques and sequence adopted in tunnelling. If the support installation is delayed, the rock mass moves into the tunnel and a stress redistribution takes place around it. Conversely, if the rock deformations are constrained, squeezing will lead to a long-term load build-up of rock support.

A comparison between squeezing and swelling phenomena by Jethwa and Dhar (1996) is given in Table 22.3. Figure 22.1 shows how radial displacements vary

Table 22.2. Method of excavation, type of supports and precautions to be adopted for different ground conditions.

S. No.	Ground conditions	Excavation method	Type of support	Precautions
1.	Self-supporting/ competent	TBM or full face drill and controlled blast	No support or spot bolting with a thin layer of shotcrete to prevent widening of joints	Look out for localised wedge/shear zone. Past experience discourages use of TBM if geological conditions change frequently
2.	Non-squeezing/ incompetent	Full face drill and controlled blast by boomers	Flexible support; shotcrete and pre-tensioned rock bolt supports of required capacity. Steel fibre reinforced shotcrete (SFRS) may or may not be required	First layer of shotcrete should be applied after some delay but within the stand-up time to release the strain energy of rock mass
3.	Ravelling	Heading and bench; drill and blast manually	Steel support with struts/ pre-tensioned rock bolts with steel fibre reinforced shotcrete (SFRS)	Expect heavy loads including side pressure
4.	Mild squeezing	Heading and bench; drill and blast	Full column grouted rock anchors and SFRS. Floor to be shotcreted to complete a support ring	Install support after each blast; circular shape is ideal; side pressure is expected; do not have a long heading which delays completion of support ring
5.	Moderate squeezing	Heading and bench; drill and blast	Flexible support; full column grouted highly ductile rock anchors and SFRS. Floor bolting to avoid floor heaving and to develop a reinforced rock frame. In case of steel ribs, these should be installed and embedded in shotcrete to withstand high support pressure	Install support after each blast; increase the tunnel diameter to absorb desirable closure; circular shape is ideal; side pressure is expected; instrumentation is essential

Table 22.2. (*continued*).

6.	High squeezing	Heading and bench in small tunnels and multiple drift method in large tunnels; use forepoling if stand-up time is low	Very flexible support; full column grouted highly ductile rock anchors and slotted SFRS; yielding steel ribs with struts when shotcrete fails repeatedly; steel ribs may be used to supplement shotcrete to withstand high support pressure; close ring by erecting invert support; encase steel ribs in shotcrete, floor bolting to avoid floor heaving; sometimes steel ribs with loose backfill are also used to release the strain energy in a controlled manner (tunnel closure more than 4 percent shall not be permitted)	Increase the tunnel diameter to absorb desirable closure; provide invert support as early as possible to mobilise full support capacity, long-term instrumentation is essential; circular shape is ideal
7.	Swelling	Full face or heading and bench; drill and blast	Full column grouted rock anchors with SFRS shall be used around the tunnel; increase 30 percent thickness of shotcrete due to weak bond of the shotcrete with rock mass; erect invert strut. The first layer of shotcrete is sprayed immediately to prevent ingress of moisture into rock mass	Increase the tunnel diameter to absorb the expected closure; prevent exposure of swelling minerals to moisture, monitor tunnel closure
8.	Running and flowing	Multiple drift with forepoles; grouting of the ground is essential; shield tunnelling may be used in soil conditions	Full column grouted rock anchors and SFRS; concrete lining up to face, steel liner in exceptional cases with shield tunnelling	Progress is very slow. Trained crew should be deployed. In reach of sudden flooding, the tunnel is realigned by-passing the same
9.	Rock burst	Full face drill and blast	Fibre reinforced shotcrete with full column resin anchors immediately after excavation	Micro-seismic monitoring is essential

Table 22.3. Comparison between squeezing and swelling phenomena (Jethwa and Dhar, 1996).

Parameter	Squeezing	Swelling
1. *Cause*	Small volumetric expansion of weak and soft ground upon stress-induced shear failure. Compaction zone can form within broken zone	Volumetric expansion due to ingress of moisture in ground containing highly swelling minerals
2. *Closure*		
* Rate of closure	1. Very high initial rate, several centimeters per day for the first 1–2 weeks of excavation	1. High initial rate for first 1–2 weeks till moisture penetrates deep into the ground
	2. Reduces with time	2. Decreases with time as moisture penetrates into the ground deeply with difficulty
* Period	3. May continue for years in exceptional cases	3. May continue for years if the moist ground is scooped out to expose fresh ground
3. *Extent*	The affected zone can be several tunnel diameters thick	The affected zone is several metres thick. Post-construction saturation may increase swelling zone significantly

with time significantly within the broken zone. The radial displacement, however, tend to converge at the interface boundary of the elastic and the broken zones. Figure 22.2 shows that a compaction zone is formed within this broken zone so that the rate of tunnel wall closure is arrested.

Various approaches for estimating the ground conditions for tunnelling on the basis of Q and modified Q, i.e., rock mass number N are dealt in Chapter 8 in book by Singh and Goel (1999).

22.3   EMPIRICAL APPROACH FOR DEGREE OF SQUEEZING

The rock mass number N is defined as modified rock mass quality Q as follows,

$$N = (RQD/J_n)(J_r/J_a)J_w \qquad (22.1)$$

where

RQD = Rock Quality Designation $\geqslant 10$,
J_n = Joint set number,
J_r = Joint roughness number for the critically oriented joint set,
J_a = Joint alteration number for the critically oriented joint set, and
J_w = Joint water reduction factor.

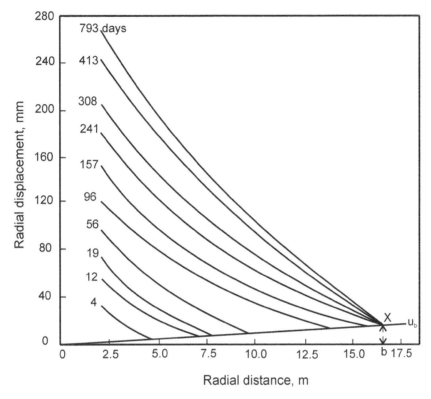

Figure 22.1. Variation of radial displacement with radial distance within slates/phyllites of Giri Tunnel, India (Dube et al., 1986).

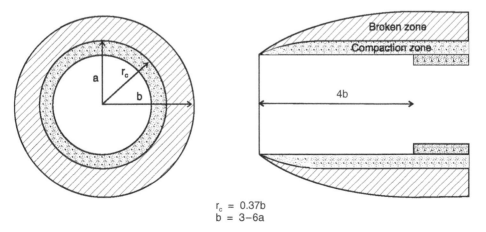

$$r_c = 0.37b$$
$$b = 3-6a$$

Figure 22.2. Compaction zone within broken zone in the squeezing ground condition (Jethwa, 1981).

Since the stress reduction factor (SRF) is often difficult to identify in rock masses, SRF has been dropped in Equation 22.1. Figure 22.3 shows zones of tunnelling hazards depending up on the values of $HB^{0.1}$ and N. Here H is the overburden in metres, B is the width of the tunnel or cavern in metres. It should be noted that B should be more than the self-supporting tunnels (Eq. 21.2 in Chapter 21).

For a squeezing ground condition

$$H \gg (275\, N^{0.33}) \cdot B^{-0.1} \text{ metres} \qquad (22.2)$$
$$\frac{J_r}{J_a} \le \frac{1}{2}$$

and

for a non-squeezing ground condition

$$H \ll (275\, N^{0.33}) \cdot B^{-0.1} \text{ metres} \qquad (22.3)$$

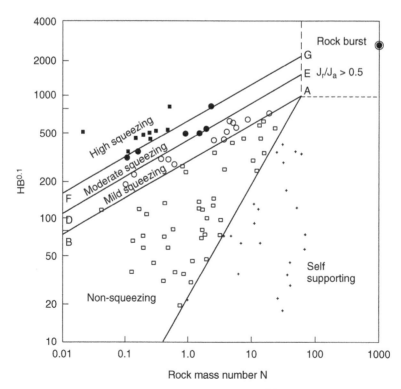

Figure 22.3. Plot between rock mass number N and $HB^{0.1}$ for predicting ground conditions (Goel, 1994).

22.4   SUDDEN FLOODING OF TUNNELS

The inclined beds of impervious rocks (shale, phyllite, schist, etc.) and pervious rocks (crushed quartzites, sandstone, limestone, fault, etc.) may be found along a tunnel alignment. The heavy rains/snow charge the beds of pervious rocks with water like an acquifer. While tunnelling through the impervious bed into a pervious bed, seepage water may gush out suddenly. The authors have studied four similar case histories of the Chhibro-Khodri, Maneri Bhali, BSL and Dulhasti hydroelectric projects in Himalaya. Experience shows that sudden flood accompanied by a huge out-wash of sand and boulders may occur ahead of the tunnel face where several shear zones exist. This flooding problem becomes dangerous where the pervious rock mass is squeezing ground also due to the excessive overburden. The machines and tunnel boring machines are buried partly.

The seepage should be monitored near the portal regularly. The discharge of water should be plotted along the R.D. of the face of the tunnel. If the peak discharge is found to increase with tunnelling, it is very likely that sudden flooding of the tunnel may take place with further tunnelling. It is suggested that international experts be consulted for tackling such situations.

22.5   CHIMNEY FORMATION

There may be local thick shear zones dipping towards a tunnel face. The soil/gouge may fall down rapidly, unless it is supported carefully and immediately. Thus, a high cavity/chimney may be formed along the thick shear zone. The chimney may be very high in a water charged rock mass.

22.6   CONCLUDING REMARKS

Rock has EGO (extraordinary geological occurrence) problems. Enormous time and money is lost due to unforeseen tunnelling hazards, particularly in Himalaya. It is said that if a shear zone is not seen within 200 m in lower Himalaya, it means that it has been missed. Thus, geological uncertainties may be managed by adopting a strategy of tunnel construction which can cope up with the most of tunnelling conditions. A hazard foreseen is a hazard controlled. The modern trend is that contractors insure their tunnelling machinery and the anticipated losses due to delays because of unexpected geological and geohydrological conditions.

REFERENCES

Barla, G. 1995. Squeezing rocks in tunnels. *ISRM News J.* 2(3&4): 44–49.
Barton, N., Lien, R. and Lunde, J. 1974. Engineering classification of rock masses for the design of tunnel support. *Rock Mechanics*, Vol. 6. Springer-Verlag, pp. 189–236.

Dube, A.K. and Singh, B. 1986. Study of squeezing pressure phenomenon in a tunnel – Part I and Part II. *Tunnelling and Underground Space Technology* Vol. 1, No.1, pp. 35–39 (Part I) and pp. 41–48 (Part II), U.S.A.

Goel, R.K. 1994. *Correlations for predicting support pressures and closures in tunnels.* Ph.D. Thesis, Nagpur University, Nagpur, India, p. 308.

ISRM–Referred the Definitions of Squeezing received from ISRM.

Jethwa, J.L. 1981. *Evaluation of rock pressures in tunnels through squeezing ground in lower Himalayas.* Ph.D. Thesis, Department of Civil Engineering, University of Roorkee, India, p. 272.

Jethwa, J.L. and Dhar, B.B. 1996. Tunnelling under squeezing ground gondition. *Proc. Recent Advances in Tunnelling Technology*, New Delhi, pp. 209–214.

Singh, Bhawani and Goel, R.K. 1999. *Rock Mass Classification – A Practical Approach in Civil Engineering.* Chapter 7. The Netherlands, Elsevier Science Ltd., p. 267.

Singh, Bhawani, Jethwa, J.L. and Dube, A.K. 1995. A classification system for support pressure in tunnels and caverns. *J. Rock Mech. Tunnell Techno India*, 1(1): 13–24.

CHAPTER 23

Deformability, strength and in situ stresses

*"Weakness is cause of strength both in materials and living beings.
So do not hate weaknesses. Nature compensates"*
Anonymous

23.1 GENERAL

The Chaos theory appears to be applicable at micro-level only in nature and mostly near surface. Furthermore, Chaos is self-organizing. For engineering use, the overall (weighted average) behaviour is all that is needed. Since there is perfect harmony in nature at macro-level, the overall behaviour should also be harmonious. Hence, in civil engineering, Chaos theory seems to find only a limited application. In fact in civil engineering practice, simple continuum characterisation is more popular for large structures. Thus, when one is talking of behaviour of jointed rock masses, one is really talking about the most probable behaviour of rock masses.

In this chapter rock mass parameters like modulus of deformation, poly-axial strength criterion and in situ stress trends are presented. However, the in situ tests must be conducted in the case of important structures in rock like underground caverns for hydroelectric projects and oil storage, etc.

23.2 MODULUS OF DEFORMATION

Dams and tunnels have to be built on, or in, natural geomaterials. An assessment of these structures for safety and stability necessitate a geotechnical characterisation of these geomaterials. In the context of this article, it implies determination of rock mass deformability which has been established by Muller (1974) as an important design parameter. In general, the presence of cracks, fissures, joints, shear zones and bedding planes in the rock mass makes the modulus function so complex that it is not possible to write a simple expression that encompasses all the important variables. Some empirical relations are, however, available which provide a first order estimate of the rock mass behaviour without having to perform time consuming and expensive field tests (Singh and Goel, 1999).

In India, a large number of hydroelectric power projects have been completed recently and several projects are under construction. These projects have generated a bulk of instrumentation data which have been analysed by Mitra (1991), Mehrotra (1992), Verman (1993), Goel (1994) and Singh (1997). These new data and their analysis have led to a revision of the existing empirical relations and formulation of new correlations which are subsequently described in this chapter.

In the first approach, the modulus of deformation of the rock mass is expressed as a function of the corresponding value for the intact rock E_r according to the following expression,

$$E_d = E_r \cdot MRF \tag{23.1}$$

where MRF = modulus reduction factor which is given as

1. Nicholson and Bieniawski (1990):

$$MRF = \frac{E_d}{E_r} = 0.0028RMR^2 + 0.9e^{(RMR/22.82)} \tag{23.2}$$

2. Mitri et al. (1994):

$$MRF = \frac{E_d}{E_r} = 0.5[1 - \cos(\pi. RMR/100)] \tag{23.3}$$

where RMR = rock mass rating according to Bieniawski (1978).

Eissa and Kazi (1988) discovered the following correlation between static modulus of elasticity (E_r) in GPa and dynamic modulus of elasticity (E_{rdyn}) in GPa from 76 ultrasonic tests on rock material.

$$Log_{10} E_r = 0.02 + 0.77 \log_{10}(\gamma E_{rdyn}) \tag{23.4}$$

where γ = unit weight of rock material (gm/cc).
The correlation coefficient of Equation 23.4 is 96 percent.

In the second approach, the E_d value is specified directly in terms of known parameters. Some typical correlations are given in Table 23.1. The following symbols are used in Table 23.1.

Q   = rock mass quality according to Barton et al. (1974),
J_w = joint water reduction factor in Q system,
q_c = uniaxial (unconfined) compressive strength of rock material at natural moisture content,
GSI = geological strength index according to Hoek and Brown (1997),
    = RMR − 5; and
H   = tunnel depth from ground surface in metres.

Mehrotra (1992) found significant effect of saturation on the modulus of deformation of water sensitive rocks (argillaceous). The state of initial stress in the rock mass depends on its depth from the ground surface and its deformation behaviour

Table 23.1. Empirical correlations for modulus of deformation of rock mass (GSI and RMR ≪ 100).

Authors	Expression for E_d (GPa)	Conditions
Bieniawski (1978)	$2RMR - 100$	$q_c > 100$ MPa and RMR > 50
Barton et al. (1980)	$25 \log Q$	$Q > 1$ and $q_c > 100$ MPa
Serafim and Pereira (1983)	$10^{(RMR-10)/40}$	$q_c \geqslant 100$ MPa
Verman (1993)	$0.3\, H^\alpha \cdot 10^{(RMR-20)/38}$	$\alpha = 0.16$ to 0.30 (higher for poor rocks) $q_c \leqslant 100$ MPa; $H \geqslant 50$ m; $J_w = 1$ Coeff. of correlation $= 0.91$
Hoek and Brown (1997)	$\dfrac{\sqrt{q_c}}{10}\, 10^{\frac{(GSI-10)}{40}}$	$q_c \leqslant 100$ Mpa
Singh (1997)	$E_d = Q^{0.36}\, H^{0.2}$	$Q < 10$; $J_w = 1$
	$E_e = 1.5 Q^{0.6}\, E_r^{0.14}$	Coeff. of correlation for $E_e = 0.96$; $J_w \leqslant 1$

Note: The above correlations are expected to provide a mean value.

should, therefore, also depend upon the depth from ground surface as reported by Verman (1993) and Singh (1997).

The field experience suggests that the mobilised strength of rock mass is very high as compared to that in the rock slopes. The following knowledge is offered for analysis of tunnels in rock masses. Hoek and Brown's criterion of strength of rock mass is too conservative and presented in Section 28.1.

23.3   STRENGTH ENHANCEMENT AND A NEW FAILURE THEORY

Consider a cube of rock mass with two or more joint sets as shown in Figure 23.1. If high intermediate principal stress is applied on the two opposite faces of the cube, then the chances of wedge failure are more than the chances of planar failure as found in the triaxial tests. The shear stress along the line of intersection of joint planes will be proportional to $\sigma_1 - \sigma_3$ because σ_3 will try to reduce shear stress. The normal stress on both the joint planes will be proportional to $(\sigma_2 + \sigma_3)/2$. Hence the criterion for peak failure at low confining stresses may be as follows (Singh et al., 1998):

$$\sigma_1 - \sigma_3 = q_{cmass} + A[(\sigma_2 + \sigma_3)/2] \tag{23.5}$$

$$q_{cmass} = q_c \left[\frac{E_d}{E_r}\right]^{0.70} \cdot \left[\frac{d}{S_{rock}}\right]^{0.20} \tag{23.6}$$

$$\Delta = \frac{\phi_p - \phi_r}{2} \tag{23.7}$$

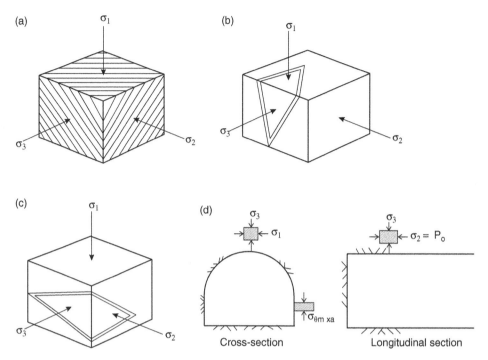

Figure 23.1. (a) Anisotropic rock material with one joint set (slate, schist, etc.). (b) Mode of failure in rock mass with 2 joint sets. (c) $p_{horizontal} \gg p_{vertical}$. (d) Direction of σ_1, σ_2, and σ_3 in the tunnel.

where

q_{cmass}	=	average uniaxial compressive strength of rock mass for various orientation of principal stresses (Ramamurthy, 1993; Roy, 1993),
$\sigma_1, \sigma_2, \sigma_3$	=	final effective principal stresses which are equal to in situ principal stress plus induced stress minus seepage pressure,
A	=	average constant for various orientation of principal stress, (value of A varies from 0.6 to 6.0),
	=	$2 \cdot \sin\phi_p/(1 - \sin\phi_p)$,
ϕ_p	=	peak angle of internal friction of rock mass,
	$\cong$	$\tan^{-1}[(J_r/J_a) + 0.10]$ at a low confining stress,
		< peak angle of internal friction of rock material,
	=	$14°$–$57°$,
S_{rock}	=	average spacing of joints,
q_c	=	average UCS of rock material for core of diameter d (for schistose rock also),
Δ	=	peak angle of dilatation of rock mass at failure,
ϕ_r	=	residual angle of internal friction of rock mass = $\phi_p - 10° > 14°$

E_d = modulus of deformation of rock mass ($\sigma_3 = 0$); and
E_r = modulus of deformation of the rock material ($\sigma_3 = 0$).

The peak angle of dilatation is approximately equal to ($\phi_p - \phi_r$)/2 for rock joints (Barton and Bandis, 1990) at low σ_3. This correlation may be assumed for jointed rock masses also. The proposed strength criterion reduces to Mohr-Coulomb criterion for triaxial conditions.

The significant rock strength enhancement in underground openings is due to σ_2 or in situ stress along tunnels and caverns which pre-stress rock wedges and prevents their failure both in the roof and the walls. However, σ_3 is released due to stress free excavation boundaries (Fig. 23.1d). In the rock slopes σ_2 and σ_3 are nearly equal and negligible. Therefore, there is an insignificant or no enhancement of the strength. As such, a block shear test on a rock mass gives realistic results for rock slopes and dam abutments only; because $\sigma_2 = 0$ in this test. Thus, Equation 23.5 may give a general criterion of jointed rock masses for underground openings, rock slopes and foundations.

Another cause of strength enhancement is higher uniaxial compressive strength of rock mass (q_{cmass}) due to higher E_d because of constrained dilatancy and restrained fracture propagation near the excavation face only in the underground structures. In rock slopes, E_d is found to be much less due to complete stress release and low confining pressure on account of σ_2 and σ_3; and long length of weathered and filled-up joints. So, q_{cmass} will also be low near rock slopes.

Through careful back analysis, both the model and its constants should be deduced (Sakurai, 1997). Thus, A, E_d and q_{cmass} should be estimated from the feedback of instrumentation data at the beginning of construction stage. With these values, forward analysis should be attempted carefully as mentioned earlier. At present, a non-linear back analysis may be difficult and it does not give unique parameters.

The proposed strength criterion is different from Mohr's strength theory which works well for homogeneous soils and isotropic materials on the scale of the specimen prepared in the laboratory. There is a basic difference in the structure of soil and rock masses. Soils generally have no pre-existing planes of weaknesses and so planar failure can occur on a typical plane with dip direction towards σ_3 in the triaxial tests. However, rocks have pre-existing planes of weaknesses like joints and bedding planes, etc. As such, failure occurs mostly along these planes of weaknesses. In the triaxial tests on rock masses, planar failure takes place along the weakest joint plane. In polyaxial stress field, a wedge type of failure may be the dominant mode of failure, if $\sigma_2 \gg \sigma_3$. Therefore, Mohr's theory needs to be modified for anisotropic and jointed rock masses and naturally inhomogeneous soils in the field.

The new strength criterion is proved by extensive polyaxial tests on anisotropic tuff (Wang and Kemeny, 1995). It is interesting to note that the constant 'A' is the same for biaxial, triaxial and polyaxial tests (Singh et al., 1998). Moreover, the effective in situ stresses on ground level in mountainous areas appear to follow Equation 23.5 ($q_{cmass} = 3$ MPa, A = 2.5) which indicates a state of failure near ground due to the tectonic stresses.

In the NJPC tunnel, India, under an abnormal overburden of 1400 m, the poly-axial strength criterion suggested that moderate to severe rock burst or squeezing condition is ruled out as $J_r/J_a \gg 0.5$. Fortunately, this prediction came to be true even at such high overburden of 1400 m in massive augen gneiss in spite of low UCS ($q_c = 27$ Mpa). The suggested hypothesis appears applicable approximately for the rock masses with three or more joint sets.

23.4   POOR ROCK MASSES

Squeezing is found to occur in tunnels in the weak rocks where overburden H is more than $350\,Q^{1/3}$ m. The tangential stress at failure shall be about $2\gamma H$ assuming hydrostatic in situ stresses. Thus, mobilised compressive strength is 2γ $350\,Q^{1/3} = 700\,\gamma Q^{1/3}$ T/m^2. In other words (Singh and Goel, 1999),

$$q_{cmass} = 0.70\gamma Q^{1/3}\,\text{MPa} \quad \text{for } Q < 10 \tag{23.8a}$$

where γ = unit weight of rock mass in kN/m^3 (22–29).
Many investigators have agreed with the above correlation (Grimstad and Bhasin, 1996; Barton, 1995; Choubey, 1998; Aydan et al., 2000). Ten case histories of tunnels in the squeezing ground have also been analysed in Chapter 25. In poor rocks, the peak angle of internal friction (ϕ_p) is back analysed and related as follows,

$$\tan\phi_p = \frac{J_r}{J_a} + 0.1 \leqslant 1.5 \tag{23.8b}$$

The addition of 0.1 accounts for interlocking of rock blocks. It may be visualized that interlocking is more in jointed rock mass due to low void ratio than in soils. Further, Kumar (2000) has shown theoretically that the internal angle of friction of rock mass is slightly higher than the sliding angle of friction of its joints.

23.5   FAILURE OF INHOMOGENEOUS GEOLOGICAL MATERIAL

In an inhomogeneous geological material, the process of failure is initiated by its weakest link (zone of loose soil and weak rock, crack, bedding plane, soft seam, etc.). Thus, natural failure surfaces are generally three-dimensional (perhaps four-dimensional) which start from this weakest link and propagate towards a free surface (or face of excavation). As such the intermediate principal stress (σ_2) plays an important role and governs the failure and the constitutive relations of the naturally inhomogeneous geological materials (both in rock masses and soils) in the field. Since micro-inhomogeneity is rather unknown, assumption of homogeneity is popular in the minds of the engineers. Therefore, the intuition is that the effective confining stress is about $[(\sigma_2 + \sigma_3)/2]$ in naturally inhomogeneous soils also.

Furthermore, the failure in an inhomogeneous geological material is progressive, whereas a homogeneous rock fails suddenly. Hence the advantage of inhomogeneous materials which is offered by nature is that it gives an advance warning of the failure process starting from the weakest zone.

23.6 RESIDUAL STRENGTH PARAMETERS

Mohr's theory will be applicable to residual failure as a rock mass would be reduced to non-dilatant soil-like condition. The mobilized residual cohesion c_r is approximately equal to 0.1 MPa and is not negligible unless tunnel closure is more than 6 percent of its diameter. The mobilized residual angle of internal friction ϕ_r is about 10° less than the peak angle of internal friction ϕ_p but more than 14°.

23.7 COEFFICIENT OF VOLUMETRIC EXPANSION

In the case of squeezing ground condition (Chapter 25), in addition to the in situ stresses the ground response (reaction) curve depends upon the peak and residual strength parameters of rock mass and also on the coefficient of volumetric expansion of rock mass (K) in the broken zone. The coefficient of volumetric expansion is defined as the ratio between increment in the volume of rock mass after failure and the initial volume of rock mass. Jethwa (1981) estimated values of K as listed in Table 23.2. It may be noted that higher degree of squeezing was associated with weak rock masses and higher K values.

Table 23.2. Coefficient of volumetric expansion of failed rock mass (K) within broken zone (Jethwa, 1981).

S.No.	Rock type	K
1	Phyllites	0.003
2	Claystones/siltstones	0.01
3	Black clays	0.01
4	Crushed sandstones	0.004
5	Crushed shales	0.005
6	Metabasics (Goel, 1994)	0.006

23.8 THE CRITICAL STRAIN OF ROCK MASS

The critical strain (ε_{mass}) is defined as the ratio between UCS (q_{cmass}) and the modulus of deformation (E_d) of rock mass (Sakurai, 1997). Hence Equation 23.6 may be rewritten to deduce ε_{mass} as follows,

$$\varepsilon_{mass} = \varepsilon_r \left[\frac{E_r}{E_d} \right]^{0.30} \left[\frac{d}{S_{rock}} \right]^{0.20} \qquad (23.9)$$

$$\geq \varepsilon_\theta$$

where

ε_r = qc/Er = critical strain of rock material,

ε_θ = tangential strain around opening,

= (observed deflection of crown in downward direction/radius of tunnel),

S_{rock} = average spacing of joints; and

q_c = UCS of rock material for core of diameter d.

The experience in Japan is that there were not many construction problems in tunnels where $\varepsilon_\theta < \varepsilon_{mass}$ or ε_r. It may be noted that critical strain appears to be size dependent according to Equation 23.9.

23.9   THE SAFE BEARING CAPACITY OF ROCK MASS

In the case of rock mass with favourable characteristics (that is, the rock surface is parallel to the base of the foundation, the load has no tangential component, the rock mass has no open discontinuities), the safe bearing pressure should be estimated from the equation suggested in the Canadian Foundation Manual,

$$q_s = q_c N_j \qquad (23.10)$$

where

q_s = safe bearing pressure (gross),

q_c = average uniaxial compressive strength of rock cores,

N_j = empirical coefficient depending on the spacing of discontinuities

$$= MRF = \frac{E_d}{E_r} = 0.0$$

δ = thickness of discontinuities in cm,

S = spacing of discontinuities in cm, and

B_f = footing width in cm.

Equation 23.10 includes a factor of safety of 3. This relationship is valid for a rock mass with a spacing of discontinuities greater than 0.3 m, aperture (opening) of discontinuities less than 10 mm (15 mm if filled with soil or rock debris) and a foundation width greater than 0.3 m. The safe bearing capacity is used to determine the size (B_f) of base plates of rock bolts, rock anchors and cable anchors to control both landslide and tunnel/cavern hazards.

In the design of foundations, the allowable bearing pressure is used. Bieniawski's (1978) rock mass rating may also be used to obtain net allowable bearing pressure as per Table 23.3 (Mehrotra, 1992). The guideline given in Table 23.3 has been developed on the basis of plate load tests at about 60 sites and calculating the allowable bearing pressure for 6 m wide raft foundation with settlement of 12 mm.

Table 23.3. Net allowable bearing pressure q_a based on RMR (Mehrotra, 1992).

Class No.	I	II	III	IV	V
Description of rock	Very good	Good	Fair	Poor	Very poor
RMR	100–81	80–61	60–41	40–21	20–0
$q_a(T/m^2)$	600–440	440–280	280–135	135–45	45–30

Note:
1. The RMR for Table 23.3 should be obtained below the foundation at depth equal to the width of the foundation provided RMR does not change with depth. If the upper part of the rock, within a depth of about one-fourth of foundation width, is of lower quality, the value of this part should be used or the inferior rock should be replaced with concrete. Since the values in Table 23.3 are based on limiting the settlement, they should not be increased if the foundation is embedded into rock.
2. In the case of (pier) foundation which will lie below the reservoir/river, the rating for ground water condition may be assumed as zero. Then the net allowable bearing pressure may be estimated from Table 23.3. Thus, the gross allowable bearing pressure is sum of net allowable bearing pressure and the submerged overburden pressure at the level of foundation. The foundation should be deeper than the unstable rock wedge, if any, in the slope.
3. During earthquake loading, the above values of allowable bearing pressure may be increased by 50 percent in view of rheological behaviour of rock masses.

23.10   IN SITU STRESSES

The in situ stresses are measured generally by a hydro-fracturing method which is economical, faster and simple than other methods. The magnitude and the orientation of in situ stresses may have major influence on planning and design of underground openings in major hydroelectric projects, mining and underground space technology. The local orientation of in situ stresses is controlled by major geological structures like fold, faults and intrusions. It is almost needless to mention that a statistically significant number of hydro-fracture tests should be conducted beforehand to know this local orientation of in situ stresses.

23.10.1 A classification of geological conditions and stress regimes

Ramsay and Hubber (1988) have shown how type of faults rotates principal in situ stresses (Fig. 23.2).

Normal fault area (Fig. 23.2a)
These are steeply dipping faults where slip is mostly along dip direction than that along its strike, and the hanging wall is moved downwards. The mechanics of failure suggests that the vertical stress (σ_v) is the major principal stress and the minimum horizontal stress (σ_h) acts along the dip-direction. As such, the order of in situ stresses is given below,

$$\sigma_v > \sigma_H > \sigma_h$$

In a sub-ducting boundary plate, normal faults are found commonly as the downward bending of this plate reduces horizontal stresses along dip direction. However, in the upper boundary plate, thrust faults are seen generally because of the tectonic thrust and thus there is an urgent need for stress analysis of interaction of plate boundaries (Nedoma, 1997).

Thrust fault area (Fig. 23.2 b)
Thrusts have mild dip with major slip along the dip direction compared to that along its strike, and the hanging wall is moved upwards. The mechanics of brittle failure indicates that the vertical stress in this case should be the minimum principal in situ stress and the horizontal stress along the dip direction is the maximum principal in situ stress. Thus, the order of the in situ stresses in the thrust fault area is as follows:

$$\sigma_H > \sigma_h > \sigma_v$$

It should be noted that the correlations developed in India refer to the geological region of upper boundary plate with frequent thrust and strike-slip faults in Himalaya.

Strike-slip fault area (Fig. 23.2c)
Such faults are steeply oriented and usually vertical. The slip is mostly along the strike than that along the dip direction. In a strike-slip fault, the major principal stress and minor principal stress are oriented as shown in Figure 23.2c. Thus, the order of the in situ stresses is given below,

$$\sigma_H > \sigma_v > \sigma_h$$

It may be noted that both magnitude and orientation of horizontal in situ stresses will change with erosion and tectonic movements, especially in hilly regions. The regional horizontal in situ stresses are relaxed in steep mountainous regions. These stresses are relaxed more with decreasing distance from the slope face. Thus, the gradient of the horizontal stress with depth (or vertical stress) may be more in steeply inclined mountainous terrain compared to that in the plane terrain. Furthermore, the vertical stress just below the valley may be much higher than the overburden pressure due to the stress concentration at the bottom of the valley.

Stephansson (1993) has reported the following trend for in situ horizontal stresses at shallow depth ($z < 1000$ m) from hydro-fracturing tests

$$\sigma_H = 2.8 + 1.48\sigma_v \text{ MPa} \tag{23.11}$$

$$\sigma_h = 2.2 + 0.89\sigma_v \text{ MPa} \tag{23.12}$$

$$\sigma_v = \gamma z \tag{23.13}$$

Hydro-fracturing tests of Sharma (1999) show that above trends apply to thrust area regime (Fig. 23.2b). It may also be noted that $\sigma_H > \sigma_h > \sigma_v$. He also showed that the measured in situ stresses depend significantly on the method of testing.

Sengupta (1998) performed a large number of hydro-fracturing tests within weak rocks in the Himalayan region. It is heartening to observe a good correlation

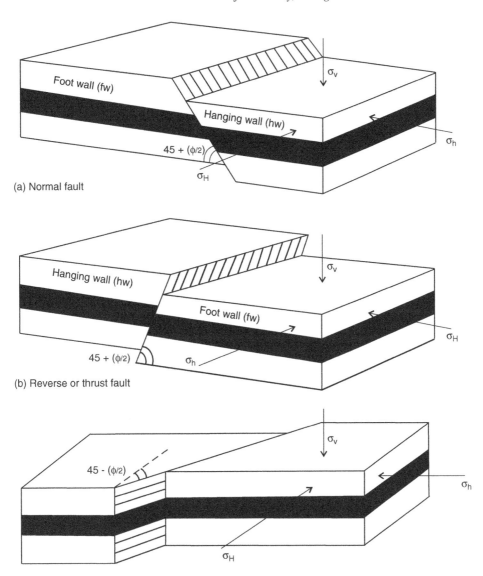

Figure 23.2. Orientation of in situ stresses in various geological conditions (Ramsay and Hubber, 1988).

between σ_H and σ_v. The correlation between σ_h and σ_v is perhaps not good due to mountainous terrain. Thus, it is inferred that for $z < 400$ metres,

$$\sigma_H = 1.5 + 1.2\sigma_v \text{ MPa} \tag{23.14}$$

$$\sigma_H = 1.0 + 0.5\sigma_v \text{ MPa} \tag{23.15}$$

Perhaps in steeply inclined mountainous terrain, Sengupta's correlations (Eqns 23.14 and 23.15) may be applicable in the strike-slip fault stress region

($\sigma_H > \sigma_v > \sigma_h$) as the in situ horizontal stresses are likely to be relaxed significantly (Fig. 23.2c).

In the computer programs, the stress ratio k is used as defined below,

$$k = \text{Horizontal stress across the opening/vertical stress} \qquad (23.16)$$

It may be noted that the stress ratio k is highest near ground surface and decreases with increasing depth of opening according to Equations 23.11 to 23.15.

REFERENCES

Aydan, Omer, Dalgic, S. and Kawamoto, T. 2000. Prediction of squeezing potential of rocks in tunnelling through a combination of an analytical method and rock mass classification. *Italian Geotech. J.* Anno XXXIV, (1): 41–45.

Barla, G. 1995. Squeezing rocks in tunnels. *ISRM News J.* 2(3&4): 44–49.

Barton, N. 1995. The influence of joint properties in modelling jointed rock masses. *Eight Int. Rock Mech. Congress*, Tokyo, Vol. 3, pp. 1023–1032.

Barton, N. and Bandis, S. 1990. Review of predictive capabilities of JRC-JCS model in engineering practice, Reprinted from: Barton, N.R. and O. Stephansson (eds), *Rock Joints Proc. of a Regional Conference of the International Society for Rock Mechanics*, Leon, 4–6.6.1990. Rotterdam, A.A. Balkema, 820 pp.

Barton, N., Lien, R. and Lunde, J. 1974. Engineering classification of rock masses for the design of tunnel support. *Rock Mechanics*, Vol. 6. Springer-Verlag, pp. 189–236.

Barton, N., Loset, F., Lien, R. and Lune, J. 1980. Application of Q-system in design decisions concerning dimensions and appropriate support for underground installations. *Subsurface Space*. Pergamon, pp. 553–561.

Bieniawski, Z.T. 1978. Determining rock mass deformability, experience from case histories. *Int. J. Rock Mech. Min. Sci. Geomech. Abstr.*, 15: 237–247.

Choubey, V.D. 1998. Potential of rock mass classification for design of tunnel supports – hydro electric projects in Himalayas. *Int. Conf. Hydro Power Development Himalayas*, Shimla, India, pp. 305–336.

Eissa, E.A. and Kazi, A. (1988). Technical note on relation between static and dynamic Young's modulus. *Int. J. Rock Mech. Min. Sci. & Geomech. Abstrs.* 25(6): 479–482.

Goel, R.K. 1994. *Correlations for predicting support pressures and closures in tunnels.* Ph.D. Thesis. Visvesvaraya Regional College of Engineering, Nagpur, India, p. 347.

Grimstad, E. and Bhasin, R. 1996. Stress strength relationship and stability in hard rock. *Proc. Conf. Recent Advances Tunnelling Technology*, New Delhi, India, Vol. 1, pp. 3–8.

Hoek, E. and Brown, E.T. 1997. Practical estimates of rock mass strength. *Int. J. Rock Mech. Min. Sci. Geomech. Abstr.* 34(8): 1165–1186.

Jethwa, J.L. 1981. *Evaluation of rock pressures in tunnels through squeezing ground in lower Himalayas.* Ph.D. Thesis. Department of Civil Engineering, University of Roorkee, India, p. 272.

Kumar, P. 2000. Mechanics of excavation in jointed underground medium. *Symp. Modern Techniques in Underground Construction*, CRRI, New Delhi, ISRMTT, pp. 49–75.

Mehrotra, V.K. 1992. *Estimation of engineering properties of rock mass.* Ph.D. Thesis. University of Roorkee, India, p. 267.

Mitra, S. 1991. *Studies on long-term behaviour of underground powerhouse cavities in soft rocks.* Ph.D. Thesis. University of Roorkee, India, p. 194.

Mitri, H.S., Edrissi, R. and Henning, J. 1994. Finite element modelling of cable bolted stopes in hard rock underground mines. Presented at the SME Annual Meeting, pp. 14–17.

Muller, L. 1974. Rock mass behaviour determination and applications in engineering practice. *Proc. 3rd Congr. Rock Mech.*, ISRM, Denver, Vol. 1, pp. 205–215.

Nedoma, J. 1997. Part I – Geodynamic analysis of the Himalayas and Part II – geodynamic analysis. *Institute of Computer Science, Academy of Sciences of the Czech Republic*, Technical Report No. 721, September, p. 44.

Nicholson, G.A. and Bieniawski, Z.T. 1990. A non-linear deformation modulus based on rock mass classification, *Int. J. Min Geol. Engg.*, (8), pp. 181–202.

Ramamurthy, T. 1993. Strength and modulus responses of anisotropic rocks. *Compr. Rock Eng.* 1, pp. 313–329.

Ramsay, G. and Hubber, M.I. 1988. The techniques of modern structural geology, Vol. 2. *Folds and Fractures*. Academic Press, pp. 564–566.

Roy, Nagendra. 1993. *Engineering behaviour of rock masses through study of jointed models*. Ph.D. Thesis. Civil Engineering Department, I.I.T., New Delhi, India, p. 365.

Sakurai, S. 1997. Lessons learned from field measurements. *Tunnel. and Underground Space Technol.* 12(4): 453–460.

Sengupta, S. 1998. *Influence of geological structures on in situ stresses*. Ph.D. Thesis. Department of Civil Engineering, Indian Institute of Technology, New Delhi, p. 275.

Serafim, J.L. and Pereira, J.P. 1983. Considerations of the geomechanics classification of Bnieniawski, *Int. Symp. Eng. Geol. Underground Constr.*, LNEC, Lisbon, Vol. 1, pp. II.33–II.42.

Sharma, S.K. 1999. *In situ stress measurements by hydro-fracturing – some case studies*. M.E. Thesis, WRDTC, University of Roorkee, p. 104.

Stephansson, O. 1993. Rock stress in the fennoscandian shield. *Comprehensive Rock Eng.* 3(17): 445–459.

Singh, Bhawani and Goel, R.K. 1999. *Rock Mass Classification – A Practical Approach in Civil Engineering*. Part of Chapters 13, 19 and 27, Permission from Elsevier Science Ltd., The Netherlands, p. 268.

Singh, Bhawani, Goel, R.K., Mehrotra, V.K., Garg, S.K. and Allu, M.R. 1998. Effect of intermediate principal stress on strength of anisotropic rock mass. *J. Tunnel. & Underground Space Technol.* 13(1): 71–79.

Singh, Suneel 1997. *Time dependent deformation modulus of rocks in tunnels*. M.E. Thesis. Dept. of Civil Engineering, University of Roorkee, India, p. 65.

Verman, Manoj 1993. *Rock mass – tunnel support interaction analysis*. Ph.D. Thesis. University of Roorkee, Roorkee, India, p. 267.

Wang, R. and Kemeny J.M. 1995. A new empirical failure criterion under Polyaxial compressive stresses. Reprinted from: Daemen, Jaak J.K. & Richard A. Schultz (eds), Rock Mechanics: *Proc. 35th U.S. Symposium*-Lake Tahoe, 4–7 June 1995. Rotterdam, A.A. Balkema, 950 pp.

CHAPTER 24

Underground wedge analysis – UWEDGE

"A hazard foreseen is hazard controlled"
Anonymous

24.1   GENERAL

The stability analysis of a rock wedge inside an underground opening was originally developed by Bray and presented in the book on *Underground Excavations in Rock* by Hoek and Brown (1980). Wedges are formed by 3 independent joint sets and the free plane of the underground opening. If there are more sets of joints, many tetrahedral wedges are created and one of them will be the most critical or unstable. The Bray's analytical model determines whether
1. the wedge tends to slide out along the line of intersection of two joint planes,
2. the wedge tends to slide out along one plane of joint only,
3. the wedge is inherently stable, and
4. the wedge floats out due to uplift seepage pressure or falls out due to the gravity.
Rotational instability is not accounted for. The analysis is further extended here to consider the support system (rock anchor system, shotcrete and welded mesh) in the program UWEDGE.

24.2   ASSUMPTIONS

There is a limitation of underground wedge theory. The joints in reality are rough and dilatant and normal stresses are generally quite high in the case of underground openings. So there is a good interlocking of rock wedges with each other. Consequently, wedge failure is rare in the arched openings which are supported well within stand-up time. It is an observed fact that the support pressures are independent of the size of opening between 2 m to 22 m in non-squeezing conditions through good rock masses (Singh et al., 1992), whereas wedge theory predicts that support pressure should increase with width of an opening. Nevertheless, UWEDGE software is recommended as wedge theory is conservative and predicted support pressures are on the safe side.

Many conventional assumptions are made in the stability analysis of wedge in a underground opening. These assumptions are the same as in the stability analysis of wedge in rock slopes. These assumptions are discussed in Section 1.2 of Chapter 1. However, in UWEDGE software, the acceleration due to an earthquake has been neglected in view of the past experiences.

24.3   A SUPPORT SYSTEM FOR CAVERNS AND TUNNELS

The philosophy of the rock bolt system is to stitch all the wedges to the adjoining rock mass. The stitching of the wedges will prevent them from sliding or falling. It is assumed that rock anchors are normal to the plane of excavation which is generally the case in engineering practice (Fig. 24.1). Further, it is assumed that input value of bolt capacity (P) is mobilised completely. So, the bolt capacity P should be reduced to the value of mobilized bolt tension as suggested in the help menu, UWEDGE.TXT.

The roof of the arched opening may be simulated to consist of three stress-free planes as shown in Figure 24.2. The wedge above the assumed central and horizontal plane of excavation appears to be the most critical. It should also be stitched to the adjoining rock masses along with other wedges.

The function of the shotcrete and welded mesh is to offer shearing resistance along the surfaces of punching by the critical rock wedge (Figs 24.2 and 24.3). Thus, the shear strength of the shotcrete is an important property. In the case of steel fibre reinforced shotcrete (SFRS), the welded mesh is not provided (input shear strength of steel to be zero) as it prevents the spraying of the shotcrete. So the modern trend is to encourage SFRS because it is ductile unlike conventional shotcrete.

The software UWEDGE has been programmed on the basis of Bray's theory (Hoek and Brown, 1982). UWEDGE enables the user to design a support system (rock anchors or rock bolts + shotcrete + welded mesh) both for roof and walls which ensures stability of all rock wedges.

It is recommended that design should also be checked by software TM as discussed in Chapter 26. Furthermore, it should be ensured that the (radial) strains in rock mass along the anchor bars do not exceed the permissible tensile strain of the anchor bar with the help of software BEM or any finite element program for rock structures (Chapter 28).

24.4   A SUPPORT SYSTEM FOR SHAFTS

It may be noted that the support system of vertical shafts of circular or square and rectangular shape may also be designed by software UWEDGE. In the case of a shaft, the plunge (SAI 5) and trend (ALPHA 5) of boundary edges of the free faces

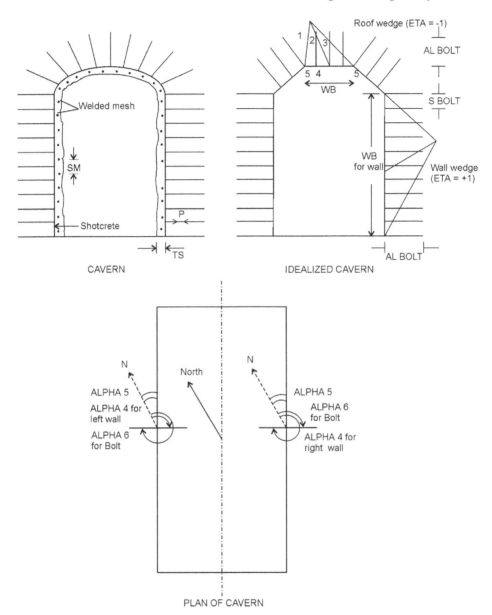

Figure 24.1. Idealization of cavern for underground wedge analysis.

are to be prescribed appropriately. It is suggested that the dip of vertical faces may be given as 89°. Similarly, the plunge of the edges of vertical faces may be assumed to be 89° for the physical understanding. It is almost needless to mention that the supports for inclined shafts of tunnels may also be designed by UWEDGE with an input of proper values of (SAI 4) and dip direction (ALPHA 4) of excavated faces; and plunge and trend of its boundary edges.

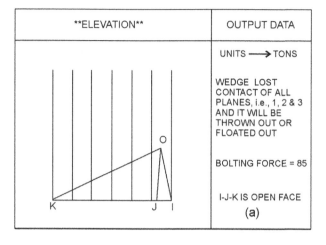

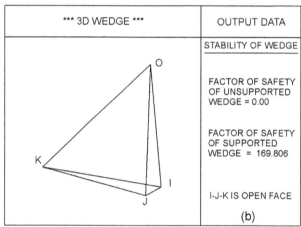

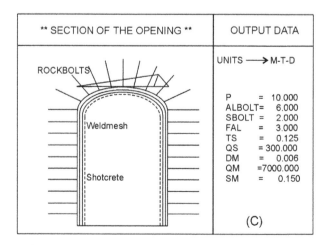

Figure 24.2. Stability analysis of reinforced wedge in roof of a cavern.

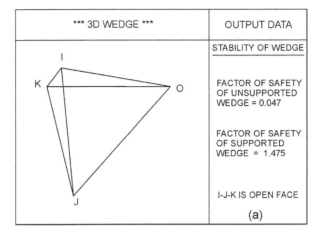

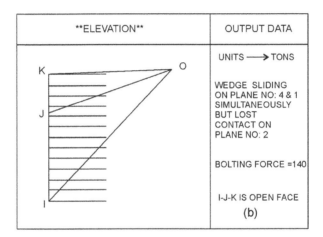

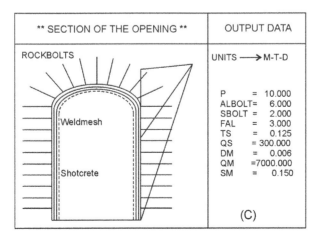

Figure 24.3. Stability analysis of reinforced wedge in wall of a cavern.

24.5　A SUPPORT SYSTEM FOR PORTALS AND HALF-TUNNELS

UWEDGE is, moreover, applicable to analyse the stability of a rock wedge in the special case of half-tunnels in the near vertical slopes in hard rocks (Singh and Goel, 1999). It is suggested that the slope face may be simulated as a frictionless and imaginary plane of rock joint with $c = \phi = u = 0$. Similarly, tunnel portals may be analysed and a support system is then designed. Of course, the slope of the tunnel portals may be analysed by software WEDGE and its rock reinforcement (reinforced rock breast wall) is designed for reliable stability of the rock slope. Unfortunately, tunnel portals fail frequently. Hence the need for cautious design of portals. A minimum rock cover equal to the width of the tunnel is recommended.

24.6　USER'S MANUAL – UWEDGE

THIS PROGRAM IS FOR STABILITY ANALYSES OF TETRAHEDRAL WEDGES IN THE UNDERGROUND OPENINGS, PORTAL, INCLINED AND VERTICAL SHAFTS IN NON-SQUEEZING GROUND (SEE FIGS 24.1, 24.2 AND 24.3)

NAME OF PROGRAM –　　　UWEDGE.FOR
UNITS USED –　　　　　　TONNE–METER–DEGREE

GIVE INPUT DATA IN THE FOLLOWING SEQUENCE

NO
TITLE OF PROBLEM IN ONE LINE (<80 CHARACTERS)
NJT
SAI(I), ALPHA(I), C(I), PHAI(I), U(I)
(REPEAT NJT TIMES)
SAI4, ALPHA4, SAI5, ALPHA5, SAI6, ALPHA6
WB, GAMA, T, ETA
P, SBOLT, ALBOLT, FAL
TS, QS, DM, QM, SM
(PLEASE REPEAT BLOCK FROM TITLE ... TO ..SM NO TIMES)

DO YOU WANT HELP REGARDING DEFINITION OF VARIABLES
ENTER　　0 FOR TERMINATION
　　　　　1 FOR FURTHER HELP
　　　　　2 FOR EXECUTION

DISCONTINUITIES ARE DENOTED BY 1,2 AND 3
EXCAVATED/FREE FACE BY 4 OR (NJT + 1), EDGE BY 5

NO	=	NUMBER OF PROBLEMS
NJT	=	NUMBER OF JOINT SETS (>2)
SAI	=	DIP OF THE JOINT PLANE(DEG.)
ALPHA	=	DIP DIRECTION OF JOINT PLANE (DEG.)
C	=	COHESION ALONG JOINT
PHAI	=	ANGLE OF FRICTION ALONG JOINT (DEG.)
	=	ARC TAN (Jr/Ja) FOR CLAY COATED JOINTS
U	=	AVERAGE SEEPAGE WATER PRESSURE ON JOINT

SAI4	=	DIP OF EXCAVATED FACE (0 FOR ROOF, 89 FOR WALLS)
ALPHA4	=	DIP DIRECTION OF EXCAVATED FACE
SAI5	=	PLUNGE OF BOUNDARY EDGES OF EXCAVATED/ FREE FACE
ALPHA5	=	TREND OF BOUNDARY EDGES OF FREE FACE
SAI6	=	PLUNGE OF BOLT FORCE (-UPWARD)
ALPHA6	=	TREND OF BOLT FORCE
WB	=	SPAN OF FREE FACE BETWEEN BOUNDARY EDGES OF ROOF
	=	HEIGHT OF FREE FACE BETWEEN BOUNDARY EDGES OF WALLS
GAMA	=	UNIT WEIGHT OF ROCK MASS (T/CUM)
ETA	=	−1. MEANS FREE FACE OVERHANGS (FOR ROOF)
ETA	=	+1. MEANS FREE FACE DOES NOT OVERHANG (FOR WALLS)
T	=	BOLT FORCE (TONNES)
	=	0 IF DESIGN OF BOLTS IS TO BE CHECKED
P	=	CAPACITY OF INDIVIDUAL BOLTS (DIVIDE BY A MOBILIZATION FACTOR UPTO 3.25(PR)^0.1 FOR PRETENSIONED BOLTS AND UPTO 9.5/PR^0.35 FOR UNTENSIONED ANCHORS RESPECTIVELY. THIS WILL ACCOUNT FOR COMPATIBILITY OF DEFORMATION OF SHOTCRETE LINING AND ROCK BOLT ARCH)
PR	=	ESTIMATED ULTIMATE SUPPORT PRESSURE (T/SQ.M)
FS	=	OVERALL FACTOR OF SAFETY OF REINFORCED WEDGE (SAFE IF FS >1.5)
SBOLT	=	SPACING OF BOLTS IN SQUARE PATTERN (<ALBOLT/2)
	=	SQUARE ROOT OF AREA OF ROCK MASS SUPPORTED BY ONE ROCK BOLT/ROCK ANCHOR
ALBOLT	=	LENGTH OF THE BOLT (>FAL)
FAL	=	FIXED ANCHOR LENGTH TO DEVELOP FULL ANCHOR CAPACITY
TS	=	THICKNESS OF SHOTCRETE
QS	=	SHEAR STRENGTH OF SHOTCRETE (DIVIDE BY(100TS)^0.05 (300 FOR SHOTCRETE AND 550 FOR STEEL FIBRE REINFORCED SHOTCRETE IN T/SQ.M)
DM	=	DIAMETER OF BARS OF WELDED MESH
QM	=	SHEAR STRENGTH OF BARS OF WELDED MESH
SM	=	SPACING OF BARS OF WELDED MESH

NOTE: (i) Half tunnel and portal slopes may be analysed by replacing slope with new joint set with C = PHI = U = 0.

  (ii) Not applicable to analysis of wedge in the floor.

(a) Input Data File: IUWEDGE.DAT

```
3
STABILITY ANALYSIS OF REINFORCED WEDGE IN ROOF OF
A CAVERN
4
80.,260.,0.,15.,0.
70.,81.,0.,15.,0.
50.,345.,0.,15.,0.
```

20.,200.,0.,15.,0.
0.,161.,0.,161.,0.,0.
12.,2.84,0.,−1.
10.,2.,6.,3.
0.125, 300.,. 006, 7000.,. 15
STABILITY ANALYSIS OF REINFORCED WEDGE IN LEFT WALL
OF A CAVERN
4
80.,260.,0.,15.,0.
70.,81.,0.,15.,0.
50.,345.,0.,15.,0.
20.,200.,0.,15.,0.
89.,251.,0.,161.,0.,0.
50.,2.84,0.,1.
10.,2.,6.,3.
0.125, 300.,. 006, 7000.,. 15
STABILITY ANALYSIS OF REINFORCED WEDGE IN RIGHT WALL
OF A CAVERN
4
80.,260.,0.,15.,0.
70.,81.,0.,15.,0.
50.,345.,0.,15.,0.
20.,200.,0.,15.,0.
89.,71.,0.,161.,0.,0.
50.,2.84,0.,1.
10.,2.,6.,3.
0.125, 300.,. 006, 7000.,.15

(b) Output Data File: OUWEDGE.DAT

STABILITY ANALYSIS OF REINFORCED WEDGE IN ROOF OF
A CAVERN

UNITS USED –	TONNE – METER – DEGREE
INPUT FILE NAME –	IUWEDGE.DAT
OUTPUT FILE NAME –	OUWEDGE.DAT

PROBLEM NO. 1
NO. OF JOINT SETS = 4
STRIKE OF JOINT NO: 1 AND BOUNDARY EDGE ARE NEARLY PARALLEL.
HENCE ALIGNMENT IS UNFAVOURABLE

STRIKE OF JOINT NO: 2 AND BOUNDARY EDGE ARE NEARLY PARALLEL.
HENCE ALIGNMENT IS UNFAVOURABLE

Case No. 1

PLANE	1	2	3	5
SAI (DEGREE)	80.0	70.0	50.0	0.0
ALPHA (DEGREE)	260.0	81.0	345.0	161.0
COHESION (T/SQM)	0.0	0.0	0.0	
PHAI (DEGREE)	15.0	15.0	15.0	
U(T/SQM)	0.0	0.0	0.0	

SAI5 = 0.0   ALPHA5 = 161.0   SAI6 = 0.0   ALPHA6 = 0.0
WB = 12.00   GAMA = 2.840   BOLT FORCE = 0.0

INDIVIDUAL BOLT CAPACITY	=	10.0
SPACING OF ROCK BOLTS	=	2.0
LENGTH OF ROCK BOLTS	=	6.0
FIXED ANCHOR LENGTH	=	3.0
THICKNESS OF SHOTCRETE	=	0.13
SHEAR STRENGTH OF SHOTCRETE	=	300.0
DIA. OF BARS OF WELDED MESH	=	0.006
SHEAR STRENGTH OF BARS OF MESH	=	7000.
SPACING OF BARS OF WELDED MESH	=	0.15

FACE OVERHANGS

WEIGHT OF THE WEDGE	=	85.
PERPENDICULAR DISTANCE FROM APEX OR ORIGIN	=	2.17

SEPARATION ON PLANE NO. 3, i.e., WEDGE WILL FALL OUT OR FLOAT
OUT BY WATER PRESSURE

Required Bolt Force	=	85.
Factor of Safety of Shotcreted/Bolted Wedge	=	very high
Required Support Pressure, kg/sq.cm	=	0.2

Case No. 2

PLANE	1	2	3	5
SAI(DEGREE)	80.0	70.0	20.0	0.0
ALPHA(DEGREE)	260.0	81.0	200.0	161.0
COHESION(T/SQM)	0.0	0.0	0.0	
PHAI(DEGREE)	15.0	15.0	15.0	
U(T/SQM)	0.0	0.0	0.0	

SAI5 = 0.0   ALPHA5 = 161.0   SAI6 = 0.0   ALPHA6 = 0.0
WB = 12.00   GAMA = 2.840   BOLT FORCE = 0.0

INDIVIDUAL BOLT CAPACITY	=	10.0
SPACING OF ROCK BOLTS	=	2.0
LENGTH OF ROCK BOLTS	=	6.0
FIXED ANCHOR LENGTH	=	3.0
THICKNESS OF SHOTCRETE	=	0.13
SHEAR STRENGTH OF SHOTCRETE	=	300.0
DIA. OF BARS OF WELDED MESH	=	0.006
SHEAR STRENGTH OF BARS OF MESH	=	7000.
SPACING OF BARS OF WELDED MESH	=	0.15

FACE OVERHANGS

WEIGHT OF THE WEDGE	=	100 Tonnes
PERPENDICULAR DISTANCE FROM APEX OR ORIGIN	=	2.50

CONTACT ON PLANE NO 4 AND 1, i.e., SEPARATION
ON PLANE NO. 2

FACTOR OF SAFETY OF UNSUPPORTED WEDGE	=	0.99
Required Bolt Force	=	1
Factor of Safety of Shotcreted/Bolted Wedge	=	very high
Required Support Pressure, kg/sq.cm	=	0.0

Case No. 3

PLANE	1	2	3	5
SAI(DEGREE)	80.0	50.0	20.0	0.0
ALPHA(DEGREE)	260.0	345.0	200.0	161.0
COHESION(T/SQM)	0.0	0.0	0.0	
PHAI(DEGREE)	15.0	15.0	15.0	
U(T/SQM)	0.0	0.0	0.0	

SAI5 = 0.0   ALPHA5 = 161.0   SAI6   = 0.0   ALPHA6 = 0.0
WB   = 12.000 GAMA   = 2.840   BOLT FORCE = 0.0

INDIVIDUAL BOLT CAPACITY	=	10.0
SPACING OF ROCK BOLTS	=	2.0
LENGTH OF ROCK BOLTS	=	6.0
FIXED ANCHOR LENGTH	=	3.0
THICKNESS OF SHOTCRETE	=	0.13
SHEAR STRENGTH OF SHOTCRETE	=	300.0
DIA. OF BARS OF WELDED MESH	=	0.006
SHEAR STRENGTH OF BARS OF MESH	=	7000.
SPACING OF BARS OF WELDED MESH	=	0.15

FACE OVERHANGS

WEIGHT OF THE WEDGE	=	92 Tonnes
PERPENDICULAR DISTANCE FROM APEX OR ORIGIN	=	2.04
CONTACT ON PLANE NO. 1, i.e., SEPARATION ON PLANE 3 AND 4		
FACTOR OF SAFETY OF UNSUPPORTED WEDGE	=	0.05
Required Bolt Force	=	88
Factor of Safety of Shotcreted/Bolted Wedge	=	30.10
Required Support Pressure, kg/sq.cm	=	0.2

Case No. 4

PLANE	1	2	3	5
SAI(DEGREE)	70.0	50.0	20.0	0.0
ALPHA(DEGREE)	81.0	345.0	200.0	161.0
COHESION(T/SQM)	0.0	0.0	0.0	
PHAI(DEGREE)	15.0	15.0	15.0	
U(T/SQM)	0.0	0.0	0.0	

SAI5 = 0.0   ALPHA5 = 161.0   SAI6   = 0.0   ALPHA6 = 0.0
WB   = 12.000 GAMA   = 2.840   BOLT FORCE = 0.0

INDIVIDUAL BOLT CAPACITY	=	10.0
SPACING OF ROCK BOLTS	=	2.0
LENGTH OF ROCK BOLTS	=	6.0
FIXED ANCHOR LENGTH	=	3.0
THICKNESS OF SHOTCRETE	=	0.13
SHEAR STRENGTH OF SHOTCRETE	=	300.0
DIA. OF BARS OF WELDED MESH	=	0.006
SHEAR STRENGTH OF BARS OF MESH	=	7000.
SPACING OF BARS OF WELDED MESH	=	0.15

FACE OVERHANGS

WEIGHT OF THE WEDGE	=	83 Tonnes
PERPENDICULAR DISTANCE FROM APEX OR ORIGIN	=	1.87

SEPARATION ON PLANE NO. 4, i.e., WEDGE WILL FALL
OUT OR FLOAT OUT BY WATER PRESSURE

Required Bolt Force	=	83.
Factor of Safety of Shotcreted/Bolted Wedge	=	14.88
Required Support Pressure, kg/sq.cm	=	0.2

STABILITY ANALYSIS OF REINFORCED WEDGE IN LEFT WALL
OF A CAVERN

UNITS USED –	TONNE – METER – DEGREE
INPUT FILE NAME –	IUWEDGE.DAT
OUTPUT FILE NAME –	OUWEDGE.DAT

PROBLEM NO.: 2

NO. OF JOINT SETS   =   4

STRIKE OF JOINT NO. 1 AND BOUNDARY EDGE ARE NEARLY PARALLEL.
HENCE ALIGNMENT IS UNFAVOURABLE

STRIKE OF JOINT NO. 2 AND BOUNDARY EDGE ARE NEARLY PARALLEL.
HENCE ALIGNMENT IS UNFAVOURABLE

UNSUPPORTED WALL MAY FAIL BY OVER-TOPPLING DUE TO
JOINT NO. 2

Case No. 1

PLANE	1	2	3	5
SAI(DEGREE)	80.0	70.0	50.0	89.0
ALPHA(DEGREE)	260.0	81.0	345.0	251.0
COHESION(T/SQM)	0.0	0.0	0.0	
PHAI(DEGREE)	15.0	15.0	15.0	
U(T/SQM)	0.0	0.0	0.0	

SAI5	=	0.0	ALPHA5	=	161.0	SAI6	= 0.0	ALPHA6	= 0.0
WB	=	50.000	GAMA	=	2.840	BOLT FORCE	= 0.0		

INDIVIDUAL BOLT CAPACITY	=	10.0
SPACING OF ROCK BOLTS	=	2.0
LENGTH OF ROCK BOLTS	=	6.0
FIXED ANCHOR LENGTH	=	3.0
THICKNESS OF SHOTCRETE	=	0.13
SHEAR STRENGTH OF SHOTCRETE	=	300.0
DIA. OF BARS OF WELDED MESH	=	0.006
SHEAR STRENGTH OF BARS OF MESH	=	7000.
SPACING OF BARS OF WELDED MESH	=	0.15

FACE DOES NOT OVERHANG

CONTACT ON PLANE NO. 1, i.e., SEPARATION ON PLANES 2 AND 3

FACTOR OF SAFETY OF UNSUPPORTED WEDGE	=	0.05
Required Bolt Force	=	1544
Factor of Safety of Shotcreted/Bolted Wedge	=	3.35
Required Support Pressure, kg/sq.cm	=	0.8

Case No. 2

PLANE	1	2	3	5
SAI(DEGREE)	80.0	70.0	20.0	89.0
ALPHA(DEGREE)	260.0	81.0	200.0	251.0
COHESION(T/SQM)	0.0	0.0	0.0	
PHAI(DEGREE)	15.0	15.0	15.0	
U(T/SQM)	0.0	0.0	0.0	

SAI5 = 0.0	ALPHA5 = 161.0	SAI6	= 0.0	ALPHA6 = 0.0
WB = 50.000	GAMA = 2.840	BOLT FORCE	= 0.0	

INDIVIDUAL BOLT CAPACITY	=	10.0
SPACING OF ROCK BOLTS	=	2.0
LENGTH OF ROCK BOLTS	=	6.0
FIXED ANCHOR LENGTH	=	3.0
THICKNESS OF SHOTCRETE	=	0.13
SHEAR STRENGTH OF SHOTCRETE	=	300.0
DIA. OF BARS OF WELDED MESH	=	0.006
SHEAR STRENGTH OF BARS OF MESH	=	7000.
SPACING OF BARS OF WELDED MESH	=	0.15

FACE DOES NOT OVERHANG
CONTACT ON PLANES 4 AND 1, i.e., SEPARATION ON PLANE NO. 2
FACTOR OF SAFETY OF UNSUPPORTED WEDGE   =   0.99
Required Bolt Force   =   126
Factor of Safety of Shotcreted/Bolted Wedge   =   very high
Required Support Pressure, kg/sq.cm   =   0.0

Case No. 3

PLANE	1	2	3	5
SAI(DEGREE)	80.0	50.0	20.0	89.0
ALPHA(DEGREE)	260.0	345.0	200.0	251.0
COHESION(T/SQM)	0.0	0.0	0.0	
PHAI(DEGREE)	15.0	15.0	15.0	
U(T/SQM)	0.0	0.0	0.0	

SAI5 = 0.0	ALPHA5 = 161.0	SAI6	= 0.0	ALPHA6 = 0.0
WB = 50.000	GAMA = 2.840	BOLT FORCE	= 0.0	

INDIVIDUAL BOLT CAPACITY	=	10.0
SPACING OF ROCK BOLTS	=	2.0
LENGTH OF ROCK BOLTS	=	6.0
FIXED ANCHOR LENGTH	=	3.0
THICKNESS OF SHOTCRETE	=	0.13
SHEAR STRENGTH OF SHOTCRETE	=	300.0
DIA. OF BARS OF WELDED MESH	=	0.006
SHEAR STRENGTH OF BARS OF MESH	=	7000.
SPACING OF BARS OF WELDED MESH	=	0.15

FACE DOES NOT OVERHANG
CONTACT ON PLANE NO. 1, i.e., SEPARATION ON PLANES 3 AND 4
FACTOR OF SAFETY OF UNSUPPORTED WEDGE   =   0.05
Required Bolt Force   =   1659

Factor of Safety of Shotcreted/Bolted Wedge   =   5.47
Required Support Pressure, kg/sq.cm   =   0.8

Case No. 4

PLANE	1	2	3	5
SAI(DEGREE)	70.0	50.0	20.0	89.0
ALPHA(DEGREE)	81.0	345.0	200.0	251.0
COHESION(T/SQM)	0.0	0.0	0.0	
PHAI(DEGREE)	15.0	15.0	15.0	
U(T/SQM)	0.0	0.0	0.0	

SAI5 = 0.0   ALPHA5 = 161.0   SAI6 = 0.0   ALPHA6 = 0.0
WB = 50.000   GAMA = 2.840   BOLT FORCE = 0.0

INDIVIDUAL BOLT CAPACITY	= 10.0
SPACING OF ROCK BOLTS	= 2.0
LENGTH OF ROCK BOLTS	= 6.0
FIXED ANCHOR LENGTH	= 3.0
THICKNESS OF SHOTCRETE	= 0.13
SHEAR STRENGTH OF SHOTCRETE	= 300.0
DIA. OF BARS OF WELDED MESH	= 0.006
SHEAR STRENGTH OF BARS OF MESH	= 7000.
SPACING OF BARS OF WELDED MESH	= 0.15

FACE DOES NOT OVERHANG
CONTACT ON PLANE NO. 4, i.e., SEPARATION ON PLANES 2 AND 3
FACTOR OF SAFETY OF UNSUPPORTED WEDGE   =   0.74
Required Bolt Force   =   1242
Factor of Safety of Shotcreted/Bolted Wedge   =   17.45
Required Support Pressure, kg/sq.cm   =   0.10

STABILITY ANALYSIS OF REINFORCED WEDGE IN RIGHT WALL OF A CAVERN

UNITS USED –   TONNE – METER – DEGREE
INPUT FILE NAME –   IUWEDGE.DAT
OUTPUT FILE NAME –   OUWEDGE.DAT

PROBLEM NO.: 3
NO. OF JOINT SETS = 4
STRIKE OF JOINT NO: 1 AND BOUNDARY EDGE ARE NEARLY PARALLEL.
HENCE ALIGNMENT IS UNFAVOURABLE

STRIKE OF JOINT NO: 2 AND BOUNDARY EDGE ARE NEARLY PARALLEL.
HENCE ALIGNMENT IS UNFAVOURABLE

UNSUPPORTED WALL MAY FAIL BY OVER-TOPPLING DUE TO JOINT NO. 1

Case No. 1

PLANE	1	2	3	5
SAI(DEGREE)	80.0	70.0	50.0	89.0
ALPHA(DEGREE)	260.0	81.0	345.0	71.0
COHESION(T/SQM)	0.0	0.0	0.0	
PHAI(DEGREE)	15.0	15.0	15.0	
U(T/SQM)	0.0	0.0	0.0	

SAI5 = 0.0	ALPHA5 = 161.0	SAI6	= 0.0	ALPHA6 = 0.0
WB = 50.000	GAMA = 2.840	BOLT FORCE	= 0.0	

INDIVIDUAL BOLT CAPACITY	= 10.0
SPACING OF ROCK BOLTS	= 2.0
LENGTH OF ROCK BOLTS	= 6.0
FIXED ANCHOR LENGTH	= 3.0
THICKNESS OF SHOTCRETE	= 0.13
SHEAR STRENGTH OF SHOTCRETE	= 300.0
DIA. OF BARS OF WELDED MESH	= 0.006
SHEAR STRENGTH OF BARS OF MESH	= 7000.
SPACING OF BARS OF WELDED MESH	= 0.15

FACE DOES NOT OVERHANG

CONTACT ON PLANES 2 AND 3, i.e., SEPARATION ON PLANE NO. 1

FACTOR OF SAFETY OF UNSUPPORTED WEDGE	= 0.32
Required Bolt Force	= 1572
Factor of Safety of Shotcreted/Bolted Wedge	= 3.58
Required Support Pressure, kg/sq.cm	= 0.50

Case No. 2

PLANE	1	2	3	5
SAI(DEGREE)	80.0	70.0	20.0	89.0
ALPHA(DEGREE)	260.0	81.0	200.0	71.0
COHESION(T/SQM)	0.0	0.0	0.0	
PHAI(DEGREE)	15.0	15.0	15.0	
U(T/SQM)	0.0	0.0	0.0	

SAI5 = 0.0	ALPHA5 = 161.0	SAI6	= 0.0	ALPHA6 = 0.0
WB = 50.000	GAMA = 2.840	BOLT FORCE	= 0.0	

INDIVIDUAL BOLT CAPACITY	= 10.0
SPACING OF ROCK BOLTS	= 2.0
LENGTH OF ROCK BOLTS	= 6.0
FIXED ANCHOR LENGTH	= 3.0
THICKNESS OF SHOTCRETE	= 0.13
SHEAR STRENGTH OF SHOTCRETE	= 300.0
DIA. OF BARS OF WELDED MESH	= 0.006
SHEAR STRENGTH OF BARS OF MESH	= 7000.
SPACING OF BARS OF WELDED MESH	= 0.15

FACE DOES NOT OVERHANG

WEDGE IS STABLE

Case No. 3

PLANE	1	2	3	5
SAI(DEGREE)	80.0	50.0	20.0	89.0
ALPHA(DEGREE)	260.0	345.0	200.0	71.0
COHESION(T/SQM)	0.0	0.0	0.0	
PHAI(DEGREE)	15.0	15.0	15.0	
U(T/SQM)	0.0	0.0	0.0	

SAI5 = 0.0	ALPHA5 = 161.0	SAI6	= 0.0	ALPHA6 = 0.0
WB = 50.000	GAMA = 2.840	BOLT FORCE	= 0.0	

INDIVIDUAL BOLT CAPACITY	=	10.0
SPACING OF ROCK BOLTS	=	2.0
LENGTH OF ROCK BOLTS	=	6.0
FIXED ANCHOR LENGTH	=	3.0
THICKNESS OF SHOTCRETE	=	0.13
SHEAR STRENGTH OF SHOTCRETE	=	300.0
DIA. OF BARS OF WELDED MESH	=	0.006
SHEAR STRENGTH OF BARS OF MESH	=	7000.
SPACING OF BARS OF WELDED MESH	=	0.15

FACE DOES NOT OVERHANG

WEDGE IS STABLE

Case No. 4

PLANE	1	2	3	5
SAI(DEGREE)	70.0	50.0	20.0	89.0
ALPHA(DEGREE)	81.0	345.0	200.0	71.0
COHESION(T/SQM)	0.0	0.0	0.0	
PHAI(DEGREE)	15.0	15.0	15.0	
U(T/SQM)	0.0	0.0	0.0	

SAI5 = 0.0 ALPHA5 = 161.0 SAI6 = 0.0 ALPHA6 = 0.0
WB = 50.000 GAMA = 2.840 BOLT FORCE = 0.0

INDIVIDUAL BOLT CAPACITY	=	10.0
SPACING OF ROCK BOLTS	=	2.0
LENGTH OF ROCK BOLTS	=	6.0
FIXED ANCHOR LENGTH	=	3.0
THICKNESS OF SHOTCRETE	=	0.13
SHEAR STRENGTH OF SHOTCRETE	=	300.0
DIA. OF BARS OF WELDED MESH	=	0.006
SHEAR STRENGTH OF BARS OF MESH	=	7000.
SPACING OF BARS OF WELDED MESH	=	0.15

FACE DOES NOT OVERHANG
CONTACT ON PLANES 2 AND 3, i.e., SEPARATION ON PLANE NO. 4

FACTOR OF SAFETY OF UNSUPPORTED WEDGE	=	0.32
Required Bolt Force	=	13202
Factor of Safety of Shotcreted/Bolted Wedge	=	1.58
Required Support Pressure, kg/sq.cm	=	1.20

KINDLY CHECK THAT MAXIMUM RADIAL STRAIN AROUND OPENING IN ROCK MASS DUE TO POST CONSTRUCTION SATURATION IS LESS THAN ONE-THIRD OF YIELD STRAIN OF ANCHOR BAR.

IN CASE NO WEDGE IS FORMED IN ROOF AND WALLS; AND OPENING IS NOT SELF-SUPPORTING, THEN A NOMINAL SUPPORT SYSTEM SHOULD BE PROVIDED TO REINFORCE TENSION ZONES AND SLIP ZONES (AS OBTAINED FROM BEM/FEM).

SPECIAL SPECIFICATIONS
1. FOR TREATMENT OF SHEAR ZONES, CROSSED ROCK ANCHORS/BOLTS SHOULD BE PROVIDED ACROSS SHEAR ZONES. FIRST GOUGE SHOULD BE CLEANED TO DESIRED EXTENT. ANCHORS ARE PROVIDED AND CONNECTED TO WELDED MESH. FINALLY, DENTAL SHOTCRETE IS BACK-FILLED. IN WIDE SHEAR ZONE,

REINFORCEMENT IN SHOTCRETE IS ALSO PLACED TO WITHSTAND HIGH SUPPORT PRESSURE. ANCHORS SHOULD BE INCLINED ACCORDING TO DIP OF SHEAR ZONE TO STOP SQUEEZING OF GOUGE AND THEREBY STABILISE DEFORMATIONS.

2. IN CASE OF STEEL RIBS IN LARGE TUNNELS AND CAVERNS, HAUNCHES SHOULD BE STRENGTHENED BY ADDITIONAL ANCHORS TO WITHSTAND HEAVY THRUST DUE TO RIBS.

3. IN CASE OF POOR ROCK MASS, SPILING BOLTS (INCLINED TOWARDS TUNNEL FACE) SHOULD BE INSTALLED BEFORE BLASTING TO INCREASE THE STAND-UP TIME OF TUNNEL ROOF. SHOTCRETE IS THEN SPRAYED ON ROOF. NEXT SPILING BOLTS ARE INSTALLED. IN FINAL CYCLE, ROOF BOLTS ARE INSTALLED.

4. IN CASES OF ARGILLACEOUS AND SWELLING ROCKS WHERE ITS BOND WITH SHOTCRETE IS POOR, THICKNESS OF SHOTCRETE MAY BE INCREASED BY ABOUT 30 PER CENT.

5. IN REACHES OF VERY POOR ROCK MASSES, STEEL RIBS SHOULD BE INSTALLED AND EMBEDDED IN SHOTCRETE TO WITHSTAND HIGH SUPPORT PRESSURES.

6. IN ROCK BURST PRONE REACHES, RESIN ANCHORS AND FIBRE REINFORCED SHOTCRETE SHOULD BE USED TO INCREASE DUCTILITY OF SUPPORT SYSTEM AND THUS CONVERT BRITTLE MODE OF FAILURE INTO DUCTILE MODE OF FAILURE.

7. IN HIGHLY SQUEEZING GROUND CONDITION (H > 350Q**0.33 M AND J_r/J_a < 0.5), STEEL RIBS WITH STRUTS SHOULD BE USED WHEN SHOTCRETE FAILS AGAIN AND AGAIN INSPITE OF MORE LAYERS. WITH STEEL RIBS, EXCAVATION BY FORE-POLING IS EASILY DONE BY PUSHING STEEL RODS INTO THE TUNNEL FACE AND WELDING OTHER ENDS TO RIBS. FLOOR HEAVING CAN BE PREVENTED BY ROCK BOLTING OF FLOOR.

8. IN CASE OF UNSTABLE PORTAL, HORIZONTAL ANCHORS OF EQUAL LENGTH SHOULD BE PROVIDED INSIDE THE CUT SLOPE, SO THAT IT ACTS AS A REINFORCED ROCK BREAST WALL.

REFERENCES

Barton, N., Lien, R. and Lunde, J. 1974. Engineering classification of rock masses for the design of tunnel support. *Rock Mechanics*, Vol. 6, pp. 189–236.

Hoek, E. and Brown, E.T. 1980. *Underground Excavations in Rock*. Institution of Mining and Metallurgy, England, Revised Edition, Appendix 6: Underground wedge analysis, p. 52.

Singh, Bhawani and Goel, R.K. 1999. *Rock Mass Classification – A Practical Approach in Civil Engineering*. Elsevier Science Ltd., Section 24.7, p. 267.

Singh, Bhawani., Jethwa, J.L., Dube, A.K. and Singh, B. 1992. Correlation between observed support pressure and rock mass quality. *J. Tunnel. and Underground Space Technol.* 7(1): 59–74.

Singh, Bhawani, Viladkar, M.N., Samadhiya, N.K. and Sandeep 1995. A semi-empirical method for the design of support systems in underground openings. *J. Tunnel. Underground Space Technol.* 10(3): 375–383.

Singh, Bhawani, Viladkar, M.N., Samadhiya, N.K. and Mehrotra, V.K. 1997. Rock mass strength parameters mobilized in tunnels, *J. Tunnel. Underground Space Technol.* 12(1): 47–54.

CHAPTER 25

Support pressure in squeezing ground condition – SQUEEZE

"Good judgement comes from experience. But where does experience come from? Experience comes from bad judgement."

Mark Twain

25.1 INITIATION OF SQUEEZING INSIDE TUNNELS

The squeezing of weak rock mass takes place due to the over-stressing around a tunnel under the high overburden. The squeezing may occur according to polyaxial strength criterion ($\sigma_3 = 0$, $\sigma_2 = P_o$ in Eq. 23.5 in Chapter 23) in the following case,

$$\sigma_{\theta max} = q_{cmass} + A(P_o/2) \tag{25.1}$$

where

$\sigma_{\theta max}$ = maximum tangential stress around periphery of a tunnel,
q_{cmass} = UCS of rock mass (Eq. 23.6, Chapter 23),
A   = $2 \sin \phi_p/(1 - \sin \phi_p)$,
ϕ_p = peak angle of internal friction of the rock mass, and
P_o = in situ stress along the tunnel axis.

The UCS of rock mass may also be found from empirical theories of Singh et al. (1997) and Grimstad and Bhasin (1996) (see Section 28.1). Equation 25.1 suggests thatsqueezing may occur where A or ϕ_p is not high, otherwise plain strain compressive strength will be enhanced considerably. Indeed the value of J_r/J_a has been found to be less than 0.5 in all the reported squeezing case histories in Himalaya and Europe (Goel, 2000). Chapter 21 reviews empirical theory of squeezing in detail. Elasto-plastic theory is also presented as a second approach for confirmation.

25.2 SOFTWARE SQUEEZE

The squeezing of rock mass inside a tunnel is a great hazard in the weak rock masses. The squeezing rock condition may be identified by observing tunnel wall closures. For example, if the tunnel closure is more than 1 or 2 percent of the tunnel width and the time of stabilisation of closure is as high as 6 to 20 months, it is the location of squeezing ground condition (Sections 22.1 and 22.2). The solution of this challenging problem is *not too stiff and too early support but in fact a flexible support system installed with calculated delay.* Consequently, the strain energy of broken rock mass is released slowly in a controlled manner until its stabilisation is achieved. Steel fibre reinforced shotcrete (SFRS) along with a rock anchor system is ideally suited to support mild to moderate squeezing grounds.

Terzaghi (1946) advocated that the phenomenon of squeezing takes place in squeezing rocks. Daemen (1975) proved that squeezing takes place in a tunnel where overburden is high such that tangential stress exceeds the strength of the rock mass. Subsequently, Singh et al. (1992) observed that squeezing occurs where overburden is more than $350 \, Q^{1/3} \, m$ and $J_r/J_a < 0.5$, where Q is rock mass quality (Chapter 21). Goel (2000) suggested criteria for mild, moderate and high squeezing based on N value (Q with SRF = 1) as discussed in Section 22.3.

A software SQUEEZE is developed to predict support pressures in tunnels based on a rational approach. It is assumed that the rock mass in the elastic zone obeys the failure criterion of Singh et al. (1998) in which confining stress is defined as average of minor and intermediate principal stress (Eq. 23.5 in Chapter 23). Figure 25.1 shows that a broken zone is developed around a tunnel due to over-stressing. Strength of rock mass within this broken zone is reduced to residual state like soil. However, it has some cohesion of the order of 0.1 MPa. Jethwa (1981) observed in more than 9 tunnels that a compaction zone is also formed within the broken zone. The radius of compaction zone is about 0.37 times the radius of the broken zone (Fig. 22.2). Hence, no compaction zone may be formed if the radius of broken zone is less than 2.7 times the radius of the tunnel, as observed in the European tunnels.

Daemen (1975) proved that the support pressure decreases rapidly with the radius of broken zone and tunnel wall closure (Fig. 25.1). The relationship between support pressure and tunnel wall closure is called the ground response (reaction) curve. It should be noted that sympathetic failure of the rock mass in the entire broken zone appears to occur after deviator strain exceeds a critical limit ($\cong 12$ percent or $u_a/a \cong 6$ percent). The sympathetic failure is observed when unstable fracture propagation takes place in the broken zone resulting in complete loss of the residual cohesion. Consequently, there is sudden rise in the support pressure where tunnel wall closure exceeds 5 to 6 percent of radius of the tunnel. Thus, the ground response curve predicted by SQUEEZE is realistic. Hence construction engineers should be advised to try to stop the tunnel wall closure beyond 4 percent of the width of a tunnel to be on safe side (see also Fig. 21.1 in Chapter 21). Computed results should be checked with empirical predictions for cross-confirmation (Singh et al., 1992; Goel, 2000).

Extensive monitoring of displacements in the broken zone (Fig. 22.1) shows that the ground response curve (Fig. 25.1) is highly time-dependent. Experience in Himalayan tunnels suggests that the ultimate support pressures are likely

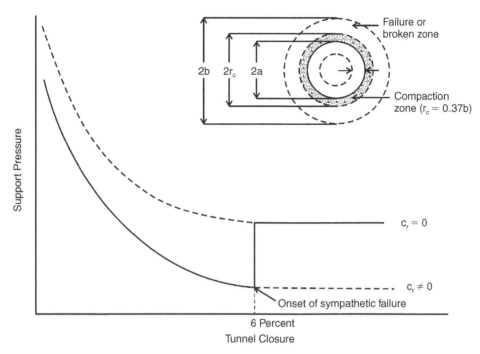

Figure 25.1. Effect of sympathetic failure of rock mass on theoretical ground response curve of squeezing ground condition.

to be 2–3 times of short-term support pressures. Accordingly, the elasto-plastic support pressure computed by SQUEEZE should be enhanced for the design of supports by program TM.

25.3   ASSUMPTIONS

The following assumptions have been made in modifying Daemen's (1975) axi-symmetric and elasto-plastic theory of squeezing in the rock tunnels.
1. The tunnel is circular, having radius a.
2. The rock mass is homogeneous, isotropic and (nearly) dry medium.
3. The broken zone is also circular with radius b and is concentric with the tunnel opening.
4. The tunnel is uniformly supported. There are no rock anchors.
5. There is hydrostatic state of in situ stresses (k = 1). The in situ stress along tunnel axis is P_o. (The modification for non-hydrostatic case is an approximation.)
6. The rock mass outside broken zone is elastic and obeys polyaxial strength criterion (Eq. 23.5 in Chapter 23) of peak strength.
7. The strength of rock mass within the broken zone is reduced to the residual state with parameters (c_r and ϕ_r) and follows Mohr's theory (Section 23.5).

8. The rock mass within the broken zone fails in a sympathetic way and suddenly when the tangential strain exceeds a critical limit.
9. There is no compaction zone within the broken zone.
10. Gravity is acting radially towards centre of the broken zone. The assumption is not serious, as the influence of gravity on support pressure is not dominating.
11. The coefficient of volumetric expansion of failed rock mass within the broken zone is known (see Table 23.2 in Chapter 23).
12. There is no rock burst or brittle failure ($J_r/J_a \leq 0.5$).

Perhaps 2D theory may also be applied in banded strata where the thickness of weak strata is more than the width of the tunnel.

25.4   VALIDATION OF THEORY IN THE FIELD

The marriage between a theory and field experiments is a difficult marriage. This marriage is a rocky one. Often the result is divorce, but on occasion, this marriage may give birth to a sweet baby of knowledge.

The output of computer program SQUEEZE shows that the predicted support pressures are of the order of those observed in 10 tunnels in the squeezing ground condition in India. There is a rather good cross-check between the proposed theory and the observations (reported by Singh et al., 1992) except in a few cases. Thus, the assumptions made in the theory are justified partially. Further, the predictions are generally conservative.

In the NJPC tunnel excavated under abnormal cover of 1400 m through massive to competent gneiss and schist gneiss, the program SQUEEZE predicted rock burst condition ($J_r/J_a = 3/4$, i.e., > 0.5). According to site geologists Pundhir, Acharya and Chadha (2000), a cracking noise was initially heard which was followed by the spalling of 5–25 cm thick rock columns/slabs and rock falls. This is a mild rock burst condition. Another cause of rock burst is the class II behaviour of gneiss according to the tests at IIT, Delhi, India. According to Mohr's theory though, most severe rock bursts or squeezing conditions were predicted under rock cover more than 300 m ($q_c = 27$ MPa and $q_{cmass} = 15.7$ MPa). Actually mild rock burst conditions were met where overburden is more than 1000 m. However, polyaxial theory (Eq. 25.1) suggested a mild rock burst condition above an overburden of 800 m. Thus, a polyaxial theory of strength is validated further by the SQUEEZE program. Fortuitously, there is a significant increase in the confidence of engineers and geologists in tackling the squeezing ground conditions.

25.5   USER'S MANUAL – SQUEEZE

PLEASE GIVE THE DATA IN THE FOLLOWING SEQUENCE:

1. NUMBER OF SQUEEZING CONDITIONS OF TUNNELS (NPROB)
2. TITLE OF THE PROJECT IN ONE LINE

HEIGHT OF OVERBURDEN IN METRE
UNIT WEIGHT OF ROCK MASS (g/cc)
ROCK MASS QUALITY Q
JOINT ROUGHNESS NUMBER
JOINT ALTERATION NUMBER
INTERMEDIATE PRINCIPAL STRESS ALONG AXIS OF TUNNEL (T/SQ.M.)
RATIO OF HORIZONTAL STRESS ACROSS TUNNEL TO VERTICAL STRESS < 1.5
AVERAGE SPACING OF JOINTS IN ROCK MASS (M)

3. RADIUS OF TUNNEL IN METRE
   RADIUS OF BROKEN ZONE IN METRE
   COEFFICIENT OF VOLUMETRIC EXPANSION OF ROCK MASS AFTER FAILURE
   MASS MODULUS OF DEFORMATION (T/SQ.M.)
   UNIAXIAL COMPRESSIVE STRENGTH OF ROCK MATERIAL (T/SQ.M.)
   MODULUS OF ELASTICITY OF ROCK MATERIAL (T/SQ.M.)

   (REPEAT THE INPUT STARTING FROM TITLE FOR NPROB TIMES)

(a) Input File: SQUEEZE.DAT

10
CHHIBRO-KHODRI TUNNEL IN THE SQUEEZING GROUND CONDITION IN RED SHALE
280 2.73 0.05 1.5 4. 765. 1. 0.05
1.5 6. 0.0025 157000. 210. 1080000.
CHHIBRO-KHODRI TUNNEL IN THE SQUEEZING GROUND CONDITION IN RED SHALE
680 2.73 0.05 1.5 4. 1850. 1. 0.05
4.5 32. 0.0025 157000. 210. 1080000.
GIRI BATA TUNNEL IN THE SQUEEZING GROUND CONDITION IN SLATES
380. 2.5 0.51 1. 2.5 950. 1. 0.10
2.3 18. 0.0025 260000. 4000. 2000000.
GIRI BATA TUNNEL IN THE SQUEEZING GROUND CONDITION IN PHYLLITES
240. 2.3 0.12 1. 3.5 550. 1. 0.06
2.3 12. 0.0025 140000. 2000. 860000.
LOKTAK TUNNEL IN THE SQUEEZING GROUND CONDITION IN SHALES
300. 2.4 0.023 1.5 4.5 720. 1. 0.05
2.4 13. 0.0043 90000. 2650. 1080000.
MANERI STAGE I TUNNEL IN THE SQUEEZING GROUND CONDITION IN CRUSHED QUARTZITE
350. 2.5 0.5 1. 4. 875. 1. 0.12
2.9 19. 0.0037 84000. 5000. 2800000.
MANERI STAGE II TUNNEL IN THE SQUEEZING GROUND CONDITION IN METABASICS
480. 2.5 0.8 1.2 3.5 1200. 1. 0.12
1.3 8. 0.0012 308000. 6000. 2200000.
MANERI STAGE II TUNNEL IN THE SQUEEZING GROUND CONDITION IN SHEARED METABASICS
410. 2.5 0.18 1.2 3.5 1025. 1. 0.06
3.5 23. 0.0012 145000. 6000. 2100000.
MANERI STAGE I TUNNEL IN THE SQUEEZING GROUND CONDITION IN WET METABASICS
800. 2.5 2.3 1.2 3.5 2000. 1. 0.15
2.4 17. 0.0037 500000. 6000. 2200000.
NJPC TUNNEL UNDER HIGH OVERBURDEN IN GNEISS

1400. 2.7 30. 0.75 1. 3000. 1. 0.75
2.5 7.5 0.004 1200000. 2700. 1500000.

(b) Output File: SQUEEZE.OUT

CHHIBRO-KHODRI TUNNEL IN THE SQUEEZING GROUND
CONDITION IN RED SHALE

HEIGHT OF OVERBURDEN IN METRE	=	280.0000
UNIT WEIGHT OF ROCK MASS (g/cc)	=	2.7300
ROCK MASS QUALITY Q	=	0.05000
JOINT ROUGHNESS NUMBER	=	1.50000
JOINT ALTERATION NUMBER	=	4.00000
PRINCIPAL STRESS ALONG AXIS OF TUNNEL	=	765.00000
RATIO OF HORIZONTAL STRESS TO VERTICAL STRESS	=	1.00000
AVERAGE SPACING OF JOINTS IN ROCK MASS (M)	=	0.05000
RADIUS OF TUNNEL IN METRE	=	1.50000
RADIUS OF BROKEN ZONE IN METRE	=	6.00000
COEFFICIENT OF VOLUMETRIC EXPANSION AFTER FAILURE	=	0.00250
MODULUS OF DEFORMATION OF ROCK MASS (T/SQ.M.)	=	157000.
UNIAXIAL COMPRESSIVE STRENGTH OF ROCK MATERIAL {ASSUMED AFTER SIZE CORRECTION FOR JOINT SPACING}	=	210.000
MODULUS OF ELASTICITY OF ROCK MATERIAL (T/SQ.M.)	=	1080000.
NORMALIZED WALL SUPPORT PRESSURE	=	2.54355
NORMALIZED ROOF SUPPORT PRESSURE	=	2.72513
TUNNEL CLOSURE IN PER CENT	=	1.89292
ELASTO PLASTIC WALL SUPPORT PRESSURE (T/SQ.M.)	=	96.89857
ELASTO PLASTIC ROOF SUPPORT PRESSURE (T/SQ.M.)	=	103.81610
BARTON'S SHORT TERM SUPPORT PRESSURE IN ROOF AND WALL	=	38.09586
RADIUS OF COMPACTION ZONE WITHIN BROKEN ZONE (M)	=	2.20800
RADIAL DISPLACEMENT AT ELASTIC PLASTIC BOUNDARY (M)	=	0.02006
CRITICAL STRAIN OF ROCK MATERIAL	=	0.00019

HEAVING MAY OCCUR AT BOTTOM OF TUNNEL

CHHIBRO-KHODRI TUNNEL IN THE SQUEEZING GROUND
CONDITION IN RED SHALE

HEIGHT OF OVERBURDEN IN METRE	=	680.0000
UNIT WEIGHT OF ROCK MASS (g/cc)	=	2.7300
ROCK MASS QUALITY Q	=	0.05000
JOINT ROUGHNESS NUMBER	=	1.50000
JOINT ALTERATION NUMBER	=	4.00000
PRINCIPAL STRESS ALONG AXIS OF TUNNEL	=	1850.00
RATIO OF HORIZONTAL STRESS TO VERTICAL STRESS = 1.00000		
AVERAGE SPACING OF JOINTS IN ROCK MASS (M)	=	0.05000
RADIUS OF TUNNEL IN METRE	=	4.50000
RADIUS OF BROKEN ZONE IN METRE	=	32.00000

COEFFICIENT OF VOLUMETRIC EXPANSION AFTER FAILURE	=	0.00250
MODULUS OF DEFORMATION OF ROCK MASS (T/SQ.M.)	=	157000.
UNIAXIAL COMPRESSIVE STRENGTH OF ROCKMATERIAL		
{ASSUMED AFTER SIZE CORRECTION FOR JOINTSPACING}	=	210.000
MODULUS OF ELASTICITY OF ROCK MATERIAL (T/SQ.M.)	=	1080000.
NORMALIZED WALL SUPPORT PRESSURE	=	5.64834
NORMALIZED ROOF SUPPORT PRESSURE	=	6.55480
TUNNEL CLOSURE IN PER CENT	=	6.40080
ELASTO PLASTIC WALL SUPPORT PRESSURE (T/SQ.M.)	=	199.33870
ELASTO PLASTIC ROOF SUPPORT PRESSURE (T/SQ.M.)	=	231.32920
BARTON'S SHORT TERM SUPPORT PRESSURE IN ROOF AND WALL	=	35.29159
RADIUS OF COMPACTION ZONE WITHIN BROKEN ZONE (M)	=	11.77600
RADIAL DISPLACEMENT AT ELASTIC PLASTIC		
BOUNDARY (M)	=	0.25243
CRITICAL STRAIN OF ROCK MATERIAL	=	0.00019

HEAVING MAY OCCUR AT BOTTOM OF TUNNEL

GIRI BATA TUNNEL IN THE SQUEEZING GROUND CONDITION IN SLATES

HEIGHT OF OVERBURDEN IN METRE	=	380.0000
UNIT WEIGHT OF ROCK MASS (g/cc)	=	2.5000
ROCK MASS QUALITY Q	=	0.51000
JOINT ROUGHNESS NUMBER	=	1.00000
JOINT ALTERATION NUMBER	=	2.50000
PRINCIPAL STRESS ALONG AXIS OF TUNNEL	=	950.00000
RATIO OF HORIZONTAL STRESS TO VERTICAL STRESS	=	1.00000
AVERAGE SPACING OF JOINTS IN ROCK MASS (M)	=	0.10000
RADIUS OF TUNNEL IN METRE	=	2.30000
RADIUS OF BROKEN ZONE IN METRE	=	18.00000
COEFFICIENT OF VOLUMETRIC EXPANSION AFTER FAILURE	=	0.00250
MODULUS OF DEFORMATION OF ROCK MASS (T/SQ.M.)	=	260000.
UNIAXIAL COMPRESSIVE STRENGTH OF ROCK		
MATERIAL		
{ASSUMED AFTER SIZE CORRECTION FOR JOINT SPACING}	=	3482.202
MODULUS OF ELASTICITY OF ROCK MATERIAL (T/SQ.M.)	=	2000000.
NORMALIZED WALL SUPPORT PRESSURE	=	1.13038
NORMALIZED ROOF SUPPORT PRESSURE	=	1.94117
TUNNEL CLOSURE IN PER CENT	=	7.83814
ELASTO PLASTIC WALL SUPPORT PRESSURE (T/SQ.M.)	=	20.45786
ELASTO PLASTIC ROOF SUPPORT PRESSURE (T/SQ.M.)	=	35.13180
BARTON'S SHORT TERM SUPPORT PRESSURE IN ROOF AND WALL	=	18.09825
RADIUS OF COMPACTION ZONE WITHIN BROKEN ZONE (M)	=	6.62400
RADIAL DISPLACEMENT AT ELASTIC PLASTIC BOUNDARY (M)	=	0.07015
CRITICAL STRAIN OF ROCK MATERIAL	=	0.00200

HEAVING MAY OCCUR AT BOTTOM OF TUNNEL

GIRI BATA TUNNEL IN THE SQUEEZING GROUND CONDITION IN PHYLLITES

HEIGHT OF OVERBURDEN IN METRE	=	240.0000
UNIT WEIGHT OF ROCK MASS (g/cc)	=	2.3000

ROCK MASS QUALITY Q	=	0.12000
JOINT ROUGHNESS NUMBER	=	1.00000
JOINT ALTERATION NUMBER	=	3.50000
PRINCIPAL STRESS ALONG AXIS OF TUNNEL	=	550.00000
RATIO OF HORIZONTAL STRESS TO VERTICAL STRESS	=	1.00000
AVERAGE SPACING OF JOINTS IN ROCK MASS (M)	=	0.06000
RADIUS OF TUNNEL IN METRE	=	2.30000
RADIUS OF BROKEN ZONE IN METRE	=	12.00000
COEFFICIENT OF VOLUMETRIC EXPANSION AFTER FAILURE	=	0.00250
MODULUS OF DEFORMATION OF ROCK MASS (T/SQ.M.)	=	140000.
UNIAXIAL COMPRESSIVE STRENGTH OF ROCK MATERIAL		
{ASSUMED AFTER SIZE CORRECTION FOR JOINT SPACING}	=	1928.385
MODULUS OF ELASTICITY OF ROCK MATERIAL (T/SQ.M.)	=	860000.
NORMALIZED WALL SUPPORT PRESSURE	=	0.41453
NORMALIZED ROOF SUPPORT PRESSURE	=	1.00786
TUNNEL CLOSURE IN PER CENT	=	3.33320
ELASTO PLASTIC WALL SUPPORT PRESSURE (T/SQ.M.)	=	8.35507
ELASTO PLASTIC ROOF SUPPORT PRESSURE (T/SQ.M.)	=	20.31386
BARTON'S SHORT TERM SUPPORT PRESSURE IN ROOF		
AND WALL	=	20.15539
RADIUS OF COMPACTION ZONE WITHIN BROKEN ZONE (M)	=	4.41600
RADIAL DISPLACEMENT AT ELASTIC PLASTIC BOUNDARY (M)	=	0.04660
CRITICAL STRAIN OF ROCK MATERIAL	=	0.00233

LOKTAK TUNNEL IN THE SQUEEZING GROUND CONDITION IN SHALES

HEIGHT OF OVERBURDEN IN METRE	=	300.0000
UNIT WEIGHT OF ROCK MASS (g/cc)	=	2.4000
ROCK MASS QUALITY Q	=	0.02300
JOINT ROUGHNESS NUMBER	=	1.50000
JOINT ALTERATION NUMBER	=	4.50000
PRINCIPAL STRESS ALONG AXIS OF TUNNEL	=	720.00000
RATIO OF HORIZONTAL STRESS TO VERTICAL STRESS	=	1.00000
AVERAGE SPACING OF JOINTS IN ROCK MASS (M)	=	0.05000
RADIUS OF TUNNEL IN METRE	=	2.40000
RADIUS OF BROKEN ZONE IN METRE	=	13.00000
COEFFICIENT OF VOLUMETRIC EXPANSION AFTER FAILURE	=	0.00430
MODULUS OF DEFORMATION OF ROCK MASS (T/SQ.M.)	=	90000.
UNIAXIAL COMPRESSIVE STRENGTH OF ROCK MATERIAL		
{ASSUMED AFTER SIZE CORRECTION FOR JOINT SPACING}	=	2650.000
MODULUS OF ELASTICITY OF ROCK MATERIAL (T/SQ.M.)	=	1080000.
NORMALIZED WALL SUPPORT PRESSURE	=	2.02555
NORMALIZED ROOF SUPPORT PRESSURE	=	2.45108
TUNNEL CLOSURE IN PER CENT	=	6.29105
ELASTO PLASTIC WALL SUPPORT PRESSURE (T/SQ.M.)	=	63.86304
ELASTO PLASTIC ROOF SUPPORT PRESSURE (T/SQ.M.)	=	77.27949
BARTON'S SHORT TERM SUPPORT PRESSURE IN ROOF		
AND WALL	=	31.52869
RADIUS OF COMPACTION ZONE WITHIN BROKEN ZONE (M)	=	4.78400
RADIAL DISPLACEMENT AT ELASTIC PLASTIC BOUNDARY (M)	=	0.09226
CRITICAL STRAIN OF ROCK MATERIAL	=	0.00245

HEAVING MAY OCCUR AT BOTTOM OF TUNNEL

MANERI STAGE I TUNNEL IN THE SQUEEZING GROUND
CONDITION IN CRUSHED QUARTZITE

HEIGHT OF OVERBURDEN IN METRE	=	350.0000
UNIT WEIGHT OF ROCK MASS (g/cc)	=	2.5000
ROCK MASS QUALITY Q	=	0.50000
JOINT ROUGHNESS NUMBER	=	1.00000
JOINT ALTERATION NUMBER	=	4.00000
PRINCIPAL STRESS ALONG AXIS OF TUNNEL	=	875.00000
RATIO OF HORIZONTAL STRESS TO VERTICAL STRESS	=	1.00000
AVERAGE SPACING OF JOINTS IN ROCK MASS (M)	=	0.12000
RADIUS OF TUNNEL IN METRE	=	2.90000
RADIUS OF BROKEN ZONE IN METRE	=	19.00000
COEFFICIENT OF VOLUMETRIC EXPANSION AFTER FAILURE	=	0.00370
MODULUS OF DEFORMATION OF ROCK MASS (T/SQ.M.)	=	84000.
UNIAXIAL COMPRESSIVE STRENGTH OF ROCK MATERIAL		
{ASSUMED AFTER SIZE CORRECTION FOR JOINT SPACING}	=	4196.892
MODULUS OF ELASTICITY OF ROCK MATERIAL (T/SQ.M.)	=	2800000.
NORMALIZED WALL SUPPORT PRESSURE	=	4.20570
NORMALIZED ROOF SUPPORT PRESSURE	=	4.91492
TUNNEL CLOSURE IN PER CENT	=	8.08280
ELASTO PLASTIC WALL SUPPORT PRESSURE (T/SQ.M.)	=	115.74330
ELASTO PLASTIC ROOF SUPPORT PRESSURE (T/SQ.M.)	=	135.26140
BARTON'S SHORT TERM SUPPORT PRESSURE IN ROOF AND WALL	=	27.52056
RADIUS OF COMPACTION ZONE WITHIN BROKEN ZONE (M)	=	6.99200
RADIAL DISPLACEMENT AT ELASTIC PLASTIC BOUNDARY (M)	=	0.13323
CRITICAL STRAIN OF ROCK MATERIAL	=	0.00179

HEAVING MAY OCCUR AT BOTTOM OF TUNNEL

MANERI STAGE-II TUNNEL IN THE SQUEEZING GROUND
CONDITION IN METABASICS

HEIGHT OF OVERBURDEN IN METRE	=	480.0000
UNIT WEIGHT OF ROCK MASS (g/cc)	=	2.5000
ROCK MASS QUALITY Q	=	0.80000
JOINT ROUGHNESS NUMBER	=	1.20000
JOINT ALTERATION NUMBER	=	3.50000
PRINCIPAL STRESS ALONG AXIS OF TUNNEL	=	1200.000
RATIO OF HORIZONTAL STRESS TO VERTICAL STRESS	=	1.00000
AVERAGE SPACING OF JOINTS IN ROCK MASS (M)	=	0.12000
RADIUS OF TUNNEL IN METRE	=	1.30000
RADIUS OF BROKEN ZONE IN METRE	=	8.00000
COEFFICIENT OF VOLUMETRIC EXPANSION AFTER FAILURE	=	0.00120
MODULUS OF DEFORMATION OF ROCK MASS (T/SQ.M.)	=	308000.
UNIAXIAL COMPRESSIVE STRENGTH OF ROCK MATERIAL		
{ASSUMED AFTER SIZE CORRECTION FOR JOINT SPACING}	=	5036.270
MODULUS OF ELASTICITY OF ROCK MATERIAL (T/SQ.M.)	=	2200000.

NORMALIZED WALL SUPPORT PRESSURE	=	0.83412
NORMALIZED ROOF SUPPORT PRESSURE	=	1.61398
TUNNEL CLOSURE IN PERCENT	=	2.23722
ELASTO PLASTIC WALL SUPPORT PRESSURE (T/SQ.M.)	=	8.93332
ELASTO PLASTIC ROOF SUPPORT PRESSURE (T/SQ.M.)	=	17.28540
BARTON'S SHORT TERM SUPPORT PRESSURE IN		
ROOF AND WALL	=	10.70982
RADIUS OF COMPACTION ZONE WITHIN BROKEN ZONE (M)	=	2.94400
RADIAL DISPLACEMENT AT ELASTIC PLASTIC BOUNDARY (M)	=	0.03378
CRITICAL STRAIN OF ROCK MATERIAL	=	0.00273

HEAVING MAY OCCUR AT BOTTOM OF TUNNEL

MANERI STAGE II TUNNEL IN THE SQUEEZING GROUND
CONDITION IN SHEARED METABASICS

HEIGHT OF OVERBURDEN IN METRE	=	410.0000
UNIT WEIGHT OF ROCK MASS (g/cc)	=	2.5000
ROCK MASS QUALITY Q	=	0.18000
JOINT ROUGHNESS NUMBER	=	1.20000
JOINT ALTERATION NUMBER	=	3.50000
PRINCIPAL STRESS ALONG AXIS OF TUNNEL	=	1025.000
RATIO OF HORIZONTAL STRESS TO VERTICAL STRESS	=	1.00000
AVERAGE SPACING OF JOINTS IN ROCK MASS (M)	=	0.06000
RADIUS OF TUNNEL IN METRE	=	3.50000
RADIUS OF BROKEN ZONE IN METRE	=	23.00000
COEFFICIENT OF VOLUMETRIC EXPANSION AFTER FAILURE	=	0.00120
MODULUS OF DEFORMATION OF ROCK MASS (T/SQ.M.)	=	145000.
UNIAXIAL COMPRESSIVE STRENGTH OF ROCK MATERIAL		
{ASSUMED AFTER SIZE CORRECTION FOR JOINT SPACING}	=	5785.155
MODULUS OF ELASTICITY OF ROCK MATERIAL (T/SQ.M.)	=	2100000.
NORMALIZED WALL SUPPORT PRESSURE	=	1.45587
NORMALIZED ROOF SUPPORT PRESSURE	=	2.90212
TUNNEL CLOSURE IN PER CENT	=	2.56389
ELASTO PLASTIC WALL SUPPORT PRESSURE (T/SQ.M.) = 23.76520		
ELASTO PLASTIC ROOF SUPPORT PRESSURE (T/SQ.M.)	=	47.37341
BARTON'S SHORT TERM SUPPORT PRESSURE IN		
ROOF AND WAL	=	16.32370
RADIUS OF COMPACTION ZONE WITHIN BROKEN ZONE (M)	=	8.46400
RADIAL DISPLACEMENT AT ELASTIC PLASTIC BOUNDARY (M)	=	0.16231
CRITICAL STRAIN OF ROCK MATERIAL	=	0.00286

HEAVING MAY OCCUR AT BOTTOM OF TUNNEL

MANERI STAGE I TUNNEL IN THE SQUEEZING GROUND
CONDITION IN WET METABASICS

HEIGHT OF OVERBURDEN IN METRE	=	800.0000
UNIT WEIGHT OF ROCK MASS (g/cc)	=	2.5000
ROCK MASS QUALITY Q	=	2.30000
JOINT ROUGHNESS NUMBER	=	1.20000
JOINT ALTERATION NUMBER	=	3.50000
PRINCIPAL STRESS ALONG AXIS OF TUNNEL	=	2000.000

RATIO OF HORIZONTAL STRESS TO VERTICAL STRESS	=	1.00000
AVERAGE SPACING OF JOINTS IN ROCK MASS (M)	=	0.15000
RADIUS OF TUNNEL IN METRE	=	2.4000
RADIUS OF BROKEN ZONE IN METRE	=	17.000
COEFFICIENT OF VOLUMETRIC EXPANSION AFTER FAILURE	=	0.00370
MODULUS OF DEFORMATION OF ROCK MASS (T/SQ.M.)	=	500000.
UNIAXIAL COMPRESSIVE STRENGTH OF ROCK MATERIAL		
{ASSUMED AFTER SIZE CORRECTION FOR JOINT SPACING}	=	4816.449
MODULUS OF ELASTICITY OF ROCK MATERIAL (T/SQ.M.)	=	2200000.
NORMALIZED WALL SUPPORT PRESSURE	=	4.68538
NORMALIZED ROOF SUPPORT PRESSURE	=	5.48897
TUNNEL CLOSURE IN PER CENT	=	9.55346
ELASTO PLASTIC WALL SUPPORT PRESSURE (T/SQ.M.)	=	99.64501
ELASTO PLASTIC ROOF SUPPORT PRESSURE (T/SQ.M.)	=	116.73510
BARTON'S SHORT TERM SUPPORT PRESSURE IN		
ROOF AND WALL	=	21.26722
RADIUS OF COMPACTION ZONE WITHIN BROKEN ZONE (M)	=	6.25600
RADIAL DISPLACEMENT AT ELASTIC PLASTIC		
BOUNDARY (M)	=	0.06742
CRITICAL STRAIN OF ROCK MATERIAL	=	0.00273

HEAVING MAY OCCUR AT BOTTOM OF TUNNEL

NJPC TUNNEL UNDER HIGH OVERBURDEN IN GNEISS

HEIGHT OF OVERBURDEN IN METRES	=	1400.0000
UNIT WEIGHT OF ROCK MASS (g/cc)	=	2.7000
ROCK MASS QUALITY Q	=	30.00000
JOINT ROUGHNESS NUMBER	=	0.75000
JOINT ALTERATION NUMBER	=	1.00000
PRINCIPAL STRESS ALONG AXIS OF TUNNEL	=	3000.000
RATIO OF HORIZONTAL STRESS TO VERTICAL STRESS	=	1.00000
AVERAGE SPACING OF JOINTS IN ROCK MASS (M)	=	0.75000
RADIUS OF TUNNEL IN METRE	=	2.50000
RADIUS OF BROKEN ZONE IN METRE	=	7.50000
COEFFICIENT OF VOLUMETRIC EXPANSION AFTER FAILURE	=	0.00400
MODULUS OF DEFORMATION OF ROCK MASS (T/SQ.M.)	=	1200000.
UNIAXIAL COMPRESSIVE STRENGTH OF ROCK MATERIAL		
{ASSUMED AFTER SIZE CORRECTION FOR JOINT SPACING}	=	1570.889
MODULUS OF ELASTICITY OF ROCK MATERIAL (T/SQ.M.)	=	1500000.

ROCK BURST MAY TAKE PLACE

REFERENCES

Barton, N. 1995. The influence of joint properties in modelling jointed rock masses. *Eight Int. Rock Mech. Congress*, Tokyo, Vol. 3, pp. 1023–1032.

Bhasin, R. and Grimstad, E. 1996. The use of stress-strength relationships in the assessment of tunnel stability. *Proc. Conf. on Recent Advances in Tunnelling Technology*, New Delhi, Vol. 1, pp. 183–196.

Daemen, J.J.K. 1975. *Tunnel support loading caused by rock failure.* Ph.D. Thesis, University of Minnesota, Minneapolis, U.S.A.

Dube, A.K., Singh, B. and Singh, Bhawani 1986. Study of squeezing pressure phenomenon in a tunnel – Part I and II. *Tunnelling Underground Space Technol.* 1(1): 35–39 (Part I) and pp. 41–48 (Part II), U.S.A.

Goel, R.K. 2000. Tunnelling in squeezing ground conditions. *Italian Geotech. J.* Anno XXXIV 1: 35–40.

Grimstad, E. and Barton, N. 1993. Updating of the Q-system for NMT. *Int. Symp. on Sprayed Concrete – Modern Use of Wet Mix Sprayed Concrete for Underground Support,* Fagernes (Editors Kompen, Opsahll and Berg. Norwegian Concrete Association, Oslo).

Grimstad, E. and Bhasin, R. 1996. Stress strength relationship and stability in hard rock. *Proc. Conf. Recent Advances in Tunnelling Technology,* New Delhi, India, Vol. 1, pp. 3–8.

Jethwa, J.L. 1981. *Evaluation of rock pressures in tunnels through squeezing ground in Lower Himalayas.* Ph.D. Thesis, Department of Civil Engineering, University of Roorkee, India, p. 272.

Pundhir, G.S., Acharya, A.K. and Chadha, A.K. 2000. Tunnelling through rock cover of more than 1000 m – a case study. *Int. Conf. Tunnelling Asia – 2000,* edited by S.P. Kaushish and T. Ramamurthy. New Delhi, India, pp. 235–240.

Singh, Bhawani, Goel, R.K., Mehrotra, V.K., Garg, S.K. and Allu, M.R. 1998. Effect of intermediate principal stress on strength of anisotropic rock mass. *J. Tunnelling Underground Space Technol.* 13(1): 71–79.

Singh, Bhawani, Jethwa, J.L., Dube, A.K. and Singh, B. 1992. Correlation between observed support pressure and rock mass quality. *Int. J. Tunnelling & Underground Space Technol.* 7(1): 59–74.

Singh, Bhawani, Viladkar, M.N., Samadhiya, N.K. and Mehrotra, V.K. 1997. Rock mass strength parameters mobilised in tunnels. *J. Tunnelling Underground Space Technol.* 12(1): 47–54.

Terzaghi, K. 1946. Rock Defects and Load on Tunnel Supports. Introduction to Rock Tunnelling with Steel Supports by Proctor, R.V. and White, T.L., Commercial Shearing and Stamping Company, Youngstown, Ohio, USA.

CHAPTER 26

Empirical design of support system for underground opening in rock mass – TM

"Out of sight is out of mind in underground space use"
Anonymous

26.1 GENERAL

The aim of supporting an underground opening is to reinforce the rock mass so that it can act as an inherently stable and robust structural system to support unstable zones in the rock mass. This objective may be achieved by constructing a reinforced rock arch, i.e., an array of (perfo/resin) rock anchors in both the roof and the side walls, according to the overall ground conditions. Welded mesh should also be used to provide local stability to rock blocks hanging in between the rock anchors.

Steel fibre reinforced shotcrete (SFRS) should be provided where squeezing ground conditions are likely to be encountered. However, in highly squeezing ground conditions (H > 350 $Q^{1/3}$m and $J_r/J_a \ll 0.5$; H is the overburden in metres and Q is Barton's rock mass quality), steel ribs with struts should be installed when the shotcrete lining begins to fail again and again, despite the addition of extra layers of shotcrete. Another advantage of steel ribs is that excavation by forepoling is easily done by pushing iron bars into the tunnel and welding their opposite ends to the ribs. The floor heaving problem in highly squeezing grounds can easily be solved by bolting the floor. Cutting the floor to maintain proper level is of no use, since heaving will redevelop. It would be wiser, however, to realign the tunnel to pass through safer zones (H < 350 $Q^{1/3}$m). Alternatively, a larger tunnel may be divided into smaller tunnels in the squeezing ground to reduce construction problems.

In the case of openings within a water-charged rock mass, steel ribs may be used to support the rock mass after it has been reinforced by Swellex rock bolts. Continuous shotcrete should not be used because it will act as a barrier to underground water. Generally, when shotcrete is applied to rock masses with well defined water bearing joints, it is important to provide drainage through the shotcrete layer in order to relieve high water pressure. Drain holes, fitted with plastic pipes are commonly used for this purpose. Water-charged rock masses may also be grouted to form a grouted arch with drainage holes.

Advantages of an integrated NATM approach to support rock masses are: a faster rate of tunnelling; flexibility with regard to construction and general strategy for a wide variety of ground conditions and weak zones; built-in stability; and greater economy. Furthermore, this type of support system can always be strengthened easily by adding more rock anchors and shotcrete layers as and when required.

In the case of pressure tunnels, however, a concrete lining may also be constructed as a permanent support and for ease of hydraulic flow. In the majority of cases, hoop reinforcement is not required (Singh et al., 1988). Shotcrete of an adequate thickness may serve the same purpose as a concrete lining, provided the tunnel diameter is increased for the same discharge to take into account the roughness of the shotcrete lining. Hoop reinforcement or a steel liner is needed in cases where the rock cover does not balance the internal water pressure (Chapter 27).

In potentially unstable portals, horizontal rock anchors of equal length should be provided inside the cut slope of the portal so that it acts as a reinforced-rock-breast wall (Section 2.8). The rock cover should be at least equal to the width of the tunnel. The aim, in a nutshell, is to construct an inherently stable and robust structural system to support a wide variety of ground conditions and weak zones, keeping in mind basic tunnel mechanics and the inherent uncertainties in the exploration, testing and behaviour of geological materials.

It has been difficult to convince Asian engineers to adopt modern methods of tunnel support. Failure of the rock bolt support system in the caverns of the Sardar Sarovar (Narmada) Project (Gujrat, India) and the tunnel of the Chamera Dam Project (Himachal Pradesh, India) have also undermined the confidence of engineers. Their hesitation is also due, in part, to confusion about the function of rock bolts. To counter the fear of trying new methods of tunnel support and to remove doubts from the mind of field engineers, a simple semi-empirical method for design of the total support system (shotcrete + rock bolt + rib + grout-arch) has been developed.

26.2 SOFTWARE TM

The advantage of steel fibre-reinforced shotcrete (SFRS) is that a thinner layer of shotcrete needs to be applied, in comparison to conventional shotcrete. Furthermore, fibre-reinforced shotcrete is especially necessary in poor rock conditions where support pressure is high.

The use of fibre-reinforced shotcrete together with resin anchors is also recommended for controlling rock burst conditions because of the high fracture toughness which especially long steel fibres provide (NGI, 1993). It appears that this type of system may also be successful in highly squeezing ground conditions, such as those encountered in the lower Himalaya. Therefore, a simple semi-empirical theory is proposed by Singh et al. (1995) to illustrate how rock bolts and shot-crete/SFRS resist the support pressures. The correlations for various types of support system have been deduced from extensive Tables and Charts of NGI. Thakur (1995) evaluated critically this semi-empirical method on the basis of over 100 case histories and found it satisfactory. Park, Kim and Lee (1997) used this design method for four food storage caverns in Korea. Samadhiya (1998) has verified the semi-empirical theory for shotcrete support by three-dimensional stress analysis of cavern with a shear zone in the Sardar Sarovar Project. The Bureau of Indian Standards (BIS) has adopted this design method for its code on Tunnelling in Rocks.

As such, the software TM is developed on the basis of the above semi-empirical method for tunnels and large caverns. It also takes into account the adverse effect of shear zones on the stability of a support system (Grimstad and Barton, 1993; Bhasin et al., 1995). The program TM has been used in three case histories. It gives realistic designs of support systems even in complex geological conditions. This program has also been used successfully for the design of rock bolt and shotcrete system at Ganwi mini hydel project, H. P., India.

The length of anchors/rock bolts in the wall of cavern is designed to prevent the buckling of reinforced rock wall column. This length depends also upon the depth of damage (d) due to blasting (Fig. 26.1). Pusch and Stranfors (1992) found the following correlation between d and weight of charge W (kg/m):

$$d = 1.94 \, W^{1.23} \text{ metres} \tag{26.1}$$

The output of TM is (i) the optimum angle of roof arch in cavern, (ii) thickness of shotcrete/SFRS and (iii) design of rock bolt system for both roof and the walls. The thickness of shotcrete (t_{sc}) may be estimated rather easily from the following equation for arched openings,

$$t_{sc} = \frac{0.6 \, BP_{roof}}{2q_{sc}} \tag{26.2}$$

where

P_{roof} = ultimate support pressure on the shotcrete in the roof,
B   = width of the opening,
0.6B = distance between vertical planes of maximum shear stresses in the shotcrete in the arched roof,
q_{sc} = shear strength of shotcrete,
    = 3 MPa in conventional shotcrete, and
    = 5.5 MPa in steel fibre reinforced shotcrete (SFRS).

It needs to be realised that shotcrete lining of adequate thickness and quality is a long-term support system. This is true for rail tunnels also. It must be ensured that there is a good bond between shotcrete and rock surface. Tensile bending stresses are not found to occur even in the irregular shotcrete lining in the roof due to a good bond between shotcrete and rock mass in an arched roof opening. Rock bolts help in better bonding. Similarly, contact grouting is essential behind the concrete lining to develop a good bond between the lining and rock mass to arrest its bending.

Rock has EGO (Extra-ordinary Geological Occurrence) problems. As such, where cracks appear in the shotcrete lining, more layers of shotcrete should be sprayed. The opening should also be monitored with the help of borehole extensometers at such locations particularly in the squeezing ground. If necessary, expert tunnel engineers should be invited to identify and solve construction problems.

In the over-stressed brittle hard rocks, rock anchors should be installed to make the reinforced rock arch a ductile arch. Thus, mode of failure is designed to be ductile from the brittle failure. Hence, failure would be slow, giving enough time for local strengthening (or retrofitting) of the existing support system.

26.3   THE EXPERIENCE IN POOR ROCK CONDITIONS

Steel fibre-reinforced shotcrete (SFRS) has proved very successful in the 6.5 km-long tunnel for the Uri Hydel Project and desilting underground chambers of NJPC in Himalaya. The main advantage is that a smaller thickness of steel fibre-reinforced shotcrete is needed. No welded mesh is required to reinforce the shotcrete. Provided that the shotcrete is graded and sprayed properly, there is less rebound, thanks to the steel fibres. This method is now economical, safer and faster than the conventional shotcrete. Controlled blasting technique is adopted to excavate the tunnel where SFRS is to be used. Furthermore, the selection of right ingredients and tight quality control over application are a key to the success of SFRS.

Experience with the use of mesh (welded mesh, etc.) has been unsatisfactory when there were overbreaks in the tunnel after blasting. In these cases, soon after the welded mesh was spreaded between bolts and shotcrete, the mesh started rebounding the shotcrete and it could not penetrate inside the mesh and fill the gap between the mesh and the overbreak. Consequently, gaps were left above the shotcrete; the sound when a hammer was struck indicated the hollow areas above the mesh. Further, loosely fitted welded wire mesh vibrates as a result of blast vibrations, causing subsequent loosening of the shotcrete.

As the overall experience with mesh-reinforced shotcrete has been unsatisfactory in handling overbreak situations, it is recommended that mesh with shotcrete should not be used where uneven surface of tunnel is available due to overbreak. In such cases, the thickness of shotcrete should be increased by perhaps 1.0 cm.

26.4   INSTRUMENTATION

Listed below are the benefits of instrumentation of caverns in general and tunnels in squeezing and swelling grounds.

1. Verifying design of support system
2. Advancing the state-of-art
3. Checking adequacy of the new construction techniques
4. Controlling quality of construction
5. Reducing construction cost and extra payments
6. Diagnosing the cause of a problem
7. Improving construction safety by providing warning system
8. Documenting as-built conditions
9. Providing legal protection
10. Enhance public relations, bonding team spirit between engineers, geologists and contractors

The World Bank therefore makes an instrumentation mandatory in its projects in every country. Most of the research carried out in India is the result of a good field database, which was prepared from tunnel instrumentation by CMRI. It should be kept in mind that the psychology of civil engineers is that they resist every effort, which reduces the momentum of enthusiasm of construction.

The displacements are measured by multipoint borehole extensometers. Extra long rock anchors may have to be installed where the rate of displacement is not decreasing rapidly. The support pressures are determined by load cells and pressure cells. The tunnel closure is obtained by tape extensometers. Displacements across cracks in shotcrete and rock mass are monitored by 3D crack meter. Grouting of cracks may be done after movements across cracks have stabilised. It is also essential to monitor rate of seepage with the help of V notch at the end of tunnel. If seepage is observed to increase with time, there is every danger of failure and flooding of tunnel within water-charged rock mass (crushed quartzite/sandstone/hard rocks, dolomite, shear zones, faults, etc.). Sometimes wide faults (>10 m) are met during tunnelling. They require the attention of experts. In case of soft ground or soil like gouge within wide faults, the tunnel lining should be designed using design method of ITA (Duddeck and Erdmann, 1985).

26.5 DRAINAGE MEASURES

The drainage system should be fully designed before the construction of tunnel and cavern. NATM (New Austrian Tunnelling Method) is very popular these days. It specifies drainage measures also. For example, radial gaps are left unshotcreted for drainage of seepage in the case of hard rock mass which is charged with water.

Very often one may observe that the seepage of water is concentrated to only one or just a few, often tubular, openings in fissures and joints. It can be worthwhile to install temporary drainage pipes in such areas before applying the shotcrete. These pipes can be plugged when the shotcrete has gained sufficient strength. Further Swellex (inflated tubular) bolts are preferred in water-charged rock masses. Cement grouted bolts are not feasible here as grout will be washed out. Resin grout may not also be reliable. It may be mentioned that the seals used in the concrete lining for preventing seepage in the road/rail tunnels may not withstand heavy water pressure.

The pressure tunnels are grouted generally alround its periphery so that the ring of grouted rock mass is able to withstand heavy ground water pressure. Polyurethane should be used as grout in rock joints under water as it swells 26 times and cements the rock mass.

26.6   SPECIAL SITUATIONS

1. The first layer of 25 mm thick shotcrete is sprayed immediately to make the opening reasonably smooth. Then welded mesh is spread and nailed into this shotcrete layer. Next rock anchors (or rock bolts) are provided. Finally, additional layers of shotcrete are sprayed according to the design. A perfect bond between shotcrete and rock mass is the trick of the trade. A strong bond results in more effective thickness of shotcrete depending on the size of rock blocks.

2. Sometimes the grouting of long bolts is not done satisfactorily because of lack of supervision and difficulties with the expanding agent (aluminium powder or expanding agent is seldom added). Therefore, pull-out tests should be conducted on at least 5 percent of the bolts to check their quality. If required, extra bolting should be done to strengthen the support system. (Pretensioned bolts must be tensioned again after initial round of blasting to excavate the tunnel face. Then they are grouted with cement mortar for long life.)

3. For deep and long tunnels in complex geological conditions, a 20 m-long probe hole should be drilled inside the tunnel face in order to obtain an accurate picture of geological conditions in advance of the tunnelling. The probe hole will also dissipate seepage pressure slowly within the water-charged rock mass, which is likely to be punctured during tunnelling. This technique will also avoid flash floods soon after blasting and consequent loss of life and damage to the support system, provided the drain is of adequate capacity.

4. In poor rock masses, spiling bolts (inclined towards the tunnel face) should be installed before blasting to increase the stand-up time of the tunnel roof (Bischoff et al., 1992). Shotcrete is then sprayed on the roof, and then the spiling bolts are installed. In the final cycle, roof bolts are installed.

5. In cases involving argillaceous rocks and swelling rocks, where the bond with shotcrete is poor, the thickness of the shotcrete should be increased by about 30 percent.

6. At the intersection of the approach tunnel and the main tunnel, the support capacity of the reinforced rock arch of the approach tunnel should be strengthened by 100 percent, up to a distance equal to say three times the bolt length of the main tunnel. This will help the reinforced rock arch of the approach tunnel to bear the thrust from the reinforced arch of the main tunnel.

7. In highly squeezing ground conditions (H $\gg$ 350 Q$^{1/3}$ m and J$_r$/J$_a$ < 0.5), steel ribs with struts should be used when the shotcrete fails repeatedly despite the addition of more layers. With steel ribs, excavation by forepoling is easily accomplished by pushing steel rods into the tunnel face and welding the opposite ends to the ribs. Floor heaving can be prevented by rock bolting of the floor. Some delay, but less than the stand-up time, is necessary to release the strain energy of the broken zone. Smaller blast holes will also be helpful. Broken zones should be instrumented. In such cases, the tunnel closure must be arrested before it reaches 4 percent of the width of the tunnel.

8. For very poor rock masses, steel ribs should be installed and embedded in the shotcrete to withstand the high support pressures.

9. In the case of steel ribs in large tunnels and caverns, haunches should be strengthened by installing more anchors to help withstand the heavy thrust due to the ribs.

10. For treatment of shear zones, crossed rock anchors/bolts should be provided across the shear zones. After the gouge has been cleaned to the desired extent, anchors are connected to the welded mesh and, finally, dental shotcrete is backfilled. In wide shear zones, shotcrete reinforcement is also placed to help withstand high support pressure. The anchors should be inclined according to the dip of the shear zone to stop squeezing of the gouge, and thereby stabilize the deformations (Fig. 2.4).

11. In the case of an unstable portal, horizontal anchors of equal length should be provided inside the cut slope, so that it acts as a reinforced rock breast wall (Fig. 2.5).

12. In rockburst-prone regions, resin anchors and steel fibre-reinforced shotcrete should be used to increase the ductility of the support system and thus convert the brittle mode of failure into the ductile mode of failure. Rock anchors should be installed immediately on the rock burst side of tunnel.

13. The concrete lining for water/pressure tunnels should be laid far away from the tunnel face within the squeezing ground where the broken zone is stabilized (i.e., about four times the radius of the broken zone). In addition, the concrete lining should be segmented within an active thrust zone to allow relative movement along the faults/thrust.

26.7 CONCLUDING REMARKS

The following conclusions are offered on the basis of modelling and design experiences of support systems in poor rock masses.

1. In a poor rock mass, the support pressure resisted by the rock bolts is small in comparison to that taken by shotcrete, which is generally the main element of the long-term support system for resisting heavy support pressure in tunnels.

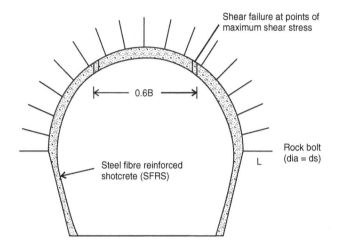

(a) Reinforced rock arch

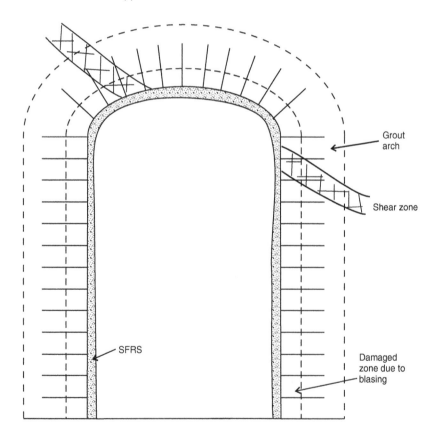

(b) Reinforced rock frame

Figure 26.1. Design of support system for underground openings.

2. The proposed semi-empirical method is based on an integrated approach in the design of shotcrete, rock bolts/anchors, steel ribs and grouted arch, taking into account both seepage pressure and support pressure. The mobilization factors for each member have been derived from NGI tables and charts of support systems for both shotcrete and fibre-reinforced shotcrete.

3. The data analysis suggests that untensioned full-column grouted bolts are likely to be more effective than the pretensioned bolts in supporting poor rock masses (Sections 2.4 and 2.7)

26.8   USER'S MANUAL – TM

PLEASE TYPE THE FOLLOWING PARAMETERS IN THE CASE OF NO SHEAR ZONE (SEE FIG. 26.1)

1. TITLE (<80 CHARACTERS)

2. SPAN OF UNDERGROUND OPENING (M)
   SLIDING ANGLE OF FRICTION ALONG JOINTS = ATAN(Jr/Ja) (DEG.)
   AVERAGE SPACING OF JOINTS APPROXIMATELY (M)
   SEEPAGE PRESSURE IN ROOF (T/SQ.M)
   MINIMUM UNIAXIAL COMPRESSIVE STRENGTH OF GROUTED ROCK MASS (T/SQ.M)
   SHEAR STRENGTH OF SHOTCRETE/FIBRE REINFORCED SHOTCRETE (T/SQ.M)
   (SAY 300 FOR SHOTCRETE AND 600 FOR FIBRE REINFORCED SHOTCRETE)
   ESTIMATED ULTIMATE SUPPORT PRESSURE IN ROOF (T/SQ.M)
   {Increase support pressure by 25 percent for seismic region}

3. LENGTH OF ROCK BOLT/ANCHOR IN ROOF (L) (M)
   FIXED ANCHOR LENGTH (100Ds/ACTUAL < L AND < 1 m FOR PRE-TENSIONED BOLT)
   SPACING OF BOLT/ANCHOR (<L/2 AND 2.25 M) (M)
   {SQUARE ROOT OF AREA OF ROCK MASS SUPPORTED BY ONE ROCK BOLT}
   BOLT/ANCHOR CAPACITY (REDUCE PROPORTIONATELY IF L < ACTUAL FAL) (T)
   CAPACITY OF STEEL RIB (T)
   SPACING OF STEEL RIBS (T)
   THICKNESS OF GROUTED ROCK MASS (M)
   {Please omit this line of data if span of opening <10 m}

4. HEIGHT OF WALL OF CAVERN (M)
   MODULUS OF DEFORMATION OF ROCK MASS (T/SQ.M)
   AVERAGE VERTICAL STRESS ABOVE HAUNCH ALONG BOLT (T/SQ.M)
   SEEPAGE PRESSURE IN WALL (T/SQ.M)
   ESTIMATED ULTIMATE WALL SUPPORT PRESSURE (T/SQ.M)
   SHEAR STRENGTH OF GROUTED ROCK MASS
   THICKNESS OF GROUTED ROCK MASS IN WALL (M)
   DEPTH OF DAMAGE TO ROCK MASS DUE TO BLASTING (1–3 M)

5. SAFE BEARING PRESSURE OF ROCK MASS IN WEAK ZONE (see Section 23.9)

6. TYPE NN EQUAL TO 1 FOR REDESIGN AT SAME LOCATION, 2 FOR REDESIGN OF WALL SUPPORT AT SAME LOCATION, 0 FOR STOP, –1 FOR NEW LOCATION AND –2 FOR AREA OF SHEAR ZONE
   {RESTART INPUT DATA FROM THE BEGINNING FOR NN = –1 OR PARA 3 FOR NN = 1 AND 2

OR
PLEASE TYPE THE FOLLOWING PARAMETERS IN CASE OF LOCATION NEAR THE
SHEAR ZONE WITH NN = −2}

TITLE

ROCK MASS QUALITY IN SHEAR ZONE
JOINT ROUGHNESS NUMBER IN SHEAR ZONE
JOINT ALTERATION NUMBER IN SHEAR ZONE
MODULUS OF DEFORMATION OF SHEAR ZONE (T/SQ.M)
WIDTH OF SHEAR ZONE (M)
STRIKE DIRECTION OF SHEAR ZONE (DEG.)
{WITH RESPECT TO AXIS OF TUNNEL/ CAVERN}

ROCK MASS QUALITY IN SURROUNDING ROCK MASS
JOINT ROUGHNESS NUMBER IN SURROUNDING ROCK MASS
JOINT ALTERATION NUMBER IN SURROUNDING ROCK
MODULUS OF DEFORMATION OF ROCK MASS (T/SQ.M)
HEIGHT OF OVERBURDEN OF ROCK MASS (M)
{PLEASE CONTINUE FROM PARA 2}

(a) Input File: TM.DAT

DESIGN OF SUPPORT SYSTEM FOR CAVERN OF TEHRI DAM PROJECT
24. 53. 0.5 0. 0. 300. 10.6
8. 3. 1. 20. 0. 0. 0.
30. 860000. 1800. 0. 1.5 0. 0. 3.
400.
2
8. 3. 3. 20. 0. 0. 0.
30. 860000. 1800. 0. 1.5 0. 0. 3.
400.
−2

DESIGN OF SUPPORT SYSTEM FOR CAVERN IN AREA OF SHEAR ZONE (SP), TEHRI
DAM (HPP)
2. 1. 2. 380000. 4. 85.
9. 1. 0.75 860000. 370.
24. 30. 0.35 0. 0. 300. 15.
8. 4. 1. 20. 0. 0. 0.
30. 470000. 1800. 0. 4.5 0. 0. 3.
400.
2
8. 4. 3. 20. 0. 0. 0.
30. 470000. 1800. 0. 4.5 0. 0. 3.
400.
−2

DESIGN OF SUPPORT SYSTEM FOR CAVERN IN AREA OF SHEAR ZONE,
TEHRI DAM (HPP)
2. 1. 2. 140000. 1.5 85.
9. 1. 0.75 860000. 370.
24. 34. 0.35 0. 0. 300. 14.
8. 4. 1. 20. 0. 0. 0.

30. 430000. 1800. 0. 4.2 0. 0. 3.
400.
2
8. 4. 3. 20. 0. 0. 0.
30. 430000. 1800. 0. 4.2 0. 0. 3.
400.
−1

DESIGN OF SUPPORT SYSTEM FOR CAVERN OF SARDAR SAROVAR PROJECT (SSP)
23. 57. 0.8 0. 0. 300. 8.8
6. 2. 1.75 32. 0. 0. 0.
52. 750000. 250. 0. 1. 0. 0. 3.
200.
−2

DESIGN OF SUPPORT SYSTEM FOR CAVERN IN AREA OF SHEAR ZONE (SSP)
1.25 1. 4. 100000. 2. 50.
9.16 1.5 0.75 750000. 70.
23. 22. 0.35 0. 0. 300. 9.8
6. 2. 1.25 32. 0. 0. 0.
52. 320000. 250. 0. 2. 0. 0. 3.
200.
−1

DESIGN OF SUPPORT SYSTEM FOR CAVITY OF UNDERGROUND POWERHOUSE,
GANWI PROJECT
14.2 35. 0.5 0. 0. 300. 13.
7. 1.5 1.5 18. 0. 0. 0.
20. 1000000. 1000. 0. 1.5 0. 0. 2.
200.
2
7. 1.5 1.75 18. 0. 0. 0.
20. 1000000. 1000. 0. 1.5 0. 0. 2.
200.
0

(b) Output File: TM.OUT

DESIGN OF SUPPORT SYSTEM FOR CAVERN OF TEHRI DAM PROJECT

SPAN OF UNDERGROUND OPENING (M)	=	24.000
SLIDING ANGLE OF FRICTION ALONG JOINTS (DEG.)	=	53.000
AVERAGE SPACING OF JOINTS (M)	=	0.500
AVERAGE VALUE OF SEEPAGE PRESSURE IN ROOF (T/SQ.M)	=	0.000
COMPRESSIVE STRENGTH OF GROUTED ROCK (T/SQ.M)	=	0.000
SHEAR STRENGTH OF SHOTCRETE (T/SQ.M)	=	300.00
ESTIMATED ULTIMATE ROOF SUPPORT PRESSURE (T/SQ.M)	=	10.600
LENGTH OF ROCK BOLT/ANCHOR IN ROOF (M)	=	8.000
FIXED ANCHOR LENGTH (M)	=	3.000
SPACING OF BOLT/ANCHOR IN ROOF (M)	=	1.000
BOLT/ANCHOR CAPACITY (T)	=	20.000

CAPACITY OF STEEL RIB (T)	=	0.000
SPACING OF STEEL RIBS (M)	=	0.000
THICKNESS OF GROUTED ROCK MASS (M)	=	0.000
CAPACITY OF SHOTCRETE LINING IN ROOF (T/SQ.M)	=	0.000
CAPACITY OF REINFORCED ROCK ARCH (T/SQ.M)	=	18.167
CAPACITY OF STEEL RIBS (T/SQ.M)	=	0.000
CAPACITY OF GROUTED ROCK ARCH (T/SQ.M)	=	0.000
THICKNESS OF SHOTCRETE IN ROOF (M)	=	0.026
RECOMMENDED ANGLE OF ARCH AT CENTRE		
(APPROXIMATE)	=	103.
HEIGHT OF WALL OF CAVERN (M)	=	30.000
MODULUS OF DEFORMATION OF ROCK MASS (T/SQ.M)	=	860000.
AVERAGE VERTICAL STRESS ABOVE HAUNCH (T/SQ.M)	=	1800.00
AVERAGE VALUE OF SEEPAGE PRESSURE IN WALL (T/SQ.M)	=	0.000
ESTIMATED WALL ULTIMATE SUPPORT PRESSURE (T/SQ.M)	=	1.500
SHEAR STRENGTH OF GROUTED ROCK (T/SQ.M)	=	0.000
THICKNESS OF GROUTED ROCK MASS IN WALL (M)	=	0.000
DEPTH OF DAMAGE TO ROCK MASS DUE TO		
BLASTING (M)	=	3.000
SUGGESTED LENGTH OF BOLT/ANCHOR FOR WALL (M)	=	8.000
SPACING OF BOLTS/ANCHORS FOR WALL (M)	=	1.000
THICKNESS OF SHOTCRETE IN WALL (M)	=	0.044

*   THE REINFORCED ROCK WALL COLUMN MAY NOT BUCKLE, SO THE SPACING OF ROCK BOLTS/ANCHORS MAY BE INCREASED AND THICKNESS OF SHOT-CRETE/ SFRS REDUCED IN WALLS ON THE BASIS OF PAST EXPERIENCE AND UNDERGROUND WEDGE ANALYSIS.

SAFE BEARING PRESSURE OF ROCK MASS (T/SQ.M)	=	400.00
SIDE OF BASE PLATE OF BOLT/ANCHOR (M)	=	0.22361
LENGTH OF ROCKBOLT/ANCHOR IN ROOF (M)	=	8.000
FIXED ANCHOR LENGTH (M)	=	3.000
SPACING OF BOLT/ANCHOR IN ROOF (M)	=	3.000
BOLT/ANCHOR CAPACITY (T)	=	20.000
CAPACITY OF STEEL RIB (T)	=	0.000
SPACING OF STEEL RIBS (M)	=	0.000
THICKNESS OF GROUTED ROCK MASS (M)	=	0.000

HEIGHT OF WALL OF CAVERN (M)	=	30.000
MODULUS OF DEFORMATION OF ROCK MASS (T/SQ.M)	=	860000.
AVERAGE VERTICAL STRESS ABOVE HAUNCH (T/SQ.M)	=	1800.000
AVERAGE VALUE OF SEEPAGE PRESSURE IN WALL (T/SQ.M)	=	0.000
ESTIMATED WALL ULTIMATE SUPPORT PRESSURE (T/SQ.M)	=	1.500
SHEAR STRENGTH OF GROUTED ROCK (T/SQ.M)	=	0.000
THICKNESS OF GROUTED ROCKMASS IN WALL (M)	=	0.000
DEPTH OF DAMAGE TO ROCK MASS DUE TO BLASTING (M)	=	3.000
SUGGESTED LENGTH OF BOLT/ANCHOR FOR WALL (M)	=	8.000
SPACING OF BOLTS/ANCHORS FOR WALL (M)	=	3.000
THICKNESS OF SHOTCRETE IN WALL (M)	=	0.044

*   THE REINFORCED ROCK WALL COLUMN MAY NOT BUCKLE, SO THE SPACING OF ROCK BOLTS/ANCHORS MAY BE INCREASED AND THICKNESS OF SHOTCRETE/

SFRS REDUCED IN WALLS ON THE BASIS OF PAST EXPERIENCE AND UNDER-
GROUND WEDGE ANALYSIS.

SAFE BEARING PRESSURE OF ROCK MASS (T/SQ.M)	=	400.00
SIDE OF BASE PLATE OF BOLT/ANCHOR (M)	=	0.22361

DESIGN OF SUPPORT SYSTEM FOR CAVERN IN AREA OF SHEAR ZONE (SP), TEHRI
DAM (HPP)

ROCK MASS QUALITY IN SHEAR ZONE	=	2.00000
JOINT ROUGHNESS NUMBER IN SHEAR ZONE	=	1.00000
JOINT ALTERATION NUMBER IN SHEAR ZONE	=	2.00000
MODULUS OF DEFORMATION OF SHEAR ZONE (T/SQ.M)	=	380000.0
WIDTH OF SHEAR ZONE (M)	=	4.00000
STRIKE DIRECTION OF SHEAR ZONE (DEG.)	=	85.00000
{WITH RESPECT TO AXIS OF TUNNEL/CAVERN}		

ROCK MASS QUALITY IN SURROUNDING ROCK MASS	=	9.00000
JOINT ROUGHNESS NUMBER IN SURROUNDING		
ROCK MASS	=	1.00000
JOINT ALTERATION NUMBER IN SURROUNDING ROCK	=	0.75000
MODULUS OF DEFORMATION OF ROCK MASS (T/SQ.M)	=	860000.0
HEIGHT OF OVERBURDEN OF ROCK MASS (M)	=	370.000

MEAN VALUES IN AREA INTERSECTED BY SHEAR OR WEAK ZONE

ROCK MASS QUALITY QM	=	2.70192
JOINT ROUGHNESS NUMBER Jrm	=	1.00000
JOINT ALTERATION NUMBER Jam	=	1.75000
MODULUS OF DEFORMATION EM (T/SQ.M)	=	476000.00
SUPPORT PRESSURE IN ROOF (T/SQ.M)	=	15.26202
SLIDING ANGLE OF FRICTION OF JOINTS (DEG.)	=	29.74481
{PLEASE USE THE ABOVE MEAN VALUES IN DESIGN IN AREA OF		
SHEAR ZONE OR WEAK ZONE IN THE NEXT ROUND}		
SUPPORT PRESSURE IN SURROUNDING ROCK (T/SQ.M)	=	10.22342

IT IS A NON-SQUEEZING GROUND

SPAN OF UNDERGROUND OPENING (M)	=	24.000
SLIDING ANGLE OF FRICTION ALONG JOINTS (DEG.)	=	30.000
AVERAGE SPACING OF JOINTS (M)	=	0.350
AVERAGE VALUE OF SEEPAGE PRESSURE IN ROOF (T/SQ.M)	=	0.000
COMPRESSIVE STRENGTH OF GROUTED ROCK (T/SQ.M)	=	0.000
SHEAR STRENGTH OF SHOTCRETE (T/SQ.M)	=	300.000
ESTIMATED ULTIMATE ROOF SUPPORT PRESSURE (T/SQ.M)	=	15.000
LENGTH OF ROCKBOLT/ANCHOR IN ROOF (M)	=	8.000
FIXED ANCHOR LENGTH (M)	=	4.000
SPACING OF BOLT/ANCHOR IN ROOF (M)	=	1.000
BOLT/ANCHOR CAPACITY (T)	=	20.000
CAPACITY OF STEEL RIB (T)	=	0.000
SPACING OF STEEL RIBS (M)	=	0.000
THICKNESS OF GROUTED ROCK MASS (M)	=	0.000

CAPACITY OF SHOTCRETE LINING IN ROOF (T/SQ.M)	=	8.634
CAPACITY OF REINFORCED ROCK ARCH (T/SQ.M)	=	6.366
CAPACITY OF STEEL RIBS (T/SQ.M)	=	0.000

CAPACITY OF GROUTED ROCK ARCH (T/SQ.M)   =   0.000
THICKNESS OF SHOTCRETE IN ROOF (M)   =   0.220
RECOMMENDED ANGLE OF ARCH AT CENTRE (APPROXIMATE)   =   103.

HEIGHT OF WALL OF CAVERN (M)   =   30.000
MODULUS OF DEFORMATION OF ROCK MASS (T/SQ.M)   =   470000.
AVERAGE VERTICAL STRESS ABOVE HAUNCH (T/SQ.M)   =   1800.000
AVERAGE VALUE OF SEEPAGE PRESSURE IN WALL (T/SQ.M)   =   0.000
ESTIMATED WALL ULTIMATE SUPPORT PRESSURE (T/SQ.M)   =   4.500
SHEAR STRENGTH OF GROUTED ROCK (T/SQ.M)   =   0.000
THICKNESS OF GROUTED ROCKMASS IN WALL (M)   =   0.000
DEPTH OF DAMAGE TO ROCK MASS DUE TO BLASTING (M)   =   3.000
SUGGESTED LENGTH OF BOLT/ANCHOR FOR WALL (M)   =   8.000
SPACING OF BOLTS/ANCHORS FOR WALL (M)   =   1.000
THICKNESS OF SHOTCRETE IN WALL (M)   =   0.140

*   THE REINFORCED ROCK WALL COLUMN MAY NOT BUCKLE, SO THE SPACING OF
      ROCK BOLTS/ANCHORS MAY BE INCREASED AND THICKNESS OF SHOTCRETE/
      SFRS REDUCED IN WALLS ON THE BASIS OF PAST EXPERIENCE AND UNDER-
      GROUND WEDGE ANALYSIS.

SAFE BEARING PRESSURE OF ROCK MASS (T/SQ.M)   =   400.00000
SIDE OF BASE PLATE OF BOLT/ANCHOR (M)   =   0.22361

LENGTH OF ROCKBOLT/ANCHOR IN ROOF (M)   =   8.000
FIXED ANCHOR LENGTH (M)   =   4.000
SPACING OF BOLT/ANCHOR IN ROOF (M)   =   3.000
BOLT/ANCHOR CAPACITY (T)   =   20.000
CAPACITY OF STEEL RIB (T)   =   0.000
SPACING OF STEEL RIBS (M)   =   0.000
THICKNESS OF GROUTED ROCK MASS (M)   =   0.000

HEIGHT OF WALL OF CAVERN (M)   =   30.000
MODULUS OF DEFORMATION OF ROCK MASS (T/SQ.M)   =   470000.
AVERAGE VERTICAL STRESS ABOVE HAUNCH (T/SQ.M)   =   1800.000
AVERAGE VALUE OF SEEPAGE PRESSURE IN WALL (T/SQ.M)   =   0.000
ESTIMATED WALL ULTIMATE SUPPORT PRESSURE (T/SQ.M)   =   4.500
SHEAR STRENGTH OF GROUTED ROCK (T/SQ.M)   =   0.000
THICKNESS OF GROUTED ROCK MASS IN WALL (M)   =   0.000
DEPTH OF DAMAGE TO ROCK MASS DUE TO BLASTING (M)   =   3.000
SUGGESTED LENGTH OF BOLT/ANCHOR FOR WALL (M)   =   8.000
SPACING OF BOLTS/ANCHORS FOR WALL (M)   =   3.000
THICKNESS OF SHOTCRETE IN WALL (M)   =   0.140

*   THE REINFORCED ROCK WALL COLUMN MAY NOT BUCKLE, SO THE SPACING OF
      ROCK BOLTS/ANCHORS MAY BE INCREASED AND THICKNESS OF SHOTCRETE/
      SFRS REDUCED IN WALLS ON THE BASIS OF PAST EXPERIENCE AND UNDER-
      GROUND WEDGE ANALYSIS.

SAFE BEARING PRESSURE OF ROCK MASS (T/SQ.M)   =   400.000
SIDE OF BASE PLATE OF BOLT/ANCHOR (M)   =   0.22361

DESIGN OF SUPPORT SYSTEM FOR CAVERN IN AREA OF SHEAR ZONE,
TEHRI DAM (HPP)

ROCK MASS QUALITY IN SHEAR ZONE	=	2.00000
JOINT ROUGHNESS NUMBER IN SHEAR ZONE	=	1.00000
JOINT ALTERATION NUMBER IN SHEAR ZONE	=	2.00000
MODULUS OF DEFORMATION OF SHEAR ZONE (T/SQ.M)	=	140000.0
WIDTH OF SHEAR ZONE (M)	=	1.50000
STRIKE DIRECTION OF SHEAR ZONE (DEG.)	=	85.00000
{WITH RESPECT TO AXIS OF TUNNEL/CAVERN}		
ROCK MASS QUALITY IN SURROUNDING ROCK MASS	=	9.00000
JOINT ROUGHNESS NUMBER IN SURROUNDING ROCK MASS	=	1.00000
JOINT ALTERATION NUMBER IN SURROUNDING ROCK	=	0.75000
MODULUS OF DEFORMATION OF ROCK MASS (T/SQ.M)	=	860000.0
HEIGHT OF OVERBURDEN OF ROCK MASS (M)	=	370.00000
MEAN VALUES IN AREA INTERSECTED BY SHEAR OR WEAK ZONE		
ROCK MASS QUALITY QM	=	3.65019
JOINT ROUGHNESS NUMBER Jrm	=	1.00000
JOINT ALTERATION NUMBER Jam	=	1.50000
MODULUS OF DEFORMATION EM (T/SQ.M)	=	428000.00
SUPPORT PRESSURE IN ROOF (T/SQ.M)	=	13.80728
SLIDING ANGLE OF FRICTION OF JOINTS (DEG.)	=	33.68999
{PLEASE USE THE ABOVE MEAN VALUES IN DESIGN IN AREA OF		
SHEAR ZONE OR WEAK ZONE IN THE NEXT ROUND}		
SUPPORT PRESSURE IN SURROUNDING ROCK (T/SQ.M)	=	10.22342
IT IS A NON-SQUEEZING GROUND		
SPAN OF UNDERGROUND OPENING (M)	=	24.000
SLIDING ANGLE OF FRICTION ALONG JOINTS (DEG.)	=	34.000
AVERAGE SPACING OF JOINTS (M)	=	0.350
AVERAGE VALUE OF SEEPAGE PRESSURE IN ROOF (T/SQ.M)	=	0.000
COMPRESSIVE STRENGTH OF GROUTED ROCK (T/SQ.M)	=	0.000
SHEAR STRENGTH OF SHOTCRETE (T/SQ.M)	=	300.000
ESTIMATED ULTIMATE ROOF SUPPORT PRESSURE (T/SQ.M)	=	14.000
LENGTH OF ROCK BOLT/ANCHOR IN ROOF (M)	=	8.000
FIXED ANCHOR LENGTH (M)	=	4.000
SPACING OF BOLT/ANCHOR IN ROOF (M)	=	1.000
BOLT/ANCHOR CAPACITY (T)	=	20.000
CAPACITY OF STEEL RIB (T)	=	0.000
SPACING OF STEEL RIBS (M)	=	0.000
THICKNESS OF GROUTED ROCK MASS (M)	=	0.000

CAPACITY OF SHOTCRETE LINING IN ROOF (T/SQ.M)	=	6.674
CAPACITY OF REINFORCED ROCK ARCH (T/SQ.M)	=	7.326
CAPACITY OF STEEL RIBS (T/SQ.M)	=	0.000
CAPACITY OF GROUTED ROCK ARCH (T/SQ.M)	=	0.000
THICKNESS OF SHOTCRETE IN ROOF (M)	=	0.168
RECOMMENDED ANGLE OF ARCH AT CENTRE (APPROXIMATE)	=	103.
HEIGHT OF WALL OF CAVERN (M)	=	30.000
MODULUS OF DEFORMATION OF ROCK MASS (T/SQ.M)	=	430000.
AVERAGE VERTICAL STRESS ABOVE HAUNCH (T/SQ.M)	=	1800.000
AVERAGE VALUE OF SEEPAGE PRESSURE IN WALL (T/SQ.M)	=	0.000
ESTIMATED WALL ULTIMATE SUPPORT PRESSURE (T/SQ.M)	=	4.200

SHEAR STRENGTH OF GROUTED ROCK (T/SQ.M)	=	0.000
THICKNESS OF GROUTED ROCKMASS IN WALL (M)	=	0.000
DEPTH OF DAMAGE TO ROCK MASS DUE TO BLASTING (M)	=	3.000
SUGGESTED LENGTH OF BOLT/ANCHOR FOR WALL (M)	=	8.000
SPACING OF BOLTS/ANCHORS FOR WALL (M)	=	1.000
THICKNESS OF SHOTCRETE IN WALL (M)	=	0.131

*   THE REINFORCED ROCK WALL COLUMN MAY NOT BUCKLE, SO THE SPACING OF ROCK BOLTS/ANCHORS MAY BE INCREASED AND THICKNESS OF SHOTCRETE/ SFRS REDUCED IN WALLS ON THE BASIS OF PAST EXPERIENCE AND UNDERGROUND WEDGE ANALYSIS.

SAFE BEARING PRESSURE OF ROCK MASS (T/SQ.M)	=	400.00000
SIDE OF BASE PLATE OF BOLT/ANCHOR (M)	=	0.22361
LENGTH OF ROCKBOLT/ANCHOR IN ROOF (M)	=	8.000
FIXED ANCHOR LENGTH (M)	=	4.000
SPACING OF BOLT/ANCHOR IN ROOF(M)	=	3.000
BOLT/ANCHOR CAPACITY (T)	=	20.000
CAPACITY OF STEEL RIB (T)	=	0.000
SPACING OF STEEL RIBS (M)	=	0.000
THICKNESS OF GROUTED ROCK MASS (M)	=	0.000

HEIGHT OF WALL OF CAVERN (M)	=	30.000
MODULUS OF DEFORMATION OF ROCK MASS (T/SQ.M)	=	430000.
AVERAGE VERTICAL STRESS ABOVE HAUNCH (T/SQ.M)	=	1800.000
AVERAGE VALUE OF SEEPAGE PRESSURE IN WALL (T/SQ.M)	=	0.000
ESTIMATED WALL ULTIMATE SUPPORT PRESSURE (T/SQ.M)	=	4.200
SHEAR STRENGTH OF GROUTED ROCK (T/SQ.M)	=	0.000
THICKNESS OF GROUTED ROCK MASS IN WALL (M)	=	0.000
DEPTH OF DAMAGE TO ROCK MASS DUE TO BLASTING (M)	=	3.000
SUGGESTED LENGTH OF BOLT/ANCHOR FOR WALL (M)	=	8.000
SPACING OF BOLTS/ANCHORS FOR WALL (M)	=	3.000
THICKNESS OF SHOTCRETE IN WALL (M)	=	0.131

*   THE REINFORCED ROCK WALL COLUMN MAY NOT BUCKLE, SO THE SPACING OF ROCK BOLTS/ANCHORS MAY BE INCREASED AND THICKNESS OF SHOTCRETE/ SFRS REDUCED IN WALLS ON THE BASIS OF PAST EXPERIENCE AND UNDERGROUND WEDGE ANALYSIS.

SAFE BEARING PRESSURE OF ROCK MASS (T/SQ.M)	=	400.000

DESIGN OF SUPPORT SYSTEM FOR CAVERN OF SARDAR SAROVAR PROJECT

SPAN OF UNDERGROUND OPENING (M)	=	23.000
SLIDING ANGLE OF FRICTION ALONG JOINTS (DEG.)	=	57.000
AVERAGE SPACING OF JOINTS (M)	=	0.800
AVERAGE VALUE OF SEEPAGE PRESSURE IN ROOF (T/SQ.M)	=	0.000
COMPRESSIVE STRENGTH OF GROUTED ROCK (T/SQ.M)	=	0.000
SHEAR STRENGTH OF SHOTCRETE (T/SQ.M)	=	300.000
ESTIMATED ULTIMATE ROOF SUPPORT PRESSURE (T/SQ.M)	=	8.800
LENGTH OF ROCK BOLT/ANCHOR IN ROOF (M)	=	6.000
FIXED ANCHOR LENGTH (M)	=	2.000
SPACING OF BOLT/ANCHOR IN ROOF (M)	=	1.750
BOLT/ANCHOR CAPACITY (T)	=	32.000
CAPACITY OF STEEL RIB (T)	=	0.000

SPACING OF STEEL RIBS (M)   =   0.000
THICKNESS OF GROUTED ROCK MASS (M)   =   0.000

CAPACITY OF SHOTCRETE LINING IN ROOF (T/SQ.M)   =   0.000
CAPACITY OF REINFORCED ROCK ARCH (T/SQ.M)   =   9.167
CAPACITY OF STEEL RIBS (T/SQ.M)   =   0.000
CAPACITY OF GROUTED ROCK ARCH (T/SQ.M)   =   0.000
THICKNESS OF SHOTCRETE IN ROOF (M)   =   0.026
RECOMMENDED ANGLE OF ARCH AT CENTRE (APPROXIMATE)   =   104.

HEIGHT OF WALL OF CAVERN (M)   =   52.000
MODULUS OF DEFORMATION OF ROCK MASS (T/SQ.M)   =   750000.
AVERAGE VERTICAL STRESS ABOVE HAUNCH (T/SQ.M)   =   250.000
AVERAGE VALUE OF SEEPAGE PRESSURE IN WALL (T/SQ.M)   =   0.000
ESTIMATED WALL ULTIMATE SUPPORT PRESSURE (T/SQ.M)   =   1.000
SHEAR STRENGTH OF GROUTED ROCK (T/SQ.M)   =   0.000
THICKNESS OF GROUTED ROCK MASS IN WALL (M)   =   0.000
DEPTH OF DAMAGE TO ROCK MASS DUE TO BLASTING (M)   =   3.000
SUGGESTED LENGTH OF BOLT/ANCHOR FOR WALL (M)   =   6.000
SPACING OF BOLTS/ANCHORS FOR WALL (M)   =   1.750
THICKNESS OF SHOTCRETE IN WALL (M)   =   0.052

*   THE REINFORCED ROCK WALL COLUMN MAY NOT BUCKLE, SO THE SPACING OF
       ROCK BOLTS/ANCHORS MAY BE INCREASED AND THICKNESS OF SHOTCRETE/
       SFRS REDUCED IN WALLS ON THE BASIS OF PAST EXPERIENCE AND UNDER-
       GROUND WEDGE ANALYSIS.

SAFE BEARING PRESSURE OF ROCK MASS (T/SQ.M)   =   200.000
SIDE OF BASE PLATE OF BOLT/ANCHOR (M)   =   0.40000

DESIGN OF SUPPORT SYSTEM FOR CAVERN IN AREA OF
SHEAR ZONE (SSP)

ROCK MASS QUALITY IN SHEAR ZONE   =   1.25000
JOINT ROUGHNESS NUMBER IN SHEAR ZONE   =   1.00000
JOINT ALTERATION NUMBER IN SHEAR ZONE   =   4.00000
MODULUS OF DEFORMATION OF SHEAR ZONE (T/SQ.M)   =   100000.0
WIDTH OF SHEAR ZONE (M)   =   2.00000
STRIKE DIRECTION OF SHEAR ZONE (DEG.)   =   50.00000
{WITH RESPECT TO AXIS OF TUNNEL/CAVERN}

ROCK MASS QUALITY IN SURROUNDING ROCK MASS   =   9.16000
JOINT ROUGHNESS NUMBER IN SURROUNDING ROCK MASS   =   1.50000
JOINT ALTERATION NUMBER IN SURROUNDING ROCK   =   0.75000
MODULUS OF DEFORMATION OF ROCK MASS (T/SQ.M)   =   750000.0
HEIGHT OF OVERBURDEN OF ROCK MASS (M)   =   70.00000

MEAN VALUES IN AREA INTERSECTED BY SHEAR OR WEAK ZONE
ROCK MASS QUALITY QM   =   2.42794
JOINT ROUGHNESS NUMBER Jrm   =   1.16667
JOINT ALTERATION NUMBER Jam   =   2.91667
MODULUS OF DEFORMATION EM (T/SQ.M)   =   316666.70
SUPPORT PRESSURE IN ROOF (T/SQ.M)   =   12.75848
SLIDING ANGLE OF FRICTION OF JOINTS (DEG.)   =   21.80136
{PLEASE USE THE ABOVE MEAN VALUES IN DESIGN IN AREA OF

SHEAR ZONE OR WEAK ZONE IN THE NEXT ROUND}
SUPPORT PRESSURE IN SURROUNDING ROCK (T/SQ.M)   =   6.37716

IT IS A NON-SQUEEZING GROUND

SPAN OF UNDERGROUND OPENING (M)	=	23.000
SLIDING ANGLE OF FRICTION ALONG JOINTS (DEG.)	=	22.000
AVERAGE SPACING OF JOINTS (M)	=	0.350
AVERAGE VALUE OF SEEPAGE PRESSURE IN ROOF (T/SQ.M)	=	0.000
COMPRESSIVE STRENGTH OF GROUTED ROCK (T/SQ.M)	=	0.000
SHEAR STRENGTH OF SHOTCRETE (T/SQ.M)	=	300.000
ESTIMATED ULTIMATE ROOF SUPPORT PRESSURE (T/SQ.M)	=	9.800
LENGTH OF ROCKBOLT/ANCHOR IN ROOF (M)	=	6.000
FIXED ANCHOR LENGTH (M)	=	2.000
SPACING OF BOLT/ANCHOR IN ROOF (M)	=	1.250
BOLT/ANCHOR CAPACITY (T)	=	32.000
CAPACITY OF STEEL RIB (T)	=	0.000
SPACING OF STEEL RIBS (M)	=	0.000
THICKNESS OF GROUTED ROCK MASS (M)	=	0.000

CAPACITY OF SHOTCRETE LINING IN ROOF (T/SQ.M)	=	6.197
CAPACITY OF REINFORCED ROCK ARCH (T/SQ.M)	=	3.603
CAPACITY OF STEEL RIBS (T/SQ.M)	=	0.000
CAPACITY OF GROUTED ROCK ARCH (T/SQ.M)	=	0.000
THICKNESS OF SHOTCRETE IN ROOF (M)	=	0.149
RECOMMENDED ANGLE OF ARCH AT CENTRE (APPROXIMATE)	=	104.00

HEIGHT OF WALL OF CAVERN (M)	=	52.000
MODULUS OF DEFORMATION OF ROCK MASS (T/SQ.M)	=	320000.
AVERAGE VERTICAL STRESS ABOVE HAUNCH (T/SQ.M)	=	250.000
AVERAGE VALUE OF SEEPAGE PRESSURE IN WALL (T/SQ.M)	=	0.000
ESTIMATED WALL ULTIMATE SUPPORT PRESSURE (T/SQ.M)	=	2.000
SHEAR STRENGTH OF GROUTED ROCK (T/SQ.M)	=	0.000
THICKNESS OF GROUTED ROCK MASS IN WALL (M)	=	0.000
DEPTH OF DAMAGE TO ROCK MASS DUE TO BLASTING (M)	=	3.000
SUGGESTED LENGTH OF BOLT/ANCHOR FOR WALL (M)	=	6.151
SPACING OF BOLTS/ANCHORS FOR WALL (M)	=	1.250
THICKNESS OF SHOTCRETE IN WALL (M)	=	0.107

SAFE BEARING PRESSURE OF ROCK MASS (T/SQ.M)	=	200.000
SIDE OF BASE PLATE OF BOLT/ANCHOR (M)	=	0.40000

DESIGN OF SUPPORT SYSTEM FOR CAVITY OF UNDERGROUND POWERHOUSE, GANWI PROJECT

SPAN OF UNDERGROUND OPENING (M)	=	14.200
SLIDING ANGLE OF FRICTION ALONG JOINTS (DEG.)	=	35.000
AVERAGE SPACING OF JOINTS (M)	=	0.500
AVERAGE VALUE OF SEEPAGE PRESSURE IN ROOF (T/SQ.M)	=	0.000
COMPRESSIVE STRENGTH OF GROUTED ROCK (T/SQ.M)	=	0.000
SHEAR STRENGTH OF SHOTCRETE (T/SQ.M)	=	300.000
ESTIMATED ULTIMATE ROOF SUPPORT PRESSURE (T/SQ.M)	=	13.000
LENGTH OF ROCKBOLT/ANCHOR IN ROOF (M)	=	7.000
FIXED ANCHOR LENGTH (M)	=	1.500
SPACING OF BOLT/ANCHOR IN ROOF (M)	=	1.500

BOLT/ANCHOR CAPACITY (T)	=	18.000
CAPACITY OF STEEL RIB (T)	=	0.000
SPACING OF STEEL RIBS (M)	=	0.000
THICKNESS OF GROUTED ROCK MASS (M)	=	0.000

CAPACITY OF SHOTCRETE LINING IN ROOF (T/SQ.M)	=	7.296
CAPACITY OF REINFORCED ROCK ARCH (T/SQ.M)	=	5.704
CAPACITY OF STEEL RIBS (T/SQ.M)	=	0.000
CAPACITY OF GROUTED ROCK ARCH (T/SQ.M)	=	0.000
THICKNESS OF SHOTCRETE IN ROOF (M)	=	0.106
RECOMMENDED ANGLE OF ARCH AT CENTRE (APPROXIMATE)	=	117.00

HEIGHT OF WALL OF CAVERN (M)	=	20.000
MODULUS OF DEFORMATION OF ROCK MASS (T/SQ.M)	=	1000000.
AVERAGE VERTICAL STRESS ABOVE HAUNCH (T/SQ.M)	=	1000.000
AVERAGE VALUE OF SEEPAGE PRESSURE IN WALL (T/SQ.M)	=	0.000
ESTIMATED WALL ULTIMATE SUPPORT PRESSURE (T/SQ.M)	=	1.500
SHEAR STRENGTH OF GROUTED ROCK (T/SQ.M)	=	0.000
THICKNESS OF GROUTED ROCK MASS IN WALL (M)	=	0.000

DEPTH OF DAMAGE TO ROCK MASS DUE TO BLASTING (M)	=	2.000
SUGGESTED LENGTH OF BOLT/ANCHOR FOR WALL (M)	=	7.000
SPACING OF BOLTS/ANCHORS FOR WALL (M)	=	1.500
THICKNESS OF SHOTCRETE IN WALL (M)	=	0.029

*   THE REINFORCED ROCK WALL COLUMN MAY NOT BUCKLE, SO THE SPACING OF ROCK BOLTS/ANCHORS MAY BE INCREASED AND THICKNESS OF SHOTCRETE/ SFRS REDUCED IN WALLS ON THE BASIS OF PAST EXPERIENCE AND UNDER-GROUND WEDGE ANALYSIS.

SAFE BEARING PRESSURE OF ROCK MASS (T/SQ.M)	=	200.000
SIDE OF BASE PLATE OF BOLT/ANCHOR (M)	=	0.30000

LENGTH OF ROCKBOLT/ANCHOR IN ROOF (M)	=	7.000
FIXED ANCHOR LENGTH (M)	=	1.500
SPACING OF BOLT/ANCHOR IN ROOF (M)	=	1.750
BOLT/ANCHOR CAPACITY (T)	=	18.000
CAPACITY OF STEEL RIB (T)	=	0.000
SPACING OF STEEL RIBS (M)	=	0.000
THICKNESS OF GROUTED ROCK MASS (M)	=	0.000

HEIGHT OF WALL OF CAVERN (M)	=	20.000
MODULUS OF DEFORMATION OF ROCK MASS (T/SQ.M)	=	1000000.
AVERAGE VERTICAL STRESS ABOVE HAUNCH (T/SQ.M)	=	1000.000
AVERAGE VALUE OF SEEPAGE PRESSURE IN WALL (T/SQ.M)	=	0.000
ESTIMATED WALL ULTIMATE SUPPORT PRESSURE (T/SQ.M)	=	1.500
SHEAR STRENGTH OF GROUTED ROCK (T/SQ.M)	=	0.000
THICKNESS OF GROUTED ROCK MASS IN WALL (M)	=	0.000
DEPTH OF DAMAGE TO ROCK MASS DUE TO BLASTING (M)	=	2.000
SUGGESTED LENGTH OF BOLT/ANCHOR FOR WALL (M)	=	7.000
SPACING OF BOLTS/ANCHORS FOR WALL (M)	=	1.750
THICKNESS OF SHOTCRETE IN WALL (M)	=	0.029

*   THE REINFORCED ROCK WALL COLUMN MAY NOT BUCKLE, SO THE SPACING OF ROCK BOLTS/ANCHORS MAY BE INCREASED AND THICKNESS OF SHOTCRETE/

SFRS REDUCED IN WALLS ON THE BASIS OF PAST EXPERIENCE AND UNDER-
GROUND WEDGE ANALYSIS.

SAFE BEARING PRESSURE OF ROCK MASS (T/SQ.M)	= 200.000
SIDE OF BASE PLATE OF BOLT/ANCHOR (M)	= 0.30000

SPECIAL SPECIFICATIONS

1. FOR TREATMENT OF SHEAR ZONES, CROSSED ROCK ANCHORS/ BOLTS SHOULD
   BE PROVIDED ACROSS SHEAR ZONES. FIRST GOUGE SHOULD BE CLEANED TO
   DESIRED EXTENT. ANCHORS ARE PROVIDED AND CONNECTED TO WELDED MESH.
   FINALLY DENTAL SHOTCRETE IS BACK-FILLED. IN WIDE SHEAR ZONE, REIN-
   FORCEMENT IN SHOTCRETE IS ALSO PLACED TO WITHSTAND HIGH SUPPORT
   PRESSURE. ANCHORS SHOULD BE INCLINED ACCORDING TO DIP OF SHEAR ZONE
   TO STOP SQUEEZING OF GOUGE AND THEREBY STABILISE DEFORMATIONS.

2. IN CASE OF POOR ROCK MASS, SPILING BOLTS (INCLINED TOWARDS TUNNEL FACE)
   SHOULD BE INSTALLED BEFORE BLASTING TO INCREASE THE STAND-UP TIME
   OF TUNNEL ROOF. SHOTCRETE IS THEN SPRAYED ON ROOF. NEXT SPILING BOLTS
   ARE INSTALLED. IN FINAL CYCLE, ROOF BOLTS ARE INSTALLED.

3. IN CASES OF ARGILLACEOUS ROCKS AND SWELLING ROCKS WHERE ITS BOND
   WITH SHOTCRETE IS POOR, THICKNESS OF SHOTCRETE MAY BE INCREASED BY
   ABOUT 30 PERCENT.

4. IN CASE OF UNSTABLE PORTALS, HORIZONTAL ANCHORS OF EQUAL LENGTH
   SHOULD BE PROVIDED INSIDE THE CUT SLOPE, SO THAT IT ACTS AS A REIN-
   FORCED ROCK BREAST WALL.

5. ALL BOLTS SHOULD BE GROUTED AS PROTECTION FROM CORROSION. AT LEAST
   5 PERCENT BOLTS SHOULD BE PULLED OUT TO CHECK THEIR PULL OUT CAPA-
   CITY PARTICULARLY NEAR SHEAR/FAULT ZONES.

REFERENCES

Barton, N., Lien, R. and Lunde, J. 1974. Engineering classification of rock masses for the
design of tunnel support. *Rock Mechanics*, Vol. 6, pp. 189–236.

Bhasin, R., Singh, R.B., Dhawan, A.K. and Sharma, V.M. 1995. Geotechnical evaluation
and a review of remedial measures in limiting deformations in distressed zones in a
Powerhouse Cavern. *Conf. Design and Construction of Underground Structures*, New
Delhi, India, pp. 145–152.

Bischoff, J.A., Klein, S.J. and Lang, T.A. 1992. Designing reinforced rock. *Civil
Engineering*, January.

Duddeck, H. and Erdmann, J. 1985. On structural design models for tunnels in soft soil.
Underground Space, Vol. 9, pp. 246–259.

Grimstad, E. and Barton, N. 1993. Updating of the Q-system for NMT. *Proc. Int. Sym. On
Sprayed Concrete – Modern Use of Wet Mix Sprayed Concrete for Underground
Support*, Fagernes, Norwegian Concrete Association, Oslo.

Hoek, E. and Brown, E.T. 1980. *Underground Excavations in Rock*. Institution of Mining
and Metallurgy, England, Revised Edition, Chapter 8: Underground excavation support
design, p. 527.

Norwegian Geological Institute. 1993. Manual for mapping of rocks, Prepared by Rock
Engineering and Reservoir Mechanics Section of NGI, for a Workshop on Norwegian
Method of Tunnelling, Sept. 1993, New Delhi, 12–54.

Park, E.S., Kim, H.Y. and Lee, H.K. 1997. A study on the design of the shallow large cavern in the Gonjiam underground storage terminal. *Proc. I Asian Rock Mechanics Sym. On Environmental and Strategy Concerns in Underground Construction*, Seoul, pp. 345–351.

Samadhiya, N.K. 1998. *Influence of anisotropy and shear zones on stability of caverns.* Ph.D. Thesis. Civil Engineering Department, University of Roorkee, India, p. 334.

Singh, Bhawani, Jethwa, J.L., Dube, A.K. and Singh, B. 1992. Correlation between observed support pressure and rock mass quality. *Tunnel. Underground Space Technol.* 7(1): 59–74.

Singh, Bhawani, Nayak, G.C, Kumar, R. and Chandra, G. 1988. Design criteria for plain concrete lining and power tunnels. *Tunnel. Underground Space Technol.* 3(2): 201–208.

Singh, Bhawani, Viladkar, M.N., Samadhiya N.K. and Sandeep 1995. A semi-empirical method for the design of support systems in underground openings. *Tunnel. Underground Space Technol.* 10(3): 375–385.

Thakur, B. 1995. *Semi-empirical method for design of supports in underground excavations.* M.E. Thesis. Civil Engineering Department, University of Roorkee, India, p. 126.

CHAPTER 27

Concrete lining in pressure tunnels – LINING

27.1 INTRODUCTION

This chapter is devoted to help the engineers and geologists in decision making on where PCC and RCC lining or steel liner are to be planned in a pressure or power tunnel. It may be mentioned that power tunnels of medium size (B = 5 to 6 m) are most economical for the generation of electricity.

No lining is required in water tunnels within massive hard rock masses as it is self-supporting (Eq. 21.2 in Chapter 21 and Fig. 22.3 in Chapter 22). However, it should be ensured that the minimum in situ principal stress is more than the internal water pressure along the entire water tunnel. (In other words, the overburden pressure of rock mass should be more than the water head.)

Most pressure (power) tunnels are lined with concrete to reduce head loss due to friction at the tunnel boundary. This reduces water loss due to seepage and also stabilises the unstable rock wedges. Plain cement concrete (PCC) lining has been used in many power tunnels in hydroelectric projects in U.P., India. No hoop reinforcement has been provided though internal water pressure is quite high. These PCC linings have been working satisfactorily since 1970 without any closure for repairs. It is heartening to know that PCC lining has worked in squeezing rock conditions also. Millions of dollars of money and construction time can be saved if unnecessary hoop reinforcement is eliminated in the conventional design of power (pressure) tunnels. Reinforcement though increases the tensile strength of the concrete and also hampers the installation of a good dense cement concrete lining. Good and compact concrete capable of withstanding high velocities and abrasion is desirable.

27.2 PCC LINING IN PRESSURE TUNNELS

A PCC lining for a water power tunnel is likely to crack radially at a number of places where the hoop tensile stress exceeds its tensile strength (Fig. 27.1). In practice six construction joints are provided while concreting. These joints are also likely to open up due to internal water pressure. In addition, cracks may also develop where the surrounding rock mass is poor. These radial cracks will be distributed nearly uniformly along circumference due to a good bond between

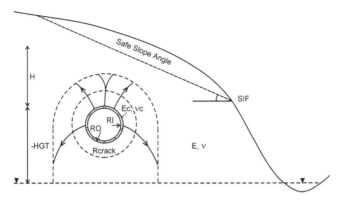

Figure 27.1. Input parameters for design of lining of power tunnel.

concrete and rock mass. Therefore, it is suggested that PCC lining should be provided in water and power tunnels. It is designed properly to ensure that the crack width is within safe limit (<3 mm) and length of segments is more than three times the thickness of the lining or 1.75 m. This would ensure self-healing of crack by precipitation of $CaCo_3$ etc. within cracks and the cracked segments will not be washed away by the water flowing with high velocity.

Nevertheless, reinforcement must be provided in the lining (i) at the tunnel intersections, (ii) at the enlargements, (iii) at inlet and outlet ends, (iv) in plug areas, (v) in the areas where the power tunnel passes through a relatively poor rock mass and (vi) where the overburden pressure due to rock cover is inadequate to counter-balance the internal water pressure.

As per field experience, the errors of surveying are higher in mountainous terrain because of many difficulties. As such, depth of rock cover cannot be estimated reliably. Re-surveying may be recommended in critical areas where overburden is not adequate.

The design methodology for PCC lining was developed by Singh et al. (1988a,b). Later, Kumar and Singh (1990) proposed a design procedure for reinforced concrete lining for water/power tunnels. The program LINING has been developed on the basis of this research work. The program also calculates seepage loss through lining using the analytical solution (Schleiss, 1998). Concrete lining tunnels are also being used as canals. The seepage loss may be estimated by the expression developed by Swamee and Kashyap (2001).

It should also be noted that the rock mass is saturated all around the lining as shown in Figure 27.1 after charging of the water conductor system. In argillaceous rocks, this saturation reduces the modulus of deformation of the rock mass significantly. Consequently, high support pressures are developed on the lining after saturation of the rock masses (Verman, 1993 and Eq. 21.7 in Chapter 21). The worst condition of design occurs when the power tunnel is empty. Thus, the PCC lining must be able to support these unusually high support pressures as well as the ground water pressure, which is nearly equal to the internal water pressure in

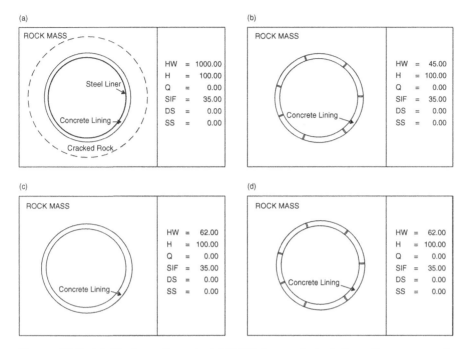

Figure 27.2. (a) Design of cracked PCC lining and steel liner for a power tunnel. (b) Ramganga river project, UP., India. (c) Maneri Bhali Project Stage I, U.P., India. (d) Yamuna Hydroelectric Project Stage II, U.P., India.

the tunnel. The recommended factor of safety in hoop compression is 3.0 for PCC/RCC (Jethwa, 1981). To make PCC lining ductile, nominal reinforcement of about 1 percent of volume of concrete is suggested so that mode of failure of lining is ductile and slow due to unexpected rock loads. Nominal reinforcement will also prevent shrinkage cracks in the concrete lining.

It may be recalled that temporary support system for a power tunnel is designed for the existing ground water condition for rock mass quality Q. Yet, it is the post-construction ground water condition around a power tunnel which will govern the long-term support pressure even in non-swelling rock masses. Hence J_w should be taken corresponding to the internal water pressure of power tunnel. There is no cause for anxiety as the extra long-term support pressure on the lining is negligible compared to the ground water pressure. Both pressures act simultaneously on the concrete lining.

Indeed, a water conductor is never charged instantly as assumed in the design. The power tunnel is pressurised slowly. Thus, seepage takes place through construction joints into the rock mass. The seepage pressure tries to counteract the internal water pressure on the concrete lining. Consequently, actual numbers of cracks are limited to construction joints mostly. The actual crack opening is much less than that predicted by theory. As such, use of PCC lining should be encouraged in good and fair rock masses where overburden is adequate. Figure 27.2 shows a typical output of number of cracks, etc. in the PCC lining.

27.3   STEEL LINEAR IN PENSTOCK

The steel liner is provided in the underground penstocks which connect the power tunnel (head race tunnel HRT) and the underground power house cavity. The steel liner can sustain very high velocities of flow of water (1.6 to 9 m/sec) and reduce the hoop tensile stresses in the surrounding PCC lining. The computer program helps in calculating the thickness of the steel liner and the spacing of stiffners. The worst condition for the steel liner is also the empty tunnel, as the seepage pressure of the order of the internal water pressure may act upon the steel liner. As such the stiffners are provided to prevent the steel liner from buckling. The steel liner is painted with anti-corrosive paint like epoxy paint. However, the minimum thickness of the liner is D/400; where D is the diameter of the penstock. This will provide adequate stiffness to the liner which is required during its fabrication and handling.

Finally, contact grouting between concrete lining and rock mass (and also between steel liner and concrete lining) is executed at low pressure. This is followed by the consolidation grouting of the surrounding ring of the rock mass under high pressure. Some experts recommend high grouting pressure to pre-stress the concrete lining. In view of the authors, pre-stressing is not needed where PCC lining is feasible. Vaidya and Gupta (1998) have reported failure of grout plugs in the steel liner due to the seepage pressure. The repair was done successfully. The above idea of PCC lining prone to cracks is also recommended for the concrete lining of the underground silt ejectors in the hydroelectric projects.

27.4   HYDRAULIC FRACTURING AT JUNCTION OF PENSTOCK

The following caution should be kept in mind. It is observed by Barton (1986) that a rock joint opens near the junction of unlined head race tunnel and the steel lined penstock. The ground water table is very high above the head race tunnel at the time of full head, but it drops suddenly above the penstock due to the impervious steel lining. Hence, consolidation of rock mass above the penstock may take place due to the drastic reduction in seepage water pressure. Consequently, this leads to development of the horizontal tensile strain in the upstream adjoining rock mass around the head race tunnel. Thus, the rock joint opens at this junction within the ungrouted rock mass and this fracture may propagate up to the top of the hill in some cases. This phenomenon of fracturing underlines the need for extensive consolidation and contact grouting of rock mass near this junction specially in the case of unlined head race tunnels.

The problem at the junction is further complicated due to the square shape of steel lining which cannot bear high outside water pressure. The buckling of steel liners of the two penstocks of the Pong Dam Project after reservoir filling may be due to the above reasons. This damage was repaired successfully (Oberoi and Gupta, 2000).

27.5 A VALIDATION OF THEORY IN THE FIELD

Finally 8 examples (including 7 case histories of PCC lining in Himalaya) of computer aided design are given to help the users. The computer analysis predicted the failure of the PCC lining of Kopli Hydel Project. This was the case in reality. Other power tunnels have been predicted to be structurally safe and are working well. Consolidation grouting has been done as suggested by the computer aided analysis.

27.6 USER'S MANUAL – LINING

THIS PROGRAM IS FOR THE DESIGN OF SOLID/CRACKED/PLAIN CEMENT CONCRETE AND RCC LINING OF PRESSURE TUNNEL AND PENSTOCK (SEE FIGS 27.1 AND 27.2)

NAME OF PROGRAM –   LINING.FOR
UNITS USED –   TONNE – METER – DEGREE – CENTIGRADE – SECOND

GIVE INPUT DATA IN THE FOLLOWING SEQUENCE
NUMBER OF PROBLEMS
TITLE OF PROBLEM IN ONE LINE (<80 CHARACTERS)
E,V,EC,VC,ES,VS
H,HW,GAMA,GAMAW,RI,RO,ALPHAC,T
FC,FT,SIGYS,DS,SS,RS,ALPHAS
K,HGWT,KO,SIF,PROOF
RSL,TSL,DELTAC
REPEAT ABOVE SIX LINES NO TIMES

DO YOU WANT HELP REGARDING DEFINITIONS OF VARIABLES USED
ENTER   0 FOR TERMINATION
           1 FOR FURTHER HELP
           2 FOR EXECUTION

E	=	MODULUS OF ELASTICITY OF ROCK MASS (MINIMUM VALUE IN ANISOTROPIC ROCK MASS)
V	=	POISSON'S RATIO OF ROCK MASS
EC	=	MODULUS OF ELASTICITY OF CONCRETE
VC	=	POISSON'S RATIO OF CONCRETE
ES	=	MODULUS OF ELASTICITY OF STEEL
GAMA	=	UNIT WEIGHT OF ROCK MASS
GAMAW	=	UNIT WEIGHT OF WATER
VS	=	POISSON'S RATIO OF STEEL
HW	=	DESIGN HEAD OF WATER
RI	=	INTERNAL RADIUS OF CONCRETE LINING
RO	=	EXTERNAL RADIUS OF CONCRETE LINING
ALPHAC	=	COEFFICIENT OF THERMAL EXPANSION OF CONCRETE
T	=	DIFFERENCE IN TEMPERATURE OF WATER AND TUNNEL ROCK
FC	=	SAFE COMPRESSIVE STRENGTH OF CONCRETE
FT	=	ULTIMATE TENSILE STRENGTH OF CONCRETE
SIGYS	=	SAFE TENSILE STRENGTH OF STEEL OR LINER {MULTIPLY BY EFFICIENCY (0.9) OF WELDED JOINT FOR STEEL LINER}

DS	=	DESIGN DIAMETER OF HOOP REINFORCEMENT BARS
		(LESS 1.5 MM TO ACCOUNT FOR CORROSION)
SS	=	SPACING OF HOOP REINFORCEMENT BARS
		(REDUCE BY HALF FOR 2 LAYERS OF HOOP REINFORCEMENT)
RS	=	AVERAGE RADIUS OF HOOP REINFORCEMENT BARS
K	=	PERMEABILITY OF ROCK MASS
HGWT	=	HEIGHT OF GROUND WATER TABLE ABOVE CROWN
	=	0 IF TUNNEL IS ABOVE GROUND WATER TABLE
KO	=	RATIO OF HORIZONTAL AND VERTICAL IN SITU STRESSES
PROOF	=	INCREASE IN SUPPORT PRESSURE ON SATURATION OF ROCK
	=	0.75*SHORT TERM SUPPORT PRESSURE (NON-SQUEEZING ROCK)
	=	(1 OR 2) SHORT TERM SUPPORT PRESSURE (SQUEEZING ROCK)
	=	5*SHORT TERM SUPPORT PRESSURE
		[WATER- CHARGED ROCK MASS WITH ERODIBLE JOINT FILLINGS]
	<	{1-Esat/Edry}*GAMA*H [POST CONSTRUCTION
		SATURATION]
ALPHAS	=	COEFFICIENT OF THERMAL EXPANSION OF STEEL LINER
H	=	HEIGHT OF OVERBURDEN OF ROCK MASS ON SLOPE SIDE > 0
SIF	=	SAFE SLOPE ANGLE OF HILL
RSL	=	INTERNAL RADIUS OF STEEL LINER
TSL	=	THICKNESS OF STEEL LINER(PAINTED FOR NO
		CORROSION)
DELTAC	=	SHRINKAGE OF CONCRETE LINING (=0.0001*RSL)

(a) Input File: ILINING.DAT

8
DESIGN OF CRACKED PCC LINING AND STEEL LINER FOR A POWER TUNNEL
200000. 0.2 2000000. 0.2 20000000. 0.3
100. 1000. 2.5 1. 3.25 3.60 0.00001 −5.
700. 200. 30000. 0.0 0.0 3.45 0.0000113
0.000001 0. 2.5 35. 10.
3.20 0.075 0.0005

RAM GANGA RIVER PROJECT (U.P.)
85000. 0.2 2000000. 0.2 20000000. 0.3
100. 45. 2.5 1. 4.5 5.25 0.00001 0.
700. 200. 45000. 0. 0. 0. 0.
0.000001 10. 2. 35. 10.
0. 0. 0.

MANERI BHALI HYDEL PROJECT STAGE-I (U.P.)
750000. 0.2 2000000. 0.2 20000000. 0.3
100. 62. 2.5 1. 2.37 2.67 0.00001 0.
700. 200. 45000. 0. 0. 0. 0.
0.000001 0. 2. 35. 10.
0. 0. 0.

YAMUNA HYDROELECTRIC PROJECT STAGE II (U.P.)
50000. 0.2 2000000. 0.2 20000000. 0.3
100. 62. 2.5 1. 3.5 4.1 0.00001 0.
700. 200. 45000. 0. 0. 0. 0.
0.000001 0. 2. 35. 10.
0. 0. 0.

MANERI BHALI HYDEL PROJECT STAGE I (U.P.)
300000. 0.2 2000000. 0.2 20000000. 0.3
100. 35. 2.5 1. 3. 3.30 0.00001 0.
700. 200. 45000. 0. 0. 0. 0.
0.000001 0. 2. 35. 10.
0. 0. 0.

TEHRI DAM PROJECT DIVERSION TUNNEL (U.P.)
300000. 0.2 2000000. 0.2 20000000. 0.3
100. 60. 2.5 1. 5.5 6.4 0.00001 0.
700. 200. 45000. 0. 0. 0. 0.
0.0000001 0. 2. 35. 10.
0. 0. 0.

TEHRI DAM PROJECT HEAD RACE TUNNEL (U.P.)
80000. 0.2 2000000. 0.2 20000000. 0.3
200. 120. 2.5 1. 4.0 4.6 0.00001 0.
700. 200. 45000. 0. 0. 0. 0.
0.0000001 0. 2. 35. 10.
0. 0. 0.

KOPLI HYDEL PROJECT (ASSAM)
57000. 0.2 2000000. 0.2 20000000. 0.3
200. 160. 2.5 1. 2.25 2.45 0.00001 0.
700. 200. 45000. 0. 0. 0. 0.
0.000001 0. 2. 35. 10.

(b) Output File: OLINING.DAT

DESIGN OF CRACKED PCC LINING AND STEEL LINER FOR A POWER TUNNEL

UNITS USED –   TONNE – METER – DEGREE – CENTIGRADE – SECOND
INPUT FILE NAME –   ILINING.DAT
OUTPUT FILE NAME –   OLINING.DAT

Case No. 1

MODULUS OF ELASTICITY OF ROCK MASS	=	200000.00
POISSON'S RATIO OF ROCK MASS	=	0.2000
MODULUS OF ELASTICITY OF CONCRETE	=	2000000.0
POISSON'S RATIO OF CONCRETE	=	0.200
MODULUS OF ELASTICITY OF STEEL/LINER	=	20000000.00
POISSON'S RATIO OF STEEL	=	0.3000000
HEIGHT OF OVERBURDEN ON SLOPE SIDE	=	100.000
DESIGN HEAD OF WATER	=	1000.000
UNIT WEIGHT OF ROCK MASS	=	2.500
UNIT WEIGHT OF WATER	=	1.000
INTERNAL RADIUS OF CONCRETE LINING	=	3.250
EXTERNAL RADIUS OF CONCRETE LINING	=	3.600
COEFFICIENT OF THERMAL EXPANSION OF CONCRETE	=	0.0000100
DIFFERENCE IN TEMPERATURE OF WATER AND ROCK	=	−5.000
SAFE COMPRESSIVE STRENGTH OF CONCRETE	=	700.000
ULTIMATE TENSILE STRENGTH OF CONCRETE	=	200.000
SAFE TENSILE STRENGTH OF STEEL/LINER	=	30000.000

DESIGN DIAMETER OF HOOP REINFORCEMENT BARS	=	0.000
RECOMMENDED SPACING OF HOOP REINFORCEMENT BARS	=	0.000
AVERAGE RADIUS OF HOOP REINFORCEMENT BARS	=	3.4500
COEFFICIENT OF THERMAL EXPANSION – STEEL/LINER	=	0.0000113
PERMEABILITY OF ROCK MASS	=	0.0000010
HEIGHT OF GROUND WATER TABLE ABOVE CROWN	=	0.00
RATIO OF HORIZONTAL AND VERTICAL IN SITU STRESSES	=	2.5000
SLOPE ANGLE OF HILL	=	35.000
ROOF SUPPORT PRESSURE AFTER SATURATION OF ROCK	=	10.000
INTERNAL RADIUS OF STEEL LINER	=	3.200
ASSUMED THICKNESS OF STEEL LINER	=	0.0750
RECOMMENDED SHRINKAGE OF CONCRETE LINING	=	0.00050
ROCK CRACKS BY HYDRAULIC FRACTURING TO RADIUS	=	4.7465400
HOOP TENSILE STRESS IN STEEL LINER	=	34755.77000
RATIO OF ROCK SHARING AND INTERNAL PRESSURES	=	0.1648104
SPACING OF STIFFNERS OF LINER	=	32.500

HOOP STRESS IN STEEL LINER EXCEEDS YIELD LIMIT

PLEASE MONITOR WATER OOZING OUT OF A VERTICAL HYDRAULIC FRACTURE WHICH IS PERPENDICULAR TO TUNNEL AXIS AND MAY OCCUR AT JUNCTION OF HEAD RACE TUNNEL AND PENSTOCK

RAMGANGA RIVER PROJECT (U.P.)

UNITS USED –	TONNE – METER – DEGREE – CENTIGRADE – SECOND
INPUT FILE NAME –	ILINING.DAT
OUTPUT FILE NAME –	OLINING.DAT

Case No. 2

MODULUS OF ELASTICITY OF ROCK MASS	=	85000.000
POISSON'S RATIO OF ROCK MASS	=	0.200
MODULUS OF ELASTICITY OF CONCRETE	=	2000000.0
POISSON'S RATIO OF CONCRETE	=	0.200
MODULUS OF ELASTICITY OF STEEL/LINER	=	20000000.
POISSON'S RATIO OF STEEL	=	0.30
HEIGHT OF OVERBURDEN ON SLOPE SIDE	=	100.00
DESIGN HEAD OF WATER	=	45.00
UNIT WEIGHT OF ROCK MASS	=	2.50

UNIT WEIGHT OF WATER	=	1.00
INTERNAL RADIUS OF CONCRETE LINING	=	4.50
EXTERNAL RADIUS OF CONCRETE LINING	=	5.250
COEFFICIENT OF THERMAL EXPANSION OF CONCRETE	=	0.000010
DIFFERENCE IN TEMPERATURE OF WATER AND ROCK	=	0.00
SAFE COMPRESSIVE STRENGTH OF CONCRETE	=	700.00
ULTIMATE TENSILE STRENGTH OF CONCRETE	=	200.00
SAFE TENSILE STRENGTH OF STEEL/LINER	=	45000.00
DESIGN DIAMETER OF HOOP REINFORCEMENT BARS	=	0.000
RECOMMENDED SPACING OF		
HOOP REINFORCEMENT BARS	=	0.00
AVERAGE RADIUS OF HOOP		
REINFORCEMENT BARS	=	0.00
COEFFICIENT OF THERMAL EXPANSION – STEEL/LINER	=	0.00
PERMEABILITY OF ROCK MASS	=	0.0000010
HEIGHT OF GROUND WATER TABLE ABOVE CROWN	=	10.00
RATIO OF HORIZONTAL AND VERTICAL		
IN SITU STRESSES	=	2.00
SLOPE ANGLE OF HILL	=	35.00
ROOF SUPPORT PRESSURE AFTER		
SATURATION OF ROCK	=	10.00
HOOP STRESS AT INNER SIDE OF SOLID LINING	=	-241.60320
HOOP STRESS AT OUTER SIDE – EMPTY SOLID LINING	=	130.76920
RATIO OF ROCK SHARING & INTERNAL PRESSURES	=	0.1551380
SIDE ROCK COVER IS ADEQUATE		
NUMBER OF CRACKS IN CRACKED PCC LINING	=	7
SPACING OF CRACKS IN CRACKED PCC LINING	=	4.03920
OPENING OF CRACKS IN CRACKED PCC LINING	=	0.0025941
COMPRESSIVE HOOP STRESS – EMPTY CRAKED LINING	=	130.76920
SEEPAGE LOSS PER UNIT LENGTH OF TUNNEL	=	0.0001843707

PLEASE GROUT ROCK MASS, SEEPAGE LOSS IS HIGH

MANERI BHALI HYDEL PROJECT STAGE I (U.P.)

UNITS USED –	TONNE – METER – DEGREE – CENTIGRADE – SECOND	
INPUT FILE NAME –	ILINING.DAT	
OUTPUT FILE NAME –	OLINING.DAT	

Case No. 3

MODULUS OF ELASTICITY OF ROCK MASS	=	750000.00
POISSON'S RATIO OF ROCK MASS	=	0.200
MODULUS OF ELASTICITY OF CONCRETE	=	2000000.
POISSON'S RATIO OF CONCRETE	=	0.200
MODULUS OF ELASTICITY OF STEEL/LINER	=	20000000.
POISSON'S RATIO OF STEEL	=	0.300
HEIGHT OF OVERBURDEN ON SLOPE SIDE	=	100.00
DESIGN HEAD OF WATER	=	62.000
UNIT WEIGHT OF ROCK MASS	=	2.50

UNIT WEIGHT OF WATER	=	1.00
INTERNAL RADIUS OF CONCRETE LINING	=	2.370
EXTERNAL RADIUS OF CONCRETE LINING	=	2.670
COEFFICIENT OF THERMAL EXPANSION OF CONCRETE	=	0.00001
DIFFERENCE IN TEMPERATURE OF WATER AND ROCK	=	0.000
SAFE COMPRESSIVE STRENGTH OF CONCRETE	=	700.00
ULTIMATE TENSILE STRENGTH OF CONCRETE	=	200.00
SAFE TENSILE STRENGTH OF STEEL/LINER	=	45000.00
DESIGN DIAMETER OF HOOP REINFORCEMENT BARS	=	0.00
RECOMMENDED SPACING OF HOOP REINFORCEMENT BARS	=	0.00
AVERAGE RADIUS OF HOOP REINFORCEMENT BARS	=	0.00
COEFFICIENT OF THERMAL EXPANSION - STEEL/LINER	=	0.00
PERMEABILITY OF ROCK MASS	=	0.000001
HEIGHT OF GROUND WATER TABLE ABOVE CROWN	=	0.00
RATIO OF HORIZONTAL AND VERTICAL IN SITU STRESSES	=	2.00
SLOPE ANGLE OF HILL	=	35.00
ROOF SUPPORT PRESSURE AFTER SATURATION OF ROCK	=	10.00
HOOP STRESS AT INNER SIDE OF SOLID LINING	=	−145.35530
HOOP STRESS AT OUTER SIDE – EMPTY SOLID LINING	=	606.94240
RATIO OF ROCK SHARING AND INTERNAL PRESSURES	=	0.6453311

SIDE ROCK COVER IS ADEQUATE

SEEPAGE LOSS PER UNIT LENGTH OF TUNNEL	=	0.0001355
RADIUS OF SATURATION ON SLOPE SIDE	=	67.74906
RADIUS OF SATURATION ON ROOF	=	14.94782

PLEASE GROUT ROCK MASS, SEEPAGE LOSS IS HIGH

YAMUNA HYDROELECTRIC PROJECT STAGE II (U.P.)

UNITS USED –	TONNE – METER – DEGREE – CENTIGRADE – SECOND
INPUT FILE NAME –	ILINING.DAT
OUTPUT FILE NAME –	OLINING.DAT

Case No. 4

MODULUS OF ELASTICITY OF ROCK MASS	=	50000.00
POISSON'S RATIO OF ROCK MASS	=	0.20
MODULUS OF ELASTICITY OF CONCRETE	=	2000000.
POISSON'S RATIO OF CONCRETE	=	0.20
MODULUS OF ELASTICITY OF STEEL/LINER	=	20000000.
POISSON'S RATIO OF STEEL	=	0.30
HEIGHT OF OVERBURDEN ON SLOPE SIDE	=	100.00
DESIGN HEAD OF WATER	=	62.00
UNIT WEIGHT OF ROCK MASS	=	2.50
UNIT WEIGHT OF WATER	=	1.00
INTERNAL RADIUS OF CONCRETE LINING	=	3.50
EXTERNAL RADIUS OF CONCRETE LINING	=	4.10
COEFFICIENT OF THERMAL EXPANSION OF CONCRETE	=	0.00001
DIFFERENCE IN TEMPERATURE OF WATER AND ROCK	=	0.00

SAFE COMPRESSIVE STRENGTH OF CONCRETE	=	700.00
ULTIMATE TENSILE STRENGTH OF CONCRETE	=	200.00
SAFE TENSILE STRENGTH OF STEEL/LINER	=	45000.00
DESIGN DIAMETER OF HOOP REINFORCEMENT BARS	=	0.00
RECOMMENDED SPACING OF HOOPREINFORCEMENT BARS	=	0.00
AVERAGE RADIUS OF HOOP REINFORCEMENT BARS	=	0.00
COEFFICIENT OF THERMAL EXPANSION – STEEL/LINER	=	0.00

PERMEABILITY OF ROCK MASS	=	0.000001
HEIGHT OF GROUND WATER TABLE ABOVE CROWN	=	0.00
RATIO OF HORIZONTAL AND VERTICAL IN SITU STRESSES	=	2.00
SLOPE ANGLE OF HILL	=	35.00
ROOF SUPPORT PRESSURE AFTER SATURATION OF ROCK	=	10.00

HOOP STRESS AT INNER SIDE OF SOLID LINING	=	-351.3531
HOOP STRESS AT OUTER SIDE – EMPTY SOLID LINING	=	458.8422
RATIO OF ROCK SHARING AND INTERNAL PRESSURES	=	0.0957331

SIDE ROCK COVER IS ADEQUATE

NUMBER OF CRACKS IN CRACKED PCC LINING	=	9
SPACING OF CRACKS IN CRACKED PCC LINING	=	2.4434670
OPENING OF CRACKS IN CRACKED PCC LINING	=	
COMPRESSIVE HOOP STRESS – EMPTY CRACKED LINING	=	458.8422

OPENING OF CRACKS IN PCC LINING EXCEEDS PERMISSIBLE LIMIT

SEEPAGE LOSS PER UNIT LENGTH OF TUNNEL	=	0.0001506
RADIUS OF SATURATION ON SLOPE SIDE	=	75.28876
RADIUS OF SATURATION ON ROOF	=	16.61134

PLEASE GROUT ROCK MASS, SEEPAGE LOSS IS HIGH

MANERI BHALI HYDEL PROJECT STAGE I (U.P.)

UNITS USED –	TONNE – METER – DEGREE – CENTIGRADE – SECOND
INPUT FILE NAME –	ILINING.DAT
OUTPUT FILE NAME –	OLINING.DAT

Case No. 5

MODULUS OF ELASTICITY OF ROCK MASS	=	300000.00
POISSON'S RATIO OF ROCK MASS	=	0.20
MODULUS OF ELASTICITY OF CONCRETE	=	2000000.
POISSON'S RATIO OF CONCRETE	=	0.20
MODULUS OF ELASTICITY OF STEEL/LINER	=	20000000.
POISSON'S RATIO OF STEEL	=	0.30

HEIGHT OF OVERBURDEN ON SLOPE SIDE	=	
DESIGN HEAD OF WATER	=	35.00
UNIT WEIGHT OF ROCK MASS	=	2.50
UNIT WEIGHT OF WATER	=	1.00
INTERNAL RADIUS OF CONCRETE LINING	=	3.00
EXTERNAL RADIUS OF CONCRETE LINING	=	3.30
COEFFICIENT OF THERMAL EXPANSION OF CONCRETE	=	0.00001
DIFFERENCE IN TEMPERATURE OF WATER AND ROCK	=	0.00

SAFE COMPRESSIVE STRENGTH OF CONCRETE	=	700.00
ULTIMATE TENSILE STRENGTH OF CONCRETE	=	200.00
SAFE TENSILE STRENGTH OF STEEL/LINER	=	45000.00
DESIGN DIAMETER OF HOOP REINFORCEMENT BARS	=	0.00
RECOMMENDED SPACING OF HOOP REINFORCEMENT BARS	=	0.00
AVERAGE RADIUS OF HOOP REINFORCEMENT BARS	=	0.00
COEFFICIENT OF THERMAL EXPANSION – STEEL/LINER	=	0.00
PERMEABILITY OF ROCK MASS	=	0.000001
HEIGHT OF GROUND WATER TABLE ABOVE CROWN	=	0.00
RATIO OF HORIZONTAL AND VERTICAL IN SITU STRESSES	=	2.00
SLOPE ANGLE OF HILL	=	35.00
ROOF SUPPORT PRESSURE AFTER SATURATION OF ROCK	=	10.00
HOOP STRESS AT INNER SIDE OF SOLID LINING	=	−161.8927
HOOP STRESS AT OUTER SIDE – EMPTY SOLID LINING	=	473.5716
RATIO OF ROCK SHARING AND INTERNAL PRESSURES	=	0.5118363

SIDE ROCK COVER IS ADEQUATE

SEEPAGE LOSS PER UNIT LENGTH OF TUNNEL	=	0.0000931
RADIUS OF SATURATION ON SLOPE SIDE	=	46.54874
RADIUS OF SATURATION ON ROOF	=	10.27029

PLEASE GROUT ROCK MASS, SEEPAGE LOSS IS HIGH

TEHRI DAM PROJECT DIVERSION TUNNEL (U.P.)

UNITS USED –	TONNE – METER – DEGREE – CENTIGRADE – SECOND
INPUT FILE NAME –	ILINING.DAT
OUTPUT FILE NAME –	OLINING.DAT

Case No. 6

MODULUS OF ELASTICITY OF ROCK MASS	=	300000.00
POISSON'S RATIO OF ROCK MASS	=	0.20
MODULUS OF ELASTICITY OF CONCRETE	=	2000000.
POISSON'S RATIO OF CONCRETE	=	0.20
MODULUS OF ELASTICITY OF STEEL/LINER	=	20000000.
POISSON'S RATIO OF STEEL	=	0.30
HEIGHT OF OVERBURDEN ON SLOPE SIDE	=	100.00
DESIGN HEAD OF WATER	=	60.00
UNIT WEIGHT OF ROCK MASS	=	2.50
UNIT WEIGHT OF WATER	=	1.00
INTERNAL RADIUS OF CONCRETE LINING	=	5.50
EXTERNAL RADIUS OF CONCRETE LINING	=	6.40
COEFFICIENT OF THERMAL EXPANSION OF CONCRETE	=	0.00001
DIFFERENCE IN TEMPERATURE OF WATER AND ROCK	=	0.00
SAFE COMPRESSIVE STRENGTH OF CONCRETE	=	700.00
ULTIMATE TENSILE STRENGTH OF CONCRETE	=	200.00
SAFE TENSILE STRENGTH OF STEEL/LINER	=	45000.00
DESIGN DIAMETER OF HOOP REINFORCEMENT BARS	=	0.00
RECOMMENDED SPACING OF HOOP REINFORCEMENT BARS	=	0.00
AVERAGE RADIUS OF HOOP REINFORCEMENT BARS	=	0.00
COEFFICIENT OF THERMAL EXPANSION – STEEL/LINER	=	0.00

PERMEABILITY OF ROCK MASS = 0.0000001
HEIGHT OF GROUND WATER TABLE ABOVE CROWN = 0.00
RATIO OF HORIZONTAL AND VERTICAL IN SITU STRESSES = 2.00
SLOPE ANGLE OF HILL = 35.00
ROOF SUPPORT PRESSURE AFTER SATURATION OF ROCK = 10.00

HOOP STRESS AT INNER SIDE OF SOLID LINING = −222.9616
HOOP STRESS AT OUTER SIDE – EMPTY SOLID LINING = 465.4247
RATIO OF ROCK SHARING AND INTERNAL PRESSURES = 0.3834392

SIDE ROCK COVER IS ADEQUATE

NUMBER OF CRACKS IN CRACKED PCC LINING = 9
SPACING OF CRACKS IN CRACKED PCC LINING = 3.839733
OPENING OF CRACKS IN CRACKED PCC LINING =
COMPRESSIVE HOOP STRESS – EMPTY CRACKED LINING = 465.4247

SEEPAGE LOSS PER UNIT LENGTH OF TUNNEL = 0.0000165
RADIUS OF SATURATION ON SLOPE SIDE = 82.4145
RADIUS OF SATURATION ON ROOF = 18.18353

TEHRI DAM PROJECT HEAD RACE TUNNEL (U.P.)

UNITS USED – TONNE – METER – DEGREE – CENTIGRADE – SECOND
INPUT FILE NAME – ILINING.DAT
OUTPUT FILE NAME – OLINING.DAT

Case No. 7
MODULUS OF ELASTICITY OF ROCK MASS = 80000.00
POISSON'S RATIO OF ROCK MASS = 0.20
MODULUS OF ELASTICITY OF CONCRETE = 2000000.
POISSON'S RATIO OF CONCRETE = 0.20
MODULUS OF ELASTICITY OF STEEL/LINER = 20000000.
POISSON'S RATIO OF STEEL = 0.30

HEIGHT OF OVERBURDEN ON SLOPE SIDE = 200.00
DESIGN HEAD OF WATER = 120.00
UNIT WEIGHT OF ROCK MASS = 2.50
UNIT WEIGHT OF WATER = 1.00
INTERNAL RADIUS OF CONCRETE LINING = 4.00
EXTERNAL RADIUS OF CONCRETE LINING = 4.60
COEFFICIENT OF THERMAL EXPANSION OF CONCRETE = 0.00001
DIFFERENCE IN TEMPERATURE OF WATER AND ROCK = 0.00

SAFE COMPRESSIVE STRENGTH OF CONCRETE = 700.00
ULTIMATE TENSILE STRENGTH OF CONCRETE = 200.00
SAFE TENSILE STRENGTH OF STEEL/LINER = 45000.00
DESIGN DIAMETER OF HOOP REINFORCEMENT BARS = 0.00
RECOMMENDED SPACING OF HOOPREINFORCEMENT BARS = 0.00
AVERAGE RADIUS OF HOOP REINFORCEMENT BARS = 0.00
COEFFICIENT OF THERMAL EXPANSION – STEEL/LINER = 0.00

PERMEABILITY OF ROCK MASS = 0.0000001
HEIGHT OF GROUND WATER TABLE ABOVE CROWN = 0.00
RATIO OF HORIZONTAL AND VERTICAL IN SITU STRESSES = 2.00

SLOPE ANGLE OF HILL	=	35.00
ROOF SUPPORT PRESSURE AFTER SATURATION OF ROCK	=	10.00
HOOP STRESS AT INNER SIDE OF SOLID LINING	=	−704.4156
HOOP STRESS AT OUTER SIDE – EMPTY SOLID LINING	=	936.2016
RATIO OF ROCK SHARING AND INTERNAL PRESSURES	=	0.1623377

COMPRESSIVE STRESS IN EMPTY LINING EXCEEDS SAFE LIMIT

SIDE ROCK COVER IS ADEQUATE

SPACING OF CRACKS IN PCC LINING IS ON UNSAFE SIDE

NUMBER OF CRACKS IN CRACKED PCC LINING	=	16
SPACING OF CRACKS IN CRACKED PCC LINING	=	1.5708
OPENING OF CRACKS IN CRACKED PCC LINING	=	0.00288
COMPRESSIVE HOOP STRESS – EMPTY CRACKED LINING	=	936.2016

COMPRESSIVE STRESS IN EMPTY LINING EXCEEDS SAFE LIMIT

SEEPAGE LOSS PER UNIT LENGTH OF TUNNEL	=	0.0000255
RADIUS OF SATURATION ON SLOPE SIDE	=	127.5433
RADIUS OF SATURATION ON ROOF	=	28.14052

KOPLI HYDEL PROJECT (ASSAM)

UNITS USED –	TONNE – METER – DEGREE – CENTIGRADE – SECOND	
INPUT FILE NAME –	ILINING.DAT	
OUTPUT FILE NAME –	OLINING.DAT	

Case No. 8

MODULUS OF ELASTICITY OF ROCK MASS	=	57000.00
POISSON'S RATIO OF ROCK MASS	=	0.20
MODULUS OF ELASTICITY OF CONCRETE	=	2000000.
POISSON'S RATIO OF CONCRETE	=	0.20
MODULUS OF ELASTICITY OF STEEL/LINER	=	20000000.
POISSON'S RATIO OF STEEL	=	0.30
HEIGHT OF OVERBURDEN ON SLOPE SIDE	=	200.00
DESIGN HEAD OF WATER	=	160.00
UNIT WEIGHT OF ROCK MASS	=	2.50
UNIT WEIGHT OF WATER	=	1.00
INTERNAL RADIUS OF CONCRETE LINING	=	2.25
EXTERNAL RADIUS OF CONCRETE LINING	=	2.45
COEFFICIENT OF THERMAL EXPANSION OF CONCRETE	=	0.00001
DIFFERENCE IN TEMPERATURE OF WATER AND ROCK	=	0.00
SAFE COMPRESSIVE STRENGTH OF CONCRETE	=	700.00
ULTIMATE TENSILE STRENGTH OF CONCRETE	=	200.00
SAFE TENSILE STRENGTH OF STEEL/LINER	=	45000.00
DESIGN DIAMETER OF HOOP REINFORCEMENT BARS	=	0.00
RECOMMENDED SPACING OF HOOP REINFORCEMENT BARS	=	0.00
AVERAGE RADIUS OF HOOP REINFORCEMENT BARS	=	0.00
COEFFICIENT OF THERMAL EXPANSION – STEEL/LINER	=	0.00
PERMEABILITY OF ROCK MASS	=	0.000001
HEIGHT OF GROUND WATER TABLE ABOVE CROWN	=	0.00
RATIO OF HORIZONTAL AND VERTICAL IN SITU STRESSES	=	2.00
SLOPE ANGLE OF HILL	=	35.00
ROOF SUPPORT PRESSURE AFTER SATURATION OF ROCK	=	10.00

HOOP STRESS AT INNER SIDE OF SOLID LINING	=	−1485.974
HOOP STRESS AT OUTER SIDE – EMPTY SOLID LINING	=	2001.117
RATIO OF ROCK SHARING AND INTERNAL PRESSURES	=	0.1944943

COMPRESSIVE STRESS IN EMPTY LINING EXCEEDS SAFE LIMIT
SIDE ROCK COVER IS ADEQUATE
SPACING OF CRACKS IN PCC LINING IS ON UNSAFE SIDE

NUMBER OF CRACKS IN CRACKED PCC LINING	=	32
SPACING OF CRACKS IN CRACKED PCC LINING	=	0.4417875
OPENING OF CRACKS IN CRACKED PCC LINING	=	0.0015158
COMPRESSIVE HOOP STRESS – EMPTY CRACKED LINING	=	2001.117

COMPRESSIVE STRESS IN EMPTY LINING EXCEEDS SAFE LIMIT

SEEPAGE LOSS PER UNIT LENGTH OF TUNNEL	=	0.0002772
RADIUS OF SATURATION ON SLOPE SIDE	=	138.6332
RADIUS OF SATURATION ON ROOF	=	30.58735

PLEASE GROUT ROCK MASS , SEEPAGE LOSS IS HIGH

REFERENCES

Barton, N.R. 1986. Deformation phenomenon in jointed rocks. *Geotechnique* 36(2): 147–163.

Jethwa, J.L. 1981. *Evaluation of rock pressures in tunnels through squeezing ground in lower Himalayas*. Ph.D. Thesis. Department of Civil Engineering, University of Roorkee, India, p. 272.

Kumar, Prabhat and Singh, Bhawani 1990. Design of reinforced concrete lining in pressure tunnels considering thermal effects and jointed rock mass. *Tunnel. Underground Space Technol.* 5(1/2): 91–101.

Oberoi, R.R. and Gupta, G.D. 2000. Tunnelling at Pong Dam – A case Study. *Int. Conf. Tunnelling Asia – 2000*, New Delhi, edited by S.P. Kaushish and T. Ramamurthy, Sponsored by CBIP and ITA, pp. 377–385.

Schleiss, A.J. 1988. Design of reinforced concrete – lined pressure tunnels. *Proc. Conf. Tunnels and Water*, Madrid, 12–15 June, pp. 1127–1133.

Singh, Bhawani, Nayak, G.C., Kumar, R. and Chandra, G. 1988a. Design criteria for plain contrete lining and power tunnels. *Tunnel. Underground Space Technol.* 3(2): 201–208.

Singh, Bhawani, Nayak, G.C., Kumar, R. and Chandra, G. 1988b. Design criteria for plain concrete lining in power and water tunnels. *Int. Symp. Underground Engineering*. New Delhi, India, edited by B. Singh, Vol. 1, pp. 281–289.

Swamee, P.K. and Kahsyap, D. 2001. Design of minimum seepage – Loss in non polygonal canal sections. *J. Irrigation Drainage Eng.*, ASCE 127(2): 113–117.

Vaidya, D.K. and Gupta, A.R. 1998. Grout Plug Failure in the Pressure Shaft of Bhabha Project. *Int. Conf. Hydro Power Development in Himalayas*. Edited by V.D. Choubey. New Delhi, Oxford and IBH Publishing Co. Pvt. Ltd., pp. 435–444.

Verman, M.K. 1993. *Rock mass-tunnel support interaction analysis*. Ph.D. Thesis. Department of Civil Engineering, University of Roorkee, India, p. 258.

CHAPTER 28

A two-dimensional boundary element method – BEM

28.1 GENERAL

The 2D boundary element analysis and computer program was developed by Bray, Hocking, Elissa and Hamett. The principle of boundary element method and program (BEM) has been reported in the book *Underground Excavations in Rock* by Hoek and Brown (1980). The following assumptions are made in the theory.
1. the medium is homogeneous, isotropic and linearly elastic,
2. the plane strain condition prevails or the opening is very long,
3. the medium is infinite,
4. the medium may contain a number of openings of any shape; and
5. the in situ stresses are uniform. Alternatively, in situ stresses increase linearly with depth.

The advantage of boundary element method is that there is no need of preparation of mesh like the finite element method. So the input data is very simple. Hoek and Brown (1980) have explained the boundary element method (BEM) for underground excavation. Results are reliable for symmetric openings. In case of asymmetric openings, only relative displacements (closures) will be accurate as there may be small rigid body displacements.

First the excavated boundary of opening is divided into a number of segments (NSEG) such as straight line, circular arc and elliptic arc (Fig. 28.1). Each segment is then sub-divided into a number of elements (NELR) of equal length. The user should specify code number/or axis of the symmetric openings.

The analysis is based on the theory of elasticity, but the computer program points out where over-stressing takes place. The output indicates no failure, shear failure and tensile failure at various grid points in the medium. The rock mass is assumed to follow the Hoek and Brown criterion of failure. However, the real zone of failure would be much less than that predicted by the elastic theory.

The following parameters in the Hoek and Brown's strength criterion are suggested for 2D plane strain stress analysis (Singh et al., 1997),

$$\sigma_1 = \sigma_3 + \sqrt{m\sigma_3\sigma_c + s\sigma_c^2} \qquad (28.1)$$

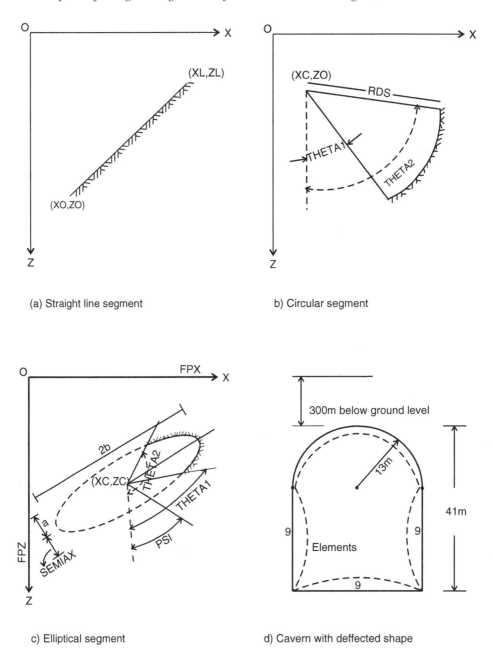

(a) Straight line segment

b) Circular segment

c) Elliptical segment

d) Cavern with deffected shape

Figure 28.1. Input data for boundary element analysis of underground opening.

where

σ_c = q_c = uniaxial compressive strength of rock material in MPa (upper bound in the case of anisotropic rock material)

$$s^{1/2} = 0.7\,\gamma Q^{1/3}/\sigma_c > 0.7\,\gamma Q^{1/3}/100 \qquad (28.2)$$

= strength reduction factor for plane strain failure (long tunnels and caverns)

$$m = m_r s^{1/3} \qquad (28.3)$$

m_r = Hoek and Brown's parameter for rock material (corresponding to the upper bound triaxial test data in the case of anisotropic rock material),

γ = unit weight of rock mass in kN/cum, and

Q = rock mass quality (Chapter 21 and Eq. 21.1).

Hoek (1994), Hoek and Brown (1997) and Hoek et al. (1998) have recommended unrealistically low values of parameters (Singh et al., 1997). Actually, the deformation of an underground opening is a case of restrained dilatancy. So its strength is likely to be close to that of rock material. Hence, there is an order of magnitude enhancement in strength in tunnels. Many investigators have also agreed with Equation 28.2 (Barla, 1995; Barton, 1995; Grimstad and Bhasin, 1996; Choubey, 1998; Aydan, 2000). It should be realised that the compressive strength in plain strain will be much higher than the test data of UCS of large blocks of rock mass. Recently, Hematian et al. (2002) have shown that previous methods for determining the rock mass parameters of the Hoek and Brown's criterion were too conservative to be applied to stability analysis of underground structures. Furthermore, the Hoek and Brown's parameter m_r and brittleness of rock is likely to increase with number of heating and cooling cycles of a rock material in the past. It is interesting to know that the ratio between uniaxial compressive strength and uniaxial tensile strength ($\cong m_r$) increases linearly with the number of heating cycles for the Carrata marble (Ortlepp and Stracey, 1998).

The help menu BEM.TXT gives the sequence of input parameters and their definitions.

The well known predictions of the theory of elasticity are as follows for any opening:

1. Stresses are independent of modulus of deformation of the medium.
2. Stresses are independent of the size of openings which are of similar shape. The stress concentrations at the boundary of opening are the same for similar openings of the same shape.
3. The flow lines of principal stress trajectories are nearly same as flow lines in the laminar flow of water. Thus the flow lines of principal stress trajectories around an opening are same as flow lines of water flowing around a pier which is of same shape as that of opening. The reason is that the differential equation for stresses and laminated flow become exactly same when Poisson's ratio is equal to 0.50. This analogy between stress flow and water flow is helpful in imagination of the stress distribution.
4. The stress distribution is independent of sequence of excavation. It, however, depends only on the final shape of the opening. (In non-linear material, it is not so.)

5. The principle of superposition is applicable in the elastic medium.

One should not be misled that stability of an opening will be independent of its size or width. In fact, there may be tendencies of wedge failures. Further, the failure of rock mass may itself take place due to high overburden.

Elastic theory may, however, be used in the design to determine the zones of tensile stresses and the stress concentration around the caverns. Unfortunately, the zones of tensile stresses are quite deep and large for caverns of large height due to release of in situ stresses along the excavated boundary. One should ensure that the maximum radial tensile strain in the rock mass does not exceed permissible tensile strain of the steel anchor bars of the full-column grouted rock bolts.

The directions of the in situ principal stresses may be inclined near the hill slopes, faults and folds. The orientation of in situ principal stresses changes the stress flow drastically (Brantmark et al., 1998). So the zones of stress concentration may be oriented towards the sharp corners at the bottom of the opening. The computer program BEM may be used to consider the orientation of in situ principal stresses indirectly by orienting the opening accordingly in the input data (see Fig. 28.1d). Plot of opening with stress contours may be re-oriented back to the original shape.

28.2   USER'S MANUAL – BEM

INPUT FILE NAME –   IBEM.DAT
OUTPUT FILE NAME –   OBEM.DAT

THIS PROBLEM IS FOR TWO DIMENSIONAL BOUNDARY ELEMENT ANALYSIS OF MULTIPLE UNDERGROUND OPENINGS (SEE FIG. 28.1)

NAME OF PROGRAM –   BEM.FOR
UNITS USED –   TONNE – METER – DEGREE

GIVE INPUT DATA IN FOLLOWING SEQUENCE

TITLE
ICODE,NSEG,KXS,KZS,NCYC,NSL,DELN,E,RNU
SIGC,RM,S
FPX,FPZ (IF ICODE = 1) OR GAMMA,FSR (IF ICODE = 2)
NELR,XO,ZO,XL,ZL,0.0 0.0 0.0 (STRAIGHT LINE SEGMENT OR)
NELR,XC,ZC,THET1,THET2,RDS,RATIO,PSI(CIRCULAR SEGMENT OR)
NELR,XC,ZC,THET1,THET2,SEMIAX(a),RATIO(b/a),PSI(of a)
{REPEAT ALL OR ANY OF ABOVE THREE LINES NSEG TIMES}
LP1,LP2,BPX,BPZ

DO YOU WANT HELP REGARDING DEFINITIONS OF VARIABLES
ENTER   0 FOR TERMINATION
             1 FOR FURTHER HELP
             2 FOR FURTHER EXECUTION

TITLE   = ANY TITLE UP TO A MAXIMUM OF 80 CHARACTERS
ICODE   = CODE FOR PROBLEM TYPE
                     = 1 FOR INFINITE MEDIUM
                     = 2 FOR GRAVITATIONAL MEDIUM

WHEN ICODE	= 1 USE INPUT LINE 8 ABOVE (LP1, ..., BPZ)
WHEN ICODE	= 2 USE INPUT LINE 9 (BLANK LINE)
NSEG	= TOTAL NUMBER OF SEGMENTS SPECIFIED
	= NUMBER OF SEGMENT LINES
SIGC	= UNIAXIAL COMPRESSIVE STRENGTH OF ROCK MATERIAL
RM	= STRENGTH PARAMETER, m (HOEK AND BROWN)
mr	= STRENGTH PARAMETER, m FOR ROCK MATERIAL
m/mr	= (S)**0.33 (HOEK, 1994)
S	= STRENGTH PARAMETER,s (HOEK AND BROWN, 1980)
	= {0.7*GAMMA(KN/cum)*Q**0.33/SIGC(MPa)}**2, Q < 10
GAMMA	= UNIT WEIGHT OF ROCK MASS
FSR	= FIELD STRESS RATIO, K = FPX/FPZ
KXS,KZS	= CODES OF SYMMETRY
	KXS = 0, KZS = 0 NO SYMMETRY
	KXS = 1, KZS = 0 SYMMETRY ABOUT X-AXIS
	KXS = 0, KZS = 1 SYMMETRY ABOUT Z-AXIS
	KXS = 1, KZS = 1 SYMMETRY ABOUT BOTH X AND Z-AXIS
NCYC	= NUMBER OF CYCLES OF ITERATION
NSL	= NUMBER OF GRID LINES PARALLEL TO THE EXCAVATION BOUNDARY
DELN	= NORMAL DISTANCE BETWEEN GRID LINE AND EXCAVATION BOUNDARY(> ELEMENT LENGTH)
RNU,E	= POISSON'S RATIO, MODULUS OF DEFORMATION (ROCK MASS)
E	= 300*ZC(m)**ALPHA*10**((RMR−20)/38) MPa, ZC > 50 m
ALPHA	= 0.16(HARD ROCKS) TO 0.35(WEAK ROCKS)[VERMAN,1993]
FPX,FPZ	= PRINCIPAL FIELD STRESSES IN X AND Z DIRECTIONS
XO,ZO	= CO-ORDINATES OF THE INITIAL POINT ON A STRAIGHT LINE SEGMENT
XL,ZL	= CO-ORDINATES OF THE FINAL POINT ON A STRAIGHT LINE SEGMENT
NELR	= NUMBERS OF ELEMENTS REQUIRED PER SEGMENT (TOTAL NUMBER OF ELEMENTS <36)
XC, ZC	= CO-ORDINATES OF THE CENTRE OF CIRCLE, ELLIPSE
THET1,THET2	= POLAR ANGLES OF INITIAL AND FINAL POINTS (WITH DOWNWARD VERTICAL Z AXIS) ON A CIRCULAR OR ELLIPTICAL SEGMENT
RDS	= RADIUS OF CIRCLE
SEMIAX	= LENGTH OF ONE SEMI-AXIS (a) OF ELLIPSE
RATIO	= b/a = RATIO OF SEMI AXES
PSI	= POLAR ANGLE OF MINOR SEMI-AXIS WITH Z AXIS
LP1,LP2	= INITIAL AND FINAL ELEMENT NUMBERS TO WHICH LOAD IS APPLIED
BPX,BPZ	= COMPONENTS OF LOAD PER UNIT OF PROJECTED AREA IN X AND Z DIRECTIONS, NOTE THAT BPX AND BPZ ARE POSITIVE INTO THE MATERIAL

OUTPUT VARIABLES IN TABLE OF BOUNDARY STRESS DISTRIBUTION

I	= ELEMENT NUMBER
CX,CZ	= X AND Z CO-ORDINATES OF ELEMENT CENTRE
SIG1,SIG3	= PRINCIPAL STRESSES AT THE CENTRE OF ELEMENT

ALPHA	= ANGLE THAT SIG1 MAKES WITH NORMAL TO BOUNDARY
UX,UZ	= DISPLACEMENTS ALONG X AND Z AXES
FOS	= FACTOR OF SAFETY
BETA	= ANGLE MADE BY FAILURE PLANE WITH SIG1 DIRECTION

OUTPUT VARIABLES IN TABLE OF INTERNAL STRESS DISTRIBUTION

I	= NORMAL GRID LINE NUMBER
CX,CZ	= X AND Z CO-ORDINATES OF INTERIOR POINTS
ALPHA	= ANGLE THAT SIG1 MAKES WITH Z-AXIS

(a) Input File: IBEM.DAT

```
UNDERGROUND CAVREN – CIRCULAR – GRAVITATIONAL FIELD
2 4 0 0 20 3 2.0 7000.0 0.25
150.0 2.5 0.004
0.027 2.0
9 0.0 341.0 26.0 341.0 0. 0. 0.
9 26.0 341.0 26.0 313.0 0. 0. 0.
9 13.0 313.0 90.0 270.0 13.0 0. 0.
9 0.0 313.0 0.0 341.0 0. 0. 0.
0 0 0 0
```

(b) Output File: OBEM.DAT

| INPUT FILE NAME – | IBEM.DAT |
| OUTPUT FILE NAME – | OBEM.DAT |

| UNITS USED – | TONNE – METER – DEGREE |

2-D STRESS ANALYSIS BY BOUNDARY ELEMENT METHOD, PLANE STRAIN
CONDITION, FICTITIOUS STRIP LOADS

UNDERGROUND CAVERN – CIRCULAR – GRAVITATIONAL FIELD

CODE OF PROBLEM TYPE   =   2

ROCK PROPERTIES, SIGC, RM, S   =   150.00, 2.50, 0.0040
NSEG,KXS,KZS,NCYC,NSL   =   4, 0, 0, 20, 3

DELN,E MODULUS,RNU   =   2.00, 7000.00, 0.250

CODE OF THE ANALYSIS IN HOMOGENEOUS GRAVITATIONAL MEDIA

GAMMA, HORZ. STRESS RATIO   =   0.027, 2.000

ELEMENTS	CENTX (FIRST-X)	CENTZ (FIRST-Z)	THET1 (LAST-X)	THET2 (LAST-Z)	RADIUS	RATIO	PSI
9	0.00	341.00	26.00	341.00			
9	26.00	341.00	26.00	313.00			
9	13.00	313.00	90.00	270.00	13.00	1.00	0.00
9	0.00	313.00	0.00	341.00			

1 NN = 0
STRESSES AND DISPLACEMENTS, AND FAILURE CRITERION AT CENTERS OF BOUNDARY ELEMENTS

I	CX	CZ	SIG1	SIG3	ALPHA	UX	UZ	F.O.S	BETA	TYPE OF FAILURE
1	1.444	341.00	57.4	-0.424E-05	-90.00	0.203E-01	-0.838E-02	0.165	25.835	SHEAR FAILURE
2	4.333	341.00	33.3	0.154E-04	-90.00	0.124E-01	-0.169E-01	0.285	21.218	SHEAR FAILURE
3	7.22	341.00	28.5	0.193E-04	-90.00	0.759E-02	-0.213E-01	0.333	19.965	SHEAR FAILURE
4	10.111	341.00	26.7	0.208E-04	-90.00	0.363E-02	-0.235E-01	0.355	19.453	SHEAR FAILURE
5	13.000	341.00	26.2	0.193E-04	90.00	0.326E-06	-0.243E-01	0.362	19.306	SHEAR FAILURE
6	15.889	341.00	26.7	0.244E-04	90.00	-0.363E-02	-0.235E-01	0.355	19.453	SHEAR FAILURE
7	18.778	341.00	28.5	0.326E-04	90.00	-0.759E-02	-0.213E-01	0.333	19.965	SHEAR FAILURE
8	21.667	341.00	33.3	0.490E-04	-90.00	-0.124E-01	-0.169E-01	0.285	21.218	SHEAR FAILURE
9	24.556	341.00	57.4	0.835E-04	-90.00	-0.203E-01	-0.836E-02	0.165	25.835	SHEAR FAILURE
10	26.000	339.444	27.2	-0.401E-04	90.00	-0.518E-01	0.926E-02	0.348	19.604	SHEAR FAILURE
11	26.000	336.333	4.86	-0.354E-04	-90.00	-0.715E-01	0.108E-01	1.954	N.A.	NO FAILURE
12	26.000	333.222	0.971	-0.356E-04	-89.99	-0.831E-01	0.970E-02	9.764	N.A.	NO FAILURE
13	26.000	330.111	0.1E-03	-0.498	-0.001	-0.906E-01	0.801E-02	0.481	0.00	SHEAR FAILURE
14	26.000	327.000	0.1E-03	-1.08	-0.001	-0.951E-01	0.611E-02	0.222	0.00	SHEAR FAILURE
15	26.000	323.889	0.1E-03	-1.15	-0.001	-0.972E-01	0.418E-02	0.208	0.00	SHEAR FAILURE
16	26.000	320.778	0.1E-03	-0.794	-0.001	-0.971E-01	0.237E-02	0.302	0.00	SHEAR FAILURE
17	26.000	317.667	0.7E-01	-0.270E-04	-89.988	-0.950E-01	0.844E-03	123.67	N.A.	NO FAILURE
18	26.000	314.556	1.49	-0.294E-04	-89.999	-0.907E-01	-0.153E-03	6.366	N.A.	NO FAILURE
19	25.608	310.777	6.72	-0.285E-04	-90.00	-0.823E-01	-0.727E-03	1.412	N.A.	NO FAILURE
20	24.087	306.599	20.3	-0.268E-04	90.00	-0.677E-01	-0.174E-03	0.468	17.343	SHEAR FAILURE
21	21.229	303.193	35.8	-0.199E-04	90.00	-0.482E-01	0.117E-02	0.265	21.832	SHEAR FAILURE
22	17.379	300.970	48.0	-0.113E-04	90.00	-0.252E-01	0.247E-02	0.198	24.299	SHEAR FAILURE
23	13.000	300.198	52.6	-0.156E-04	90.00	-0.592E-01	0.300E-02	0.180	25.082	SHEAR FAILURE
24	8.621	300.970	48.0	0.825E-04	90.00	-0.252E-01	0.247E-02	0.198	24.299	SHEAR FAILURE
25	4.771	303.193	35.8	0.156E-04	90.00	-0.482E-01	0.117E-02	0.265	21.832	SHEAR FAILURE
26	1.913	306.599	20.3	0.536E-05	-90.00	0.677E-01	-0.174E-03	0.468	17.343	SHEAR FAILURE
27	0.392	310.777	6.72	-0.859E-06	-90.00	0.823E-01	-0.727E-03	1.412	N.A.	NO FAILURE
28	0.00	314.556	1.49	-0.503E-05	-89.999	0.907E-01	-0.153E-03	6.368	N.A.	NO FAILURE
29	0.00	317.667	0.7E-01	0.149E-05	-89.989	0.950E-01	0.844E-03	123.92	N.A.	NO FAILURE
30	0.00	320.778	0.1E-03	-0.794	-0.001	0.971E-01	0.237E-02	0.302	0.00	TENSILE FAILURE
31	0.00	323.889	0.1E-03	-1.15	-0.001	0.972E-01	0.418E-02	0.208	0.00	TENSILE FAILURE
32	0.00	327.000	0.1E-03	-1.08	-0.001	0.951E-01	0.611E-02	0.222	0.00	TENSILE FAILURE
33	0.00	330.111	0.1E-03	-0.498	-0.002	0.906E-01	0.801E-02	0.482	0.00	TENSILE FAILURE
34	0.00	333.222	0.972	-0.633E-05	-89.99	0.831E-01	0.970E-02	9.761	N.A.	NO FAILURE
35	0.00	336.333	4.86	-0.812E-05	-90.00	0.715E-01	0.108E-01	1.954	N.A.	NO FAILURE
36	0.00	339.444	27.2	0.00	90.00	0.518E-01	0.926E-02	0.348	19.604	SHEAR FAILURE

1 NN = 1
STRESSES AND DISPLACEMENTS, AND FAILURE CRITERION AT CENTERS OF BOUNDARY ELEMENTS

I	CX	CZ	SIG1	SIG3	ALPHA	UX	UZ	F.O.S	BETA	TYPE OF FAILURE
1	1.444	341.00	41.2	7.98	68.632	0.224E−01	−0.552E−02	1.541	N.A.	NO FAILURE
2	4.333	343.00	33.7	1.62	82.683	0.152E−01	−0.135E−01	0.832	20.907	SHEAR FAILURE
3	7.222	343.00	29.3	0.810	86.808	0.933E−02	−0.181E−01	0.705	19.962	SHEAR FAILURE
4	10.111	343.00	27.5	0.609	88.691	0.447E−02	−0.205E−01	0.672	19.497	SHEAR FAILURE
5	13.000	343.00	27.0	0.562	−90.000	0.313E−06	−0.212E−01	0.664	19.358	SHEAR FAILURE
6	15.889	343.00	27.5	0.609	−88.691	−0.446E−02	−0.205E−01	0.672	19.497	SHEAR FAILURE
7	18.778	343.00	29.3	0.810	−86.808	−0.933E−02	−0.181E−01	0.705	19.962	SHEAR FAILURE
8	21.667	343.00	33.7	1.62	−82.683	−0.152E−01	−0.135E−01	0.832	20.907	SHEAR FAILURE
9	24.556	343.00	41.2	7.98	−68.632	−0.224E−01	−0.552E−02	1.541	N.A.	NO FAILURE
10	28.000	339.444	29.7	1.94	−41.873	−0.479E−01	−0.134E−02	1.028	N.A.	NO FAILURE
11	28.000	336.333	11.5	0.284	−21.864	−0.671E−01	−0.255E−02	1.247	N.A.	NO FAILURE
12	28.000	333.222	5.35	0.222	−16.312	−0.790E−01	−0.413E−02	2.501	N.A.	NO FAILURE
13	28.000	330.111	2.88	0.222	−12.535	−0.866E−01	0.438E−02	4.645	N.A.	NO FAILURE
14	28.000	327.000	1.81	0.232	−5.277	−0.912E−01	0.403E−01	7.496	N.A.	NO FAILURE
15	28.000	323.889	1.55	0.191	7.037	−0.933E−01	0.346E−01	8.332	N.A.	NO FAILURE
16	28.000	320.778	1.94	0.900E−01	15.720	−0.933E−01	0.289E−02	5.772	N.A.	NO FAILURE
17	28.000	317.667	2.96	−0.761E−02	19.034	−0.911E−01	0.257E−02	3.147	N.A.	NO FAILURE
18	28.000	314.556	4.85	−0.247	19.439	−0.867E−01	0.273E−02	0.971	0.00	TENSILE FAILURE
19	28.000	310.430	9.71	−0.359	27.542	−0.772E−01	0.362E−02	0.667	0.00	TENSILE FAILURE
20	27.578	305.599	19.8	0.748	42.514	−0.600E−01	0.412E−02	1.009	N.A.	NO FAILURE
21	25.819	301.661	31.0	2.69	57.655	−0.398E−01	0.239E−02	1.157	N.A.	NO FAILURE
22	22.515	299.090	39.6	4.23	73.644	−0.195E−01	−0.157E−02	1.139	N.A.	NO FAILURE
23	18.063	298.198	42.9	4.81	90.000	0.630E−01	−0.136E−01	1.127	N.A.	NO FAILURE
24	13.000	299.090	39.6	4.23	−73.644	0.195E−01	−0.157E−02	1.139	N.A.	NO FAILURE
25	7.937	301.661	31.0	2.69	−57.655	0.398E−01	0.239E−02	1.157	N.A.	NO FAILURE
26	3.458	305.599	19.8	0.748	−42.515	0.600E−01	0.412E−02	1.009	N.A.	NO FAILURE
27	0.181	310.430	9.71	−0.359	−27.543	0.772E−01	0.362E−02	0.667	0.00	TENSILE FAILURE
28	−1.578	314.556	4.85	−0.424E−02	−19.440	0.867E−01	0.273E−02	0.971	0.00	TENSILE FAILURE
29	−2.000	317.667	2.96	0.900E−01	−19.036	0.911E−01	0.257E−02	3.147	N.A.	NO FAILURE
30	−2.000	320.778	1.94	0.191	−15.722	0.933E−01	0.289E−02	5.772	N.A.	NO FAILURE
31	−2.000	323.889	1.55	0.232	−7.039	0.933E−01	0.346E−01	8.334	N.A.	NO FAILURE
32	−2.000	327.000	1.81	0.222	5.266	0.912E−01	0.403E−01	7.487	N.A.	NO FAILURE
33	−2.000	330.111	2.88	0.222	12.535	0.866E−01	0.438E−02	4.644	N.A.	NO FAILURE
34	−2.000	333.222	5.35	0.284	16.311	0.790E−01	−0.413E−02	2.501	N.A.	NO FAILURE
35	−2.000	336.333	11.5	0.284	21.864	0.671E−01	−0.255E−02	1.248	N.A.	NO FAILURE
36	−2.000	339.444	29.7	1.94	41.873	0.479E−01	−0.134E−02	1.028	N.A.	NO FAILURE

1 NN = 2
STRESSES AND DISPLACEMENTS, AND FAILURE CRITERION AT CENTERS OF BOUNDARY ELEMENTS

I	CX	CZ	SIG1	SIG3	ALPHA	UX	UZ	F.O.S	BETA	TYPE OF FAILURE
1	1.444	345.000	33.0	9.33	68.130	0.202E−01	−0.521E−02	2.101	N.A.	NO FAILURE
2	4.333	345.000	32.6	3.71	77.654	0.153E−01	−0.108E−01	1.297	N.A.	NO FAILURE
3	7.222	345.000	30.2	1.89	83.771	0.993E−01	−0.150E−01	0.997	19.916	SHEAR FAILURE
4	10.111	345.000	28.8	1.33	87.344	0.484E−01	−0.174E−01	0.888	19.667	SHEAR FAILURE
5	13.000	345.000	28.3	1.19	−90.00	0.348E−01	−0.181E−01	0.859	19.575	SHEAR FAILURE
6	15.889	345.000	28.8	1.33	−87.771	−0.484E−01	−0.174E−01	0.888	19.667	SHEAR FAILURE
7	18.779	345.000	30.2	1.89	−83.771	−0.993E−01	−0.150E−01	0.998	19.916	SHEAR FAILURE
8	21.667	345.000	32.6	3.71	−77.654	−0.153E−01	−0.108E−01	1.297	N.A.	NO FAILURE
9	24.556	345.000	33.0	9.33	−68.130	−0.202E−01	−0.521E−02	2.101	N.A.	NO FAILURE
10	30.000	339.444	26.5	2.18	−51.004	−0.468E−01	−0.508E−02	1.218	N.A.	NO FAILURE
11	30.000	336.333	16.2	0.468	−33.198	−0.630E−01	−0.265E−02	1.034	N.A.	NO FAILURE
12	30.000	333.222	9.71	0.382	−23.338	−0.746E−01	−0.239E−03	1.612	N.A.	NO FAILURE
13	30.000	330.111	6.41	0.463	−15.95	−0.823E−01	0.126E−02	2.608	N.A.	NO FAILURE
14	30.000	327.000	4.79	0.524	−7.605	−0.870E−01	0.211E−02	3.646	N.A.	NO FAILURE
15	30.000	323.889	4.22	0.504	2.507	−0.892E−01	0.266E−02	4.076	N.A.	NO FAILURE
16	30.000	320.778	4.49	0.387	12.110	−0.892E−01	0.313E−02	3.503	N.A.	NO FAILURE
17	30.000	317.667	5.53	0.209	19.324	−0.870E−01	0.375E−02	2.387	N.A.	NO FAILURE
18	30.000	314.556	7.51	−0.636E−0	25.136	−0.826E−01	0.472E−02	1.074	N.A.	NO FAILURE
19	29.547	310.082	12.10	−0.253	35.629	−0.725E−01	0.636E−02	0.950	0.00	TENSILE FAILURE
20	27.551	304.599	20.0	1.10	51.326	−0.541E−01	0.661E−02	1.175	N.A.	NO FAILURE
21	23.800	300.129	27.7	4.03	65.176	−0.339E−01	0.311E−02	1.590	N.A.	NO FAILURE
22	18.747	297.211	33.3	6.70	77.863	−0.158E−01	−0.163E−02	1.734	N.A.	NO FAILURE
23	13.000	296.198	35.3	7.78	90.00	0.622E−06	−0.380E−02	1.772	N.A.	NO FAILURE
24	7.253	297.211	33.3	6.70	−77.863	0.158E−01	−0.163E−02	1.734	N.A.	NO FAILURE
25	2.200	300.129	27.7	4.03	−65.176	0.339E−01	0.311E−02	1.590	N.A.	NO FAILURE
26	−1.551	304.599	20.0	1.10	−51.326	0.541E−01	0.661E−02	1.175	N.A.	NO FAILURE
27	−3.547	310.082	12.1	−0.253	−35.629	0.725E−01	0.636E−02	0.949	0.00	TENSILE FAILURE
28	−4.00	314.556	7.51	−0.637E−01	−25.136	0.826E−01	0.472E−02	1.074	N.A	NO FAILURE
29	−4.00	317.667	5.53	0.209	−19.324	0.870E−01	0.375E−02	2.387	N.A.	NO FAILURE
30	−4.00	320.778	4.49	0.387	−12.110	0.892E−01	0.313E−02	3.503	N.A.	NO FAILURE
31	−4.00	323.889	4.22	0.504	−2.508	0.893E−01	0.266E−02	4.077	N.A.	NO FAILURE
32	−4.00	327.000	4.79	0.524	7.603	0.870E−01	0.212E−02	3.646	N.A.	NO FAILURE
33	−4.00	330.111	6.41	0.463	15.950	0.823E−01	0.126E−02	2.608	N.A.	NO FAILURE
34	−4.00	333.222	9.71	0.382	23.338	0.746E−01	−0.239E−03	1.612	N.A.	NO FAILURE
35	−4.00	336.333	16.2	0.468	33.198	0.630E−01	−0.265E−02	1.034	N.A.	NO FAILURE
36	−4.00	339.444	26.5	2.18	51.004	0.468E−01	−0.508E−02	1.218	N.A.	NO FAILURE

1 NN = 3
STRESSES AND DISPLACEMENTS, AND FAILURE CRITERION AT CENTERS OF BOUNDARY ELEMENTS

I	CX	CZ	SIG1	SIG3	ALPHA	UX	UZ	F.O.S	BETA	TYPE OF FAILURE
1	1.444	347.000	29.2	9.48	70.708	0.179E−01	−0.511E−02	2.390	N.A.	NO FAILURE
2	4.333	347.000	30.4	5.49	76.239	0.142E−01	−0.888E−02	1.707	N.A.	NO FAILURE
3	7.222	347.000	29.8	3.34	82.059	0.963E−02	−0.123E−01	1.343	N.A.	NO FAILURE
4	10.111	347.000	29.0	2.46	86.420	0.481E−02	−0.145E−01	1.181	N.A.	NO FAILURE
5	13.000	347.000	28.8	2.23	−90.000	0.317E−06	−0.152E−01	1.135	N.A.	NO FAILURE
6	15.889	347.000	29.0	2.46	−86.420	−0.481E−02	−0.145E−01	1.181	N.A.	NO FAILURE
7	18.779	347.000	39.8	3.34	−82.059	−0.963E−02	−0.123E−01	1.343	N.A.	NO FAILURE
8	21.667	347.000	30.4	5.49	−76.239	−0.142E−01	−0.888E−02	1.707	N.A.	NO FAILURE
9	24.556	347.000	29.2	9.48	−70.708	−0.179E−01	−0.511E−02	2.390	N.A.	NO FAILURE
10	32.000	339.444	24.1	2.66	−54.407	−0.463E−01	−0.710E−02	1.480	N.A.	NO FAILURE
11	32.000	336.333	17.8	1.21	−39.823	−0.597E−01	−0.560E−02	1.381	N.A.	NO FAILURE
12	32.000	333.222	12.4	0.960	−28.909	−0.705E−01	−0.326E−02	1.790	N.A.	NO FAILURE
13	32.000	330.111	9.00	1.01	−19.795	−0.781E−01	−0.117E−02	2.515	N.A.	NO FAILURE
14	32.000	327.000	7.13	1.07	−10.218	−0.828E−01	0.493E−03	3.263	N.A.	NO FAILURE
15	32.000	323.889	6.37	1.06	0.574	−0.851E−01	0.187E−02	3.628	N.A.	NO FAILURE
16	32.000	320.778	6.51	0.923	11.198	−0.851E−01	0.314E−02	3.348	N.A.	NO FAILURE
17	32.000	317.667	7.47	0.707	20.152	−0.830E−01	0.445E−02	2.616	N.A.	NO FAILURE
18	32.000	314.556	9.26	0.463	27.912	−0.787E−01	0.591E−02	1.802	N.A.	NO FAILURE
19	31.517	309.735	13.3	0.284	40.00	−0.684E−01	0.801E−02	1.072	N.A.	NO FAILURE
20	29.283	303.599	20.0	1.51	56.468	−0.495E−01	0.805E−02	1.360	N.A.	NO FAILURE
21	25.086	298.596	25.7	4.57	70.398	−0.297E−01	0.353E−02	1.831	N.A.	NO FAILURE
22	19.431	295.331	29.3	7.76	81.358	−0.132E−01	−0.244E−02	2.132	N.A.	NO FAILURE
23	13.000	294.198	30.5	9.13	90.000	0.563E−06	−0.515E−02	2.242	N.A.	NO FAILURE
24	6.569	295.331	29.3	7.76	−81.358	0.132E−01	−0.244E−02	2.132	N.A.	NO FAILURE
25	0.914	298.596	25.7	4.57	−70.398	0.297E−01	0.353E−02	1.831	N.A.	NO FAILURE
26	−3.283	303.599	20.0	1.51	−56.468	0.495E−01	0.805E−02	1.360	N.A.	NO FAILURE
27	−5.517	309.735	13.3	0.284	−40.00	0.684E−01	0.805E−02	1.070	0.00	TENSILE FAILURE
28	−6.00	314.556	9.26	0.463	−27.912	0.787E−01	0.591E−02	1.802	N.A.	NO FAILURE
29	−6.00	317.667	7.47	0.707	−20.153	0.830E−01	0.445E−02	2.616	N.A.	NO FAILURE
30	−6.00	320.778	6.51	0.923	−11.198	0.851E−01	0.314E−02	3.348	N.A.	NO FAILURE
31	−6.00	323.889	6.37	1.06	−0.575	0.851E−01	0.187E−02	3.629	N.A.	NO FAILURE
32	−6.00	327.000	7.13	1.07	10.218	0.828E−01	0.494E−03	3.263	N.A.	NO FAILURE
33	−6.00	330.111	9.00	1.01	19.795	0.781E−01	−0.117E−02	2.515	N.A.	NO FAILURE
34	−6.00	333.222	12.4	0.960	28.909	0.705E−01	−0.326E−02	1.791	N.A.	NO FAILURE
35	−6.00	336.333	17.8	1.21	39.823	0.597E−01	−0.560E−02	1.382	N.A.	NO FAILURE
36	−6.00	339.444	24.1	2.66	54.407	0.463E−01	−0.710E−02	1.480	N.A.	NO FAILURE

REFERENCES

Aydan, Omer, Dalgic, S. and Kawamoto, T. 2000. Prediction of squeezing potential of rocks in tunnelling through a combination of an analytical method and rock mass classification. *Italian Geotech J.* Anno XXXIV(1): 41–45.

Barla, G. 1995. Squeezing rocks in tunnels. *ISRM News J.* 2(3&4): 44–49.

Barton, N. 1995. The influence of joint properties in modelling jointed rock masses. *Eight Int. Rock Mech. Congress*, Tokyo, Vol. 3, pp. 1023–1032.

Brantmark, J., Heiner, A., Martana, J. and Stille, H. 1998. Stabilisation of the Machinery Hall, Uri Hydroelectric Power Project, India. *Int. Conf. Hydro Power Development in Himalayas*. Edited by V.D. Choubey. New Delhi, India, Oxford 7 IBH Publishing Co. Pvt. Ltd., pp. 178–192.

Choubey, V.D. 1998. Potential of rock mass classification for design of tunnel supports – hydro electric projects in Himalayas. *Int. Conf. on Hydro Power Development in Himalayas*, Shimla, India, pp. 305–336.

Grimstad, E. and Bhasin, R. 1996. Stress strength relationship and stability in hard rock. *Proc. Conf. on Recent Advances in Tunnelling Technology*, New Delhi, India, Vol. 1, pp. 3–8.

Hematian, J., Porter, I. and Singh, R.N. 2002. Application of the conventional and a new failure criteria in the stability analysis of underground structures. *J. Rock Mech. Tunnel. Tech.*, Indian Society of Rock Mechanics and Tunnelling Technology, New Delhi, Vol. 8, No. 1 (in print).

Hoek, E. 1994. Strength of rock and rock Mases. *ISRM News J.* 2(2): 4–16.

Hoek, E. and Brown, E.T. 1980, *Underground Excavations in Rock*. Revised Edition. Appendix 4: Two-dimensional boundary element stress analysis. England, Institution of Mining and Metallurgy, p. 527.

Hoek, E. and Brown, E.T. 1997. Practical estimates of rock mass strength. *Int. J. Rock Mech. Mining Sci. Geomech. Abstr.* 34(8): 1165–1186.

Hoek, E., Marinos, P. and Benissi, M. 1998, Applicability of the geological strength index (GSI) classification for very weak and sheared rock masses – the case of the Athens Schist formation. *Bull. Eng. Geo. Environ.* 15: 151–160.

Ortlepp, W.D. and Stracey, T.R. 1998. Rock Burst Mechanism in tunnels and shafts. *Rock Mech. Rock Eng.* 31(3).

Singh, Bhawani, Viladkar, M.N., Samadhiya N.K. and Mehrotra, V.K. 1997. Rock mass strength parameters mobilised in tunnels. *Tunnel. Underground Space Technol.* 12(1): 47–54.

Verman, M.K. 1993. *Rock mass – tunnel support interaction analysis*. Ph.D. Thesis. University of Roorkee, India.

CHAPTER 29

Subsidence due to mining – MSEAMS

29.1   GENERAL

Subsidence is an unwanted phenomenon. It may cause damage to surface structures. Subsidence control measures may be taken in stages; (i) prediction, (ii) prevention and (iii) protection. The effectiveness of preventive and protective measures depends greatly on the accuracy of the prediction. The civil and mining engineers should join forces to reduce damage to buildings which are worth of billion dollars all over the world, particularly in the over-populated regions.

Subsidence due to mining is quite high. Other causes of subsidence are (i) compaction of aquifer system, (ii) consolidation of the clayey soils and rocks due to load of structures or lowering of ground water table, (iii) pumping of oil through oil wells, (iv) pumping out of ground water from the aquifer between the clayey layers, (v) drainage and subsequent oxidation of organic soils; and (vi) collapse of the susceptible rocks. Here only subsidence due to mining is analysed.

The two-dimensional program MSEAMS developed by Crouch and Starfield (1983) is based on Displacement Discontinuity Method (DDM) and is suitable for stress analysis of excavation in seams which are nearly horizontal and tabular ore deposits. The program assumes the overburden rock mass as homogeneous orthotropic or transversely isotropic and linearly elastic medium. It has been shown that layered overburden rock mass generally found above the coal mines can be simulated by transversely isotropic medium of very low shear modulus (Singh, 1973).

The program is specially suitable for the study of subsidence. With proper selection of input rock properties the program may produce the desired magnitude and profile of the subsidence movements. Aston et al. (1987) found MSEAMS producing results virtually identical to those obtained with non-linear FEM programs GEOROC and ADINA. The program has been found to yield best predictions among those obtained from other numerical modelling programs viz. MINAP, THREED and VEDICA (Gurtunca and Bhatacharyya, 1988). McNabb (1987) compared the results of MSEAMS with three dimensional DDM program MINLAY and found MSEAMS giving more realistic profiles. The 3D program MULSIM/BM of USBM and MULSIM of Sinha (1979) assumed rock mass to be isotropic.

29.2   CAPABILITIES OF THE PROGRAM

The program MSEAMS is suitable for two-dimensional idealisation of multi-seam workings. Since the effect of the ground surface is considered here, the program is specially suitable for subsidence studies. The coal seam elements need to be defined as mined or unmined and depth of excavation is to be specified. Both, longwall and board and pillar mines may be modelled with this program. Because of its capability of modelling board and pillar mines, this program has been used for specific studies like effects of subsidence on service decline in mines having partial extraction (Bhattacharyya et al., 1988).

This program written in Fortran 77 is capable of modelling up to 201 elements in each seam. Any consistent set of units may be used in the program. The program is simple to use and at the same time versatile as well and takes into account the anisotropy of rock mass. The program computes the stresses, displacements and strains induced by mining of horizontally lying seams. These values may be computed along a desired profile line at the seam level and/or at the specified off-seam planes. As many as ten off-seam planes in the overburden rock mass including the surface of the ground ($y = 0$) can be specified. The program cannot be used for dipping seams and the time factor is not taken into account. Proper selection of rock mass properties is of great significance in numerical modelling of a subsidence problem. This may also be done by calibrating a model with known field data.

In the transversely anisotropic material model MSEAMS, Young's modulus in horizontal and vertical directions, E_h and E_v, shear modulus G and Poisson's ratio v are required to be defined. Therefore, five independent constants required are Young's modulus in vertical and horizontal directions, the two Poisson's ratios v_v, v_h and independent shear modulus. It may be noted that unloading of stresses takes place in the vertical direction and so E_v is taken as modulus of elasticity in the unloading condition. For the isotropic seam material, Young's modulus E_c and shear modulus G_c are given.

Bahuguna (1993) has made an extensive study of subsidence due to coal mines in 65 Indian cases. He concluded that the anisotropic model gives more realistic predictions than the isotropic model. The back analysed values of G/E_v are given in MSEAMS.TXT (help menu) for various classifications of the rock masses. The more disturbed the rock mass is, the lesser is the ratio between shear modulus and Young's modulus. Further, the anisotropic model used in MSEAMS simulates realistically the effect of seam width and overburden on the subsidence. Crouch's program has been extensively used worldwide with satisfaction. Computed results should be checked with empirical predictions for cross-confirmation (Bahuguna et al., 1993).

29.3   THE TOLERANCE LIMIT OF STRUCTURES

It is equally important to check the tolerance limits of the civil engineering structures (Fig. 29.1). Table 29.1 lists the safe limits of subsidence. The horizontal

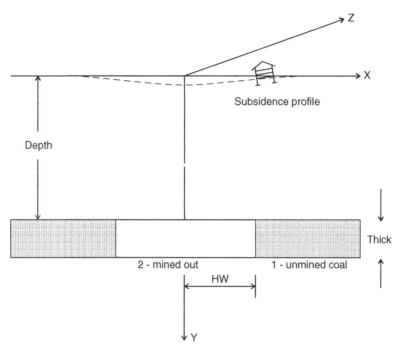

Figure 29.1. Input data for calculation of subsidence above horizontal mines.

Table 29.1. Safe limits of subsidence movement.

S.No.	Structure	Safe limits
1	Railway lines of jointed construction	
	Maximum permissible strain	3 mm/m
	Limiting operating long gradient	1 in 100
2	Railway lines of welded construction	No movement permitted
3	Water bodies	
	Maximum permissible tensile strain	4.5 mm/m
4	Buildings	
	Maximum permissible total compression or elongation for slight repairable damages	see Table 10.2
5	High tension pylons and aerial ropeway trestles	
	Maximum permissible strain	3 mm/m
	Maximum permissible displacement of the top most point	1/3 of the radius of the base

strain in the foundation of existing building induced by subsidence is most damaging (Table 29.2). According to Mair, Taylor and Burland (1996) the permissible horizontal strain is 0.075 percent for buildings of brick masonry. They have also given a chart for permissible subsidence for any horizontal strain. Table 29.3 gives the classification of damage to buildings.

Table 29.2. Relationship between category of damage and limiting tensile strain ε_{lim} (Mair, Taylor and Burland, 1996).

Category of damage	Normal degree of severity	Limiting tensile strain (ε_{lim}), in percent
0	Negligible	0–0.05
1	Very slight	0.05–0.075
2	Slight	0.075–0.15
3	Moderate	0.15–0.30
4–5	Severe to very severe	>0.30

Table 29.3. Classification of visible damage to walls with particular reference to ease of repair of plaster and brick work or masonry (Mair, Taylor and Burland, 1996).

Category	Normal degree of severity	Description of typical damage (ease of repair is in italics)
0	Negligible	Hairline cracks less than about 0.1 mm
1	Very slight	*Fine cracks which are easily treated during normal decoration.* Damage generally restricted to internal wall finishes. Close inspection may reveal some cracks in external brick work or masonry. Typical crack widths up to 1 mm
2	Slight	*Cracks easily filled up. Re-decoration probably required. Recurrent cracks can be masked by suitable lining.* Cracks may be visible externally and *some repointing may be required to ensure weathering tightness.* Doors and windows may stick slightly. Typical crack widths up to 5 mm.
3	Moderate	*The cracks require some opening up and can be patched by a mason. Repointing of external brick work and possibly a small amount of brick work to be replaced.* Doors and windows sticking. Service pipes may fracture. Weather tightness often impaired. Typical crack widths are 5 to 15 mm or several greater than 3 mm.
4	Severe	*Extensive repair work involving breaking out and replacing sections of walls, especially over doors and windows.* Windows and door frames distorted, floor sloping noticeably.* Walls leaning or bulging noticeably, some loss of bearing in beams. Service pipes disrupted. Typical crack widths are 15 to 25 mm but also depends on the number of cracks.
5	Very severe	*This requires a major repair job involving partial or complete rebuilding.* Beams lose bearing, walls lean badly and require shoring. Windows broken with distortion. Danger of instability. Typical crack widths are greater than 25 mm but depends on the number of cracks.

*Local deviation of slope from the horizontal or vertical of more than 1/100 will normally be clearly visible. Overall deviations in excess of 1/150 are undesirable.

Note: Crack width is only one factor in assessing category of damage and should not be used on its own as a direct measure of it. In the case of RCC frames of multi-storeyed buildings with isolated footings, the vertical subsidence will be more critical than the horizontal strain.

29.4 USER'S MANUAL – MSEAMS

THIS PROGRAM IS FOR COMPUTING THE DISPLACEMENTS AND STRESSES INDUCED BY MINING OF MULTIPLE AND PARALLEL SEAM TYPE DEPOSITS (USING DISPLACEMENT DISCONTINUITY METHOD FOR ANISOTROPIC ROCK MASS) (SEE FIG. 29.1)

NAME OF PROGRAM –	MSEAMS.FOR
UNITS USED –	TONNE – METER – DEGREE

GIVE INPUT DATA IN THE FOLLOWING SEQUENCE

TITLE OF PROBLEM (<81 CHARACTERS)
NSEAMS,NSEG,NITER,NUMOS
EXX,EYY,EZZ,GXY,VYX,VZX,VYZ
DENSE,OMEGA,SRATIO,HW
THICK,DEPTH,ESEAM,GSEAM
MIN (I),I = 1,NSEG
REPEAT ABOVE TWO LINES NSEAMS TIMES
YP,ISEGB,ISEGE
REPEAT ABOVE LINE NUMOS TIMES

DO YOU WANT HELP REGARDING DEFINITIONS OF VARIABLES USED
ENTER   0 FOR TERMINATION
         1 FOR FURTHER HELP
         2 FOR EXECUTION

NSEAMS	=	NO OF SEAMS(<6)
NSEG	=	NUMBER OF SEAM ELEMENTS IN EACH SEAM(<201)
NITER	=	MAXIMUM NUMBER OF ITERATIONS TO BE PERFORMED FOR SOLVING EQUATIONS BY THE METHOD OF SUCCESSIVE OVER-RELAXATION (<1000)
NUMOS	=	NUMBER OF SEPARATE LINES (Y = CONSTANT,NOT ON SEAM<11) WHERE DISPLACEMENTS AND STRESSES ARE TO BE COMPUTED
EXX	=	MODULUS OF DEFORMATION ALONG X AXIS (HORIZ.)
EYY	=	MODULUS OF DEFORMATION ALONG Y AXIS (VERTICAL) IN MPa = 500 + 195*(% OF HARD ROCK LAYERS IN OVERBURDEN)
EZZ	=	MODULUS OF DEFORMATION ALONG Z AXIS
GXY	=	SHEAR MODULUS FOR ANISOTROPIC ROCK MASS ON X–Y PLANE (VALUE FOR NEARLY ISOTROPIC CASE WILL GIVE ERROR)
GXY/EYY	=	0.07–0.38 FOR MASSIVE ROCK MASS WITH FEW PLANES OF WEAKNESSES
	=	0.03–0.07 FOR NO PREVIOUS MINING BUT FEW NATURAL DISCONTINUITIES
	=	0.016–0.03 FOR PARTING BETWEEN MINED SEAMS MORE THAN 5 TIMES SEAM THICKNESS OR CASES WITH NO PREVIOUS MINING BUT MANY NATURAL DISCONTINUITIES
	=	0.005–0.016 FOR PARTING BETWEEN MINED SEAMS NOT MORE THAN 5 TIMES SEAM THICKNESS
		<0.005 FOR HIGHLY FRAGMENTED ROCK MASS WITH REPETITIVE WORKINGS OR HAVING VERY THICK SEAMS MINED IN DESCENDING SLICING
VYX	=	POISSON'S RATIO ON Y–X PLANE
VZX	=	POISSON'S RATIO ON Z–X PLANE
VYZ	=	POISSON'S RATIO ON Y–Z PLANE
DENSE	=	UNIT WEIGHT OF ROCK MASS
OMEGA	=	OVER-RELAXATION FACTOR FOR SOLVING EQUATIONS (=0)
SRATIO	=	RATIO OF IN SITU (PRIMITIVE) STRESSES ALONG X/Y AXES
HW	=	HALF WIDTH OF SEAM ELEMENTS

THICK   = SEAM THICKNESS
DEPTH   = DEPTH OF SEAM
ESEAM   = MODULUS OF DEFORMATION OF SEAM
GSEAM   = SHEAR MODULUS OF SEAM
MIN   = 2 IF ELEMENT IS UNMINED
         = 1 IF ELEMENT IS MINED
YP   = Y-CO-ORDINATE OF OFF SEAM LINE FOR CALCULATION OF STRESSES
           FROM ISEGB TO ISEGE ELEMENTS

(a) Input File: IMSEAMS.DAT

Crouch anisotropic test problem – 168 m wide, 600 m deep
1 80 999 1
6250. 25000. 6250. 1250. 0.0 0.0 0.0
0.026 1.9 1.00 4.2
1.2 600. 3500. 1200.
0. 41 80

(b) Output File: OMSEAMS.DAT

INPUT FILE NAME –   IMSEAMS.DAT
OUTPUT FILE NAME –   OMSEAMS.DAT
UNITS USED –   KN, TONNE – METER – DEGREE

Crouch anisotropic test problem – 168 m wide, 600 m deep
NUMBER OF SEPARATE SEAMS   =   1
NUMBER OF SEGMENTS IN EACH SEAM   =   80
HALF-WIDTH OF SEGMENTS   =   4.20
NUMBER OF LINES Y = CONSTANT (OTHER THAN A SEAM) ALONG
WHICH DISPLACEMENTS AND STRESSES ARE TO BE COMPUTED   =   1
MAXIMUM NUMBER OF ITERATIONS TO BE PERFORMED   =   999
OVER-RELAXATION FACTOR   =   1.90
UNIT WEIGHT OF ROCK   =   0.0260
RATIO OF PRE-EXISTING STRESSES SIGXX/SIGYY   =   1.0000

ELASTIC CONSTANTS
EXX   = 0.6250E + 04
EYY   = 0.2500E + 05
EZZ   = 0.6250E + 04
GXY   = 0.1250E + 04
VYX   = 0.00
VZX   = 0.00
VYZ   = 0.00

MINING PATTERN (1 = MINED, 2 = UNMINED)

ELASTIC PARAMETERS
K1   = 2.00
K2   = 10.00
GAM1   = 4.45
GAM2   = 0.45
Q1   = 97.99
Q2   = 0.01

NUMBER OF ITERATIONS   =   133
MAXIMUM DIFFERENCE BETWEEN SUCCESSIVE ITERATES   =   0 .730E−06

STRESSES, DISPLACEMENTS AND DISPLACEMENT DISCONTINUITY COMPONENTS AT SEAM LEVEL(S)

SEAM	SEG	UX (POS)	UX (NEG)	DX (=DS)	UY (POS)	UY (NEG)	DY (=DN)	SIGXX	SIGYY
1	1	−0.015344	−0.015253	0.000091	−0.021076	−0.020909	0.000167	0.1	16.1
1	2	−0.015798	−0.015698	0.000099	−0.021794	−0.021606	0.000188	0.1	16.1
1	3	−0.016278	−0.016173	0.000104	−0.022534	−0.022328	0.000206	0.1	16.2
1	4	−0.016784	−0.016676	0.000109	−0.023298	−0.023073	0.000225	0.1	16.3
1	5	−0.017319	−0.017206	0.000113	−0.024088	−0.023843	0.000245	0.1	16.3
1	6	−0.017884	−0.017767	0.000117	−0.024904	−0.024638	0.000266	0.1	16.4
1	7	−0.018483	−0.018362	0.000121	−0.025747	−0.025456	0.000290	0.1	16.4
1	8	−0.019118	−0.018993	0.000125	−0.026616	−0.026299	0.000317	0.1	16.5
1	9	−0.019793	−0.019664	0.000129	−0.027512	−0.027166	0.000346	0.1	16.6
1	10	−0.020512	−0.020379	0.000133	−0.028434	−0.028057	0.000378	0.1	16.7
1	11	−0.021282	−0.021145	0.000137	−0.029382	−0.028969	0.000413	0.1	16.8
1	12	−0.022106	−0.021966	0.000140	−0.030356	−0.029904	0.000453	0.1	16.9
1	13	−0.022994	−0.022850	0.000144	−0.031355	−0.030858	0.000497	0.1	17.0
1	14	−0.023952	−0.023805	0.000147	−0.032378	−0.031831	0.000546	0.1	17.2
1	15	−0.024991	−0.024841	0.000150	−0.033423	−0.032821	0.000602	0.1	17.4
1	16	−0.026122	−0.025969	0.000153	−0.034489	−0.033824	0.000665	0.2	17.5
1	17	−0.027361	−0.027206	0.000156	−0.035576	−0.034838	0.000738	0.2	17.8
1	18	−0.028726	−0.028568	0.000158	−0.036680	−0.035859	0.000821	0.2	18.0
1	19	−0.030237	−0.030078	0.000159	−0.037800	−0.036882	0.000918	0.2	18.3
1	20	−0.031926	−0.031765	0.000160	−0.038935	−0.037903	0.001032	0.2	18.6
1	21	−0.033827	−0.033666	0.000161	−0.040083	−0.038914	0.001169	0.2	19.0
1	22	−0.035991	−0.035830	0.000161	−0.041243	−0.039908	0.001334	0.2	19.5
1	23	−0.038483	−0.038322	0.000161	−0.042416	−0.040875	0.001541	0.2	20.1
1	24	−0.041398	−0.041237	0.000161	−0.043605	−0.041800	0.001804	0.2	20.9
1	25	−0.044872	−0.044712	0.000160	−0.044818	−0.042664	0.002155	0.2	21.9
1	26	−0.049119	−0.048962	0.000159	−0.046077	−0.043433	0.002643	0.2	23.3
1	27	−0.054498	−0.054338	0.000160	−0.047426	−0.044049	0.003377	0.2	25.5
1	28	−0.061692	−0.061528	0.000164	−0.048988	−0.044372	0.004616	0.2	29.1
1	29	−0.072291	−0.072112	0.000179	−0.051181	−0.043961	0.007220	0.2	36.7

STRESSES, DISPLACEMENTS AND DISPLACEMENT DISCONTINUITY COMPONENTS AT SEAM LEVEL(S)

SEAM	SEG	UX (POS)	UX (NEG)	DX (=DS)	UY (POS)	UY (NEG)	DY (=DN)	SIGXX	SIGYY
1	30	−0.091811	−0.091555	0.000256	−0.056694	−0.040083	0.016612	0.3	64.0
1	31	−0.101522	−0.099807	0.001716	−0.161671	0.063275	0.224946	0.0	0.0
1	32	−0.092727	−0.090542	0.002185	−0.211294	0.111266	0.322560	0.0	0.0
1	33	−0.082627	−0.080288	0.002338	−0.245693	0.144169	0.389862	0.0	0.0
1	34	−0.072058	−0.069755	0.002303	−0.271717	0.168851	0.440569	0.0	0.0
1	35	−0.061242	−0.059107	0.002136	−0.291963	0.187923	0.479886	0.0	0.0
1	36	−0.050271	−0.048402	0.001869	−0.307721	0.202686	0.510407	0.0	0.0
1	37	−0.039193	−0.037667	0.001526	−0.319725	0.213883	0.533608	0.0	0.0
1	38	−0.028043	−0.026914	0.001129	−0.328423	0.221969	0.550392	0.0	0.0
1	39	−0.016844	−0.016152	0.000692	−0.334089	0.227224	0.561313	0.0	0.0
1	40	−0.005618	−0.005384	0.000233	−0.336885	0.229813	0.566698	0.0	0.0
1	41	0.005618	0.005384	−0.000233	−0.336885	0.229813	0.566699	0.0	0.0
1	42	0.016844	0.016152	−0.000692	−0.334089	0.227224	0.561313	0.0	0.0
1	43	0.028043	0.026914	−0.001129	−0.328423	0.221969	0.550392	0.0	0.0
1	44	0.039193	0.037667	−0.001526	−0.319725	0.213883	0.533609	0.0	0.0
1	45	0.050271	0.048402	−0.001869	−0.307721	0.202686	0.510407	0.0	0.0
1	46	0.061242	0.059107	−0.002136	−0.291963	0.178923	0.479886	0.0	0.0
1	47	0.072058	0.069755	−0.002303	−0.271717	0.168851	0.440569	0.0	0.0
1	48	0.082627	0.080288	−0.002338	−0.245693	0.144169	0.389862	0.0	0.0
1	49	0.092727	0.090542	−0.002185	−0.211295	0.111266	0.322561	0.0	0.0
1	50	0.101522	0.099807	−0.001716	−0.16167	0.063275	0.224946	0.0	0.0
1	51	0.091811	0.091555	−0.000256	−0.056694	−0.040083	0.016611	−0.3	64.1
1	52	0.072291	0.072112	−0.000179	−0.051181	−0.043961	0.007220	−0.2	36.7
1	53	0.061692	0.061528	−0.000164	−0.048988	−0.044372	0.004616	−0.2	29.1
1	54	0.054498	0.054338	−0.000160	−0.047426	−0.044049	0.003377	−0.2	25.5
1	55	0.049119	0.048960	−0.000159	−0.046077	−0.043433	0.002643	−0.2	23.3

STRESSES, DISPLACEMENTS AND DISPLACEMENT DISCONTINUITY COMPONENTS AT SEAM LEVEL(S)

SEAM	SEG	UX (POS)	UX (NEG)	DX (=DS)	UY (POS)	UY (NEG)	DY (=DN)	SIGXX	SIGYY
1	56	0.044872	0.044712	−0.000160	−0.044818	−0.042664	0.002155	−0.2	21.9
1	57	0.041398	0.041237	−0.000161	−0.043605	−0.041800	0.001804	−0.2	20.9
1	58	0.038483	0.038322	−0.000161	−0.042416	−0.040875	0.001541	−0.2	20.1
1	59	0.035991	0.035830	−0.000161	−0.041243	−0.039908	0.001334	−0.2	19.5
1	60	0.033827	0.033666	−0.000161	−0.040083	−0.038914	0.001169	−0.2	19.0
1	61	0.031926	0.031765	−0.000160	−0.038935	−0.037903	0.001032	−0.2	18.6
1	62	0.030237	0.030078	−0.000159	−0.037800	−0.036882	0.000918	−0.2	18.3
1	63	0.028726	0.028568	−0.000158	−0.036680	−0.035859	0.000821	−0.2	18.0
1	64	0.027361	0.027206	−0.000156	−0.035576	−0.034838	0.000738	−0.2	17.8
1	65	0.026122	0.025969	−0.000153	−0.034489	−0.033824	0.000665	−0.2	17.5
1	66	0.024991	0.024841	−0.000150	−0.033423	−0.032821	0.000602	−0.2	17.4
1	67	0.023952	0.023805	−0.000147	−0.032378	−0.031831	0.000546	−0.1	17.2
1	68	0.022994	0.022850	−0.000144	−0.031355	−0.030858	0.000497	−0.1	17.0
1	69	0.022106	0.021966	−0.000140	−0.030356	−0.029904	0.000453	−0.1	16.9
1	70	0.021282	0.021145	0.000137	−0.029382	−0.028969	0.000413	−0.1	16.8
1	71	0.020512	0.020379	0.000133	−0.028434	−0.028057	0.000378	−0.1	16.7
1	72	0.019793	0.019664	0.000129	−0.027512	−0.027166	0.000346	−0.1	16.6
1	73	0.019118	0.018993	0.000125	−0.026616	−0.026299	0.000317	−0.1	16.5
1	74	0.018483	0.018362	0.000121	−0.025747	−0.025456	0.000290	−0.1	16.4
1	75	0.017884	0.017767	0.000117	−0.024904	−0.024638	0.000266	−0.1	16.4
1	76	0.017319	0.017206	0.000113	−0.024088	−0.023843	0.000245	−0.1	16.3
1	77	0.016784	0.016676	0.000109	−0.023292	−0.023073	0.000225	−0.1	16.3
1	78	0.016278	0.016173	0.000104	−0.022534	−0.022328	0.000206	−0.1	16.2
1	79	0.015798	0.015698	0.000099	−0.021794	−0.021606	0.000188	−0.1	16.1
1	80	0.015344	0.015253	0.000091	−0.021076	−0.020909	0.000167	−0.1	16.1

STRESSES AND DISPLACEMENTS AT SPECIFIED OFF-SEAM LOCATIONS IN HALF-SPACE Y LE O

SEG	Y-COORD	X-DISP	Y-DISP	XX-STRAIN	YY-STRAIN	XY-STRAIN	SIGXX	SIGYY	SIGXY
41	0.00	−0.002153	−0.183905	0.000535	0.000	0.000	3.3	0.00	0.00
42	0.00	−0.006433	−0.182974	0.000528	0.000	0.000	3.3	0.00	0.00
43	0.00	−0.010630	−0.181132	0.000514	0.000	0.000	3.2	0.00	0.00
44	0.00	−0.014693	−0.178417	0.000495	0.000	0.000	3.1	0.00	0.00
45	0.00	−0.018572	−0.174885	0.000469	0.000	0.000	2.9	0.00	0.00
46	0.00	−0.022225	−0.170608	0.000438	0.000	0.000	2.7	0.00	0.00
47	0.00	−0.025614	−0.165672	0.000404	0.000	0.000	2.5	0.00	0.00
48	0.00	−0.028709	−0.160173	0.000366	0.000	0.000	2.3	0.00	0.00
49	0.00	−0.031488	−0.154216	0.000325	0.000	0.000	2.0	0.00	0.00
50	0.00	−0.033938	−0.147909	0.000284	0.000	0.000	1.8	0.00	0.00
51	0.00	−0.036055	−0.141359	0.000243	0.000	0.000	1.5	0.00	0.00
52	0.00	−0.037842	−0.134670	0.000202	0.000	0.000	1.3	0.00	0.00
53	0.00	−0.039309	−0.127938	0.000163	0.000	0.000	1.0	0.00	0.00
54	0.00	−0.040474	−0.121248	0.000127	0.000	0.000	0.8	0.00	0.00
55	0.00	−0.041357	−0.114674	0.000094	0.000	0.000	0.6	0.00	0.00
56	0.00	−0.041984	−0.108278	0.000063	0.000	0.000	0.4	0.00	0.00
57	0.00	−0.042380	−0.102105	0.000036	0.000	0.000	0.2	0.00	0.00
58	0.00	−0.042574	−0.096191	0.000013	0.000	0.000	0.1	0.00	0.00
59	0.00	−0.042591	−0.090560	−0.000007	0.000	0.000	0.0	0.00	0.00

STRESSES, DISPLACEMENTS AT SPECIFIED OFF-SEAM LOCATIONS IN HALF-SPACE Y LE O

SEAM	SEG	UX (POS)	UX (NEG)	DX (=DS)	UY (POS)	UY (NEG)	DY (=DN)	SIGXX	SIGYY
60	0.00	−0.042458	−0.085226	−0.000025	0.000	0.000	−0.2	0.00	0.00
61	0.00	−0.042196	−0.080194	−0.000039	0.000	0.000	−0.2	0.00	0.00
62	0.00	−0.041828	−0.075464	−0.000051	0.000	0.000	−0.3	0.00	0.00
63	0.00	−0.041373	−0.071029	−0.000061	0.000	0.000	−0.4	0.00	0.00
64	0.00	−0.040847	−0.066880	−0.000069	0.000	0.000	−0.4	0.00	0.00
65	0.00	−0.040266	−0.063005	−0.000075	0.000	0.000	−0.5	0.00	0.00
66	0.00	−0.039642	−0.059389	−0.000080	0.000	0.000	−0.5	0.00	0.00
67	0.00	−0.03898	−0.056018	−0.000083	0.000	0.000	−0.5	0.00	0.00
68	0.00	−0.038304	−0.052877	−0.000085	0.000	0.000	−0.5	0.00	0.00
69	0.00	−0.037607	−0.049950	−0.000087	0.000	0.000	−0.5	0.00	0.00
70	0.00	−0.036901	−0.047223	−0.000088	0.000	0.000	−0.5	0.00	0.00
71	0.00	−0.036190	−0.044682	−0.000088	0.000	0.000	−0.6	0.00	0.00
72	0.00	−0.035479	−0.042312	−0.000088	0.000	0.000	−0.5	0.00	0.00
73	0.00	−0.034771	−0.040102	−0.000087	0.000	0.000	−0.5	0.00	0.00
74	0.00	−0.034069	−0.038039	−0.000086	0.000	0.000	−0.5	0.00	0.00
75	0.00	−0.033375	−0.036111	−0.000085	0.000	0.000	−0.5	0.00	0.00
76	0.00	−0.032691	−0.034310	−0.000084	0.000	0.000	−0.5	0.00	0.00
77	0.00	−0.032018	−0.032625	−0.000083	0.000	0.000	−0.5	0.00	0.00
78	0.00	−0.031357	−0.031047	−0.000081	0.000	0.000	−0.5	0.00	0.00
79	0.00	−0.030709	−0.029569	−0.000079	0.000	0.000	−0.5	0.00	0.00
80	0.00	−0.030075	−0.028182	−0.000078	0.000	0.000	−0.5	0.00	0.00

REFERENCES

Aston, T.R.C., Tammemagi, H.Y. and Poon, A.W. 1987. Review and evaluation of empirical and analytical subsidence prediction techniques. *Mining Sci. Technol.* 5: 59–69.

Bahuguna, P.P. 1993. *Development of mine subsidence prediction model for indian coalfields.* Ph.D. Thesis. Department of Civil Engineering, University of Roorkee, p. 177.

Bhauguna, P.P., Singh, Bhawani, Srivastava, A.M.C. and Saxena, N.C. 1993. Semiempirical method for calculation of maximum subsidence in coal mines. *Geotech. Geol. Eng.* 11: 249–261.

Bhattacharyya, A.K., Shu, D.M. and Lama, R.D. 1988. Prediction of subsidence effects on a service decline from the partial extraction of an underground protective pillars. *Proc. Conf. on Buildings and Structures Subjected to Mine Subsidence*, Maitland, pp. 72–78.

Crouch, S.L. and Starfield, A.M. 1983. *Boundary Element Methods in Solid Mechanics.* London, George Allen and Unwin Ltd., p. 322.

Gurtunca, R.G. and Bhattacharyya, A.K. 1988. Modelling of surface subsidence in the southern coalfield of New South Wales. *5th Australia–New Zealand Conf. Geomechanics*, Sydney, August 22–23, pp. 346–350.

Mair, R.J., Taylor, R.N. and Burland, J.B. 1996. Prediction of ground movements and assessment of risk of building damages due to bored tunnelling, edited by Mair, R.J. and Taylor, R.N., *Conf Geotechnical Aspects of Underground Construction in Soft Ground*, pp. 713–718.

McNabb, K.E. 1987. Three Dimensional Numerical Modelling of Surface Subsidence Induced by Underground Mining, Technical Report No. 146, Division of Geomechanics, CSIRO, Victoria, Australia, p. 20.

Singh, Bhawani 1973. Continuum Characterisation of Jointed Rock Mass – Part II: Significance of Low Shear Modulus. *Int. J. Rock Mech. Mining Sci. Geomech. Abst.* 10: 337–349.

Singh, Bhawani, Viladkar, M.N., Samadhiya, N.K. and Mehrotra, V.K. 1997. Rock mass strength parameters mobilised in tunnels. *Tunnel. Underground Space Technol.* 12(1): 47–54.

Sinha, K.P. 1979. *Displacement discontinuity technique for analysing stresses and displacements due to mining in seam deposits.* Ph.D. Thesis. University of Minnesota, p. 311.

CHAPTER 30

Web sites in rock engineering

30.1 INTRODUCTION

In 1969 ARPANET, a group of three American Universities, started research on computer networking, which ultimately led to the birth of Internet in the year 1982. Since then development has taken place at a rapid pace in the so-called 'online world'. Today we have millions of public databases available online all over the world. With so many online services, and large variety of access methods, it is often difficult to find out way through the maze of offerings.

Information is abundant, anyone navigating the online world needs to know what is available, and how to find and use it. Getting there takes time, but the potential rewards are immense for researchers, designers and consultants. Therefore, half the battle is figuring out where to look. One will also discover that using the online resource can be quite fun and entertaining.

30.2 THE WORLD WIDE WEB

The World Wide Web (WWW, or just Web) allows one to connect to all services provided by the InterNet or Internet. Currently there are more than 30 million users connected to the Internet, and each month about half a million more start surfing the Web. About 10 percent of all users provide a Web-Server, to which all users in the world can connect. And naturally all these servers provide information or services, which may be of importance for you. Via the InterNet one may e.g. check the library of a University or even of the American Congress to see if they carry the book one is interested in. One can also discuss a specific problem in special forums with people all over the world. For this one needs to search the Web, and these searches involve using a 'search engine' available on the Internet as Web directories.

Finding relevant data not only depends on search engines but also on the type of keywords one searches for. For example if one wants to search for information on Geomechanics, one can simply type Geomechanics in the space provided for keywords on the search engine. The search engine will list the URL (Universal Resource Locator or simply address for the page) of sites containing 'geomechanics'. To do an effective search, it is helpful to know how search engines operate. There are many ways to narrow down your search. Boolean logic is one of the most common and effective way and involves using AND, OR and NOT in the key words. One can also use the '+' sign in your statements to give more focused results. Similarly, you can use the '−' sign to exclude certain words. Most of the search engines offer their 'advance search' where one have all these options and more listed in pull down menus. Some of the useful search engines available on the net are *http://www.altavista.com, http://www.infoseek.com, http://www.excite. com, http://www.yahoo.com, http://www.lycos.com, www.google.com.*

Suggested key words: tunnel, cavern, rock + engineering, geomechanics, rock + mechanics, landslide, rock + slope.

30.3   USENET

Another option is to search the Newsgroups – an online discussion group or a virtual community of people with similar interests. Usenet News (User Network News), as newsgroups are officially known as it is a network of networks and computers all of which have made bilateral agreements with other members of Usenet to share and exchange news. To be able to surf the Usenet one needs to have access to a NNTP (Network News Transfer Protocol) Servers through the ISP (Internet Service Provider). Usenet is divided into newsgroups and today, there are over 20,000 different newsgroups organised around every topic imaginable e.g., computers, books, music, science, research, environment, business etc. In order to make them easily recognisable, newsgroups follow a standard naming convention. The name of a newsgroup starts with a category type, followed by a dot and a subject, which can be followed by any number of subcategories, each separated by a dot. Some of the common categories of newsgroups are 'alt' for alternative, 'biz' for business, 'comp' related to computers, 'sci' related to science etc. More information on Usenet can be found at the web site – *http://vlib.stanford. edu/Overview2.html.* Some newsgroups of specific interest to rock engineers are *sci.engr.geomechanics, sci.engr.civil, sci.geo.geology, sci.geo.earthquakes.*

30.4   THE URLs OF SOME INTERESTING SITES

A list of web addresses, also called URL (Uniform Resource Locator), of some of the related sites for rock engineers are given below (Swarup and Prasad, 2000).

Societies
- *http://www.lnec.pt/ISRM/* – Site for ISRM
- *http://www.ita-aites.org* – Site for ITA

- *http://www.cgs.ca/* – The Canadian Geotechnical Society's Home Pages
- *http://www.tunnelcanada.ca* – Tunnelling Association of Canada
- *http://www.auca.org* – American Underground Construction Association
- *http://www.armarocks.org* – American Rock Mechanics Association
- *http://www.tucss.com* – Tunnelling and Underground Construction Society, Singapore

Software directories
- *http://www.ggsd.com* – Geotechnical and Geoenvironmental Software Directory. A comprehensive site for freeware and shareware software for download.
- *http://members.tripod.com/~mclean/software.htm*
- *http://grc.laurentian.ca/resource/resource/html*
- *www.tunnel.no, www.mrtunnel.com, www.bigdig.com* – For tunnel enquiry.

Library, Journals, Magazines, Publications, Personal Home Page, etc.
- *http://geotech.civen.okstate.ed/index.htm* – Virtual Library of Geotechnical Engineering. The site provides links to related universities from all over world apart from other relevant sites. It also provides link to *Electronic Journal of Geotechnical Engineering* (EDJE).
- *http://www.rockscience.com* – This site contains 'Practical Rock Engineering' a book published on Internet by Prof. E. Hoek. It also has a collection of Software available for download free of cost.
- *http://isrm.luns.net/index.htm* – Home page of Prof. J.A. Hudson.
- *http://www.arcat.com/arcatcos/cos08/arc08914.cfm* – Underground Space Center, University of Minnesota Profile Page.
- *http://www.pubs.asce.org* – Journals published by ASCE.
- *http://www.usace.army.mil/inet/usace-docs/eng-manuals/* – Civil works engineering manuals.
- *http://www.sciam.com* – Scientific American.
- *http://www.hbz-nrw.de/elsevier/08867798*
- *http://www.elsevier.nl/inca/publications/store/2/5/6/* – Journals published by Elsevier Science.
- *http://www.worldtunnelling.com*
- *http://www.enr.com*
- *http://www.sunet.se/sweden/science_geomechanics.html* – XYZ of Geomechanics
- *http://www.elsevier.nl/locate/isbn/0080430139* – Information of book on 'Rock Mass Classification: A Practical Approach in Civil Engineering'.
- *http://jrmtt.in.freeservers.com* – Site of *Indian Journal of Rock Mechanics and Tunnelling Technology* and to order computer programs with graphics at low cost.
- *http://rkgoel.cmri.freeservers.com* – Home page of Dr. R.K. Goel, Scientist, Central Mining Research Institute Regional Centre, CBRI campus, Roorkee, India.

- *http://rockmass.net* – Web page of Dr. Arild Palmstrom. Connections to download his Ph.D. thesis are also given.

Many similar sites are available on the internet and many more are being added regularly catering to various aspects of geomechanics. One has to search for them through the various search engines listed above.

REFERENCE

Swarup, Anil and Prasad, V.V.R. 2000. A note on internet for rock engineers. *J. Rock Mech. Tunnel. Technol.* 6(1): 61–63.

APPENDIX 1

Units used in the computer programs

The units used in computer programs for the landslide and tunnel hazards are generally – Tonne (T), Meter (m), and Degree (°). SI units may also be used.

Let us imagine a tank of 1m × 1m × 1m which is filled with water completely. The weight of the water in tank is one tonne (excluding the weight of the empty tank). The pressure/stress at the bottom of tank is 1.0 T/m^2, as unit weight of water is 1.0 T/m^3. So the suggested units make a physical sense. Their relationship with SI units and other units are as follows:

1 m	= 100 cm	= 1000 mm	= 3.28 ft	= 39.36 inches
1 T	= 10 kN	= 1000 kgf(force)	= 2240 lb	
1 T/m^2	= 0.01 MPa	= 10 kPa	= 0.1 kgf/cm^2	= 1.5 psi
1 T/m^3	= 0.1 kN/m^3	= 1.0 g/cc	= 62.5 lb/cft	
1 kN	= 0.1 T	= 100 kgf	= 224 lb	
1 kgf	= 0.001 T	= 0.01 kN	= 10 N	
1 kPa	= 0.1 T/m^2			
1 MPa	= 100 T/m^2			
1 kN/m^3	= 10 T/m^3	= 10 g/cc		
1 g/cc	= 1 T/m^3			
1 lb	= (1/2240) T	= (1/2.24) kgf	= (10/2240) kN	
1 psi	= (1/15) kgf/cm^2	= (1/1.5) T/m^2		
1 tsf	= (1/10.8) T/m^2			
1 ft	= 30.48 cm			

It may be noted that g, kg and tonne are units of mass in SI units. However, tonne is used extensively as unit of weight and force in Geotechnical engineering because it is easy to understand. As such, tonne has been suggested as unit of force and weight in all programs of landslide and tunnels. It should be kept in mind that the unit weight of soil/rock is T/m^3 or g/cc which is one-tenth of kN/m^3.

APPENDIX 2

Do's and don'ts in case of a landslide
(Ranjan et al., 1998)

- Do not let people to get panicky, as it will upset entire relief operations.
- Use media (television and newspapers) to attract the attention of leaders in case of major landslides.
- Do not remove the debris due to a landslide, where it provides a toe support for the slope materials above.
- With the help of the experts, assess the causes of failures and identify a set of short and long terms measures.
- Give mobile phones to rail/ road engineers and geologists for co-ordinating the actions on landslide clearance.
- Removal of debris at the toe may lead to widening of tension cracks in addition to producing more cracks. This will cause excessive seepage of surface water into the ground.
- Buried persons should be removed very carefully within 48 hrs.
- Since water spring generally plays a major role in the instability, provide drainage to clear the surface water away from the slide zone. Choked drains should be cleaned.
- Divert the traffic through alternative routes.
- Erect warning boards on the road on many locations indicating the location of the landslide as well as where the road has been closed.
- Locally available bamboo sticks or tree twigs can be driven into the slide debris near the toe to increase the shear strength of the slope materials so as to control further sliding.
- Local people should be educated about the ill effects of landslides and should be involved in the rehabilitation activities.
- The debris should not be dumped on the slopes below, but should be disposed off at selected places so that it does not interfere with the existing drainage system.
- A systematic investigation of the entire failure phenomenon will help to evolve a meaningful result for the environmental regeneration of the degraded landslide slope. The long-term measures based on such studies will have effective impacts on the restoration process.

- Movements of curious onlookers on the landslide debris before the stabilization of the landslide may be very harmful. In case of rock fall area, do not move up or down the slope but run laterally.
- Cattle grazing should not be permitted in the landslide prone areas, which are indicated by tilted trees.
- Rebuilding of damaged towns due to debris flow and mudflows, etc. must be allowed on safe areas.

REFERENCE

Ranjan, G., Anbalagan, R., Singh, Bhawani and Singh, Bhoop 1998. Landslide and sustainable development in mountainous regions. Sponsored by Department of Science and Technology, India, p. 23.

APPENDIX 3

Shear and normal stiffness of rock joints

The normal stiffness k_N of a unweathered rock joint is estimated as follows:

$$k_N = \frac{\text{normal stress } (\sigma)}{\text{joint closure } (\delta)}$$

$$\delta \propto \frac{\sigma}{E_r}$$

or

$$k_N \propto E_r$$

$$\therefore \frac{E_r}{k_N} = \text{constant for a given joint profile (A)} \qquad (A3.1)$$

where E_r = modulus of elasticity of asperities/rock material. The physical significance of the parameter A is that k_N is equal to the stiffness of rock layer of the thickness A.

The manual of U.S. Corps of Engineers (1973) and Singh and Goel (1982) summarize the typical values of parameter A on the basis of results of uniaxial jacking tests in USA and India, which are marked by * in Table A3.1.

On the basis of experiences of the back analysis of underground powerhouses at Sardar Sarovar and Tehri Dam Projects, Samadhiya (1999) suggested values for the normal and shear stiffness of joints which are summarised in Table A3.1 for various kinds of joints. It may be noted that normal stiffness during unloading (relaxation of normal stresses) is much higher than that during loading as expected.

Pressure dependent modulus
In highly jointed rock masses, the modulus of deformation is significantly dependent upon the confining pressure. In fact in case of soft rock materials like shales, slates, claystones, etc. the effect of confining pressure on modulus of deformation is very significant (Janbu, 1963). The effect of pressure dependency

Table A3.1. Back analysis values of normal and shear stiffnesses of rock joints.

S.No.	Joint type	A = E_r/K_n		K_s/K_n loading
		Loading (cm)	Unloading (cm)	
1	Continuous joint or loose bedding plane in weathered rock mass	115–125	16–18	1/10
2	Continuous joint or loose bedding plane in unweathered rock mass	60*	12	1/10
3	Discontinuous joints in unweathered rock mass	15–25*	5–7	1/10
4	Unweathered cleavage planes but separated	5*	2	1/10

*Adopted from Singh and Goel (1982) and US Corps of Engineers Manual (1973).

has been taken into account by the following relation in which effect of intermediate principal stress σ_2 has been included in addition to σ_3:

$$E_r = E_o \left[\frac{\sigma_2 + \sigma_3}{2\,P_a} \right]^\alpha > E_o \qquad (A3.2)$$

where

E_r = pressure dependent modulus of elasticity of a rock material in triaxial condition,

E_o = modulus of deformation corresponding to unit confining pressure (which may be taken to be approximately equal to the modulus of deformation from uni-axial compressive strength tests),

σ_2, σ_3 = effective intermediate and minor principal stresses,

P_a = atmospheric pressure and

α = the modulus exponent obtained from triaxial tests conducted at different confining pressures,

= 0.15 for hard rocks,

= 0.30 for medium rocks; and

= 0.50 for very soft rocks.

It may be noted that the increase in modulus of elasticity due to confining pressure shall also result in a corresponding increase in the joints stiffness. The stiffness k_n and k_s may also be increased in the same proportion as the modulus of elasticity of a rock material.

REFERENCES

Janbu, N. 1963. Soil compressibility as determined by odometer and triaxial tests, *European Conf. on Soil Mechanics and Foundation Engineering*, Wiesbaden, Vol. 1, pp. 19–25.

Manual of the US Army Corps of Engineers, EM1110-2-2901, 1973.

Samadhiya, N.K. 1999. *Influence of anisotropy and shear zones on Stability of Caverns*. Ph.D. Thesis. Department of Civil Engineering, University of Roorkee, India, p. 334.

Singh, Bhawani and Goel, P.K. 1982. *Estimation of Elastic Modulus of Jointed Rock Masses from Field Wave Velocity*, R.S. Mithal Commemorative Volume on Engineering Geosciences, Edited by B.B.S. Singhal, Sarita Prakashan, India, pp. 156–172.

APPENDIX 4

Instructions for using the computer program/software CD in the WINDOWS environment

All the computer programs of Landslide and Tunnelling are available in Folder SOFTWARE/SLOPE and SOFTWARE/TUNNEL respectively in the enclosed computer disc (CD). The Fortran source programs are also given in the folder SOFTWARE/SLOPE/SOURCE and SOFTWARE/TUNNEL/SOURCE. Before running the programs, these should be copied on your PC.

The attached CD has user's manual file for almost all the programs. User's manual is also presented in each chapter discussing a particular program.

For more clarifications of users, typical input data files are also given beginning with I. Similarly, corresponding output files are given, beginning with O. File details, for example, are as follows (Considering Program TOPPLE):

File Name	Details
Topple.txt	Users Manual
Itopple.dat	Typical input data file
Otopple.dat	Typical output data file

User may use simple commands in Windows Explorer. Figure A4.1 shows the printout of the Window Explorer screen. The computer programs are in Directory SOFTWARE. The program on landslide are in Folder 'Slope', whereas programs on Tunnels are in Folder 'Tunnel'. Both the Folders have different sub-folder like Circular, Debris, Footing, Planar, etc. as shown in Figure A4.1. Programs like Asc, Basc, and Sarc related to the circular landslide problem (discussed in Chapter 12) are in this sub-folder. Each program has users manual (.txt), input, output and exe files. Double click on exe file for executing the program. Programs Qult, Squeeze and Tm (Chapters 17, 25 and 26 respectively) are user-friendly programs.

Users may use other file names for different sites. New files say Ixyz.DAT may be opened using simple command in Windows Explorer. Then the input data is

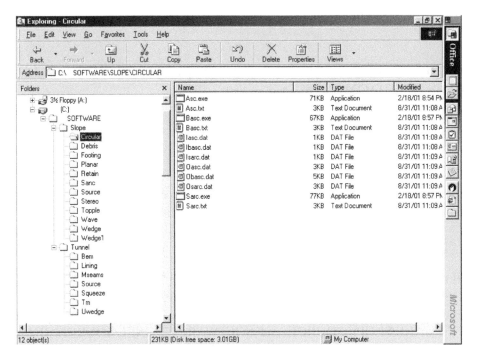

Figure A4.1. Window explorer showing various files in circular sub-folder.

typed as per the sequence shown in users manual. Similarly, new output file Oxyz.DAT should be created and empty file is closed. Finally, program Xyz is run with commands in above paragraphs. After execution of the program, a temporary empty output file is created which should be deleted.

Author index

Subject index

Printed and bound by CPI Group (UK) Ltd, Croydon, CR0 4YY

23/10/2024

01777679-0005